Quantum Theory Handbook

Quantum Theory Handbook

Edited by **Richard Burrows**

New York

Published by NY Research Press,
23 West, 55th Street, Suite 816,
New York, NY 10019, USA
www.nyresearchpress.com

Quantum Theory Handbook
Edited by Richard Burrows

International Standard Book Number: 978-1-63238-384-6 (Hardback)

Printed in the United States of America.

Contents

Preface

Over the recent decade, advancements and applications have progressed exponentially. This has led to the increased interest in this field and projects are being conducted to enhance knowledge. The main objective of this book is to present some of the critical challenges and provide insights into possible solutions. This book will answer the varied questions that arise in the field and also provide an increased scope for furthering studies.

This book presents an updated account of the research and development undergoing in the domain of quantum theory. Quantum theory is the primary theoretical approach that aids one to favorably understand the atomic and sub-atomic worlds which are too far from the cognition on the basis of general intuitions or experiences of daily-life. It is a very rational theory in which a balanced system of hypotheses and relevant mathematical methods facilitate exact explanation of the dynamics of quantum systems whose measurements are systematically influenced by objective uncertainties. Quantum theory has enabled us to apply and regulate novel quantum devices and technologies in quantum optics and lasers, quantum electronics, quantum computing and modern area of nanotechnologies.

I hope that this book, with its visionary approach, will be a valuable addition and will promote interest among readers. Each of the authors has provided their extraordinary competence in their specific fields by providing different perspectives as they come from diverse nations and regions. I thank them for their contributions.

Editor

Part 1

New Concepts in Quantum Theory

Quantum Theory of Multi-Local Particle

Takayuki Hori
Teikyo University
Japan

1. Introduction

In those days before the development of the gauge field theories there were many attempts to construct multi-local field theories of hadrons. The motivation for that was in the existence of wide variety of hadrons which may be categorized by quantum numbers indebted to the presumed internal structures. The success of QFD and/or QCD, however, impressed us the power of the local field theories of quarks and leptons, and swept away almost all alternative attempts describing the low energy physics. On the other hand the concept of the multi-locality (or non-locality) was promoted to the string model which is now regarded as one of candidates of the quantum gravity.

Although the realm of validity of the local field theory may be extended to the Planck scale the conceptual gap between the string and the local field theory is so large that we cannot treat them on an equal footing. Is there no room for the multi-local field theory in describing the phenomena near the Planck scale?

We have sought the theoretical possibility of the multi-local objects, consisting of N particles, which stay in an intermediate position between the local particle and the string. In the papers (Hori, 1992)–(Hori, 2009) we have constructed the models with $N = 2$, which have resembling properties as the string, though extremely simple in structure. The simplest model with $N = 2$ is a system of two relativistic particles with specific interaction among them. We called the object as a bilocal particle. We have found a hidden gauge symmetry in the bilocal model (Hori, 1992), which reveals $SL(2, \mathbb{R})$ in the canonical theory. This causes the pathological property that the amount of the gauge invariance does not match with the number of the first class constraints in the canonical theory. This means breakdown of Dirac's conjecture (Dirac, 1950).

The BRST analysis of the bilocal model shows the existence of spacetime critical dimensions, $D = 2$ or $D = 4$ (Hori, 1996). But the quantum theory of the model can not be treated in the similar way as the ordinary gauge theories, since the ghost numbers of the physical states are not zero. Because the reason of the difficulty is in the constraint structure we have constructed an improved version of $N = 2$ model based on the object called complex particle (Hori, 2009).

The coordinates of a complex particle are complex numbers and depend on the internal time. In the lagrangian formulation the gauge degrees of freedom is two in the ordinary sense. This causes the breakdown of Dirac's conjecture as in the bilocal model. We argued that a modification of the definition of the physical equivalence remedies the situation, and the system has all of the three gauge freedom of $SL(2, \mathbb{R})$. The constraint structure is different from that of the bilocal model in such a way that two of three constraints are hermitian conjugate to each other, and a natural quantization scheme can be applied similar to the string theory. The physical state conditions are fulfilled in the ghost number zero sector, and the requirement that the momentum eigenstate should be physical restricts the dimension of the spacetime to be two or four.

In the present paper we achieve the complete first quantization of the complex particle, supplementing the results obtained in ref.(Hori, 2009). We also propose the field theory action of the Chern-Simons type. The action is shown to be invariant under gauge transformations in the field theoretical sense only if $D = 4$.

Finally we extend the previous results to $N \geq 3$ particle system. Especially we define an open N-particle and closed N-particle systems. We restrict ourselves, however, to the open cases because the constraint structure in the canonical theory is much complicated in the closed cases compared with the open cases.

2. Preliminary remark

The notion of gauge invariance or physical equivalence in the models considered in the present paper is so subtle that one may easily fall into confusion. Therefore let us consider first the ordinary relativistic particle, and count the number of gauge as well as physical degrees of freedom. The spacetime coordinates of the particles $x^\mu, (\mu = 0, 1, 2, .., D-1)$ are functions of internal time τ, where D is the dimension of the spacetime. The action is written as [1]

$$I_0 = \int d\tau \, \frac{\dot{x}^\mu \dot{x}_\mu}{2g},$$ (1)

where g is the einbein needed for the reparametrization of the internal time. The action is invariant under the transformations

$$\delta x^\mu = \epsilon \dot{x}^\mu + \epsilon^{\mu\nu} x_\nu + a^\mu, \qquad \delta g = \frac{d}{d\tau}(\epsilon g), \qquad (\epsilon^{\mu\nu} = -\epsilon^{\nu\mu})$$ (2)

where infinitesimal constant parameters $\epsilon^{\mu\nu}, a^\mu$ are those of Lorenz transformations and translations, respectively, and the parameter ϵ depends on τ, corresponding to the reparametrization of τ.

Now what is the gauge freedom, by which the τ development of variables is not determined uniquely? The existence of the invariance of the action, with τ dependent parameter, ϵ, leads to redundant variables, because of which the Euler-Lagrange(EL) equations have not unique solutions even if one chooses suitable initial conditions.

We get the answer by first choosing gauge fixing conditions and by ascertaining consistency of the solutions to EL equations. The first integral of EL equations is

$$\dot{x}^\mu = c^\mu g, \qquad \dot{x}^\mu \dot{x}_\mu = 0,$$ (3)

---

[1] The metric convension is $\eta_{\mu\nu} = \text{diag}(-1, 1, 1, ..1)$.

where c's are arbitrary constants. By using the freedom $\epsilon(\tau)$ we can fix the gauge as

$$x^0(\tau) = \tau. \tag{4}$$

The remaining freedom (of finite degrees) is $\epsilon^{0i}, \epsilon^{ij}, a^i$, which is counted $\frac{1}{2}(D-1)(D+2)$.

Setting the initial conditions as

$$x^i(0) = x^i_0, \qquad \dot{x}^i(0) = c^i g_0, \qquad g(0) = g_0, \qquad (c^i)^2 = 1/g_0^2, \tag{5}$$

we get the unique solution

$$x^i(\tau) = c^i g_0 \tau + x^i_0, \qquad g(\tau) = g_0. \tag{6}$$

That is, if the spacial coordinates and the spacial direction of the particle both at $\tau = 0$ and the value $g(0)$ are given, the whole orbit of the particle moving with velocity of light is determined. The number of the physical degrees of freedom must be the number of degrees of freedom to put the independent initial condition, i.e., $2(D-1)$. On the other hand, among the $\frac{1}{2}(D-1)(D+2)$ degrees of freedom of the remaining symmetry in the gauge (4) the number of freedom which does not move c^i is $\frac{1}{2}(D-1)(D-2)$. Hence the net degrees of freedom for changing the initial condition is $\frac{1}{2}(D-1)(D+2) - \frac{1}{2}(D-1)(D-2) = 2(D-1)$. This coincidence implies that the gauge freedom corresponds to the transformation with the parameter $\epsilon(\tau)$, by which one can fix one variable for all τ.

Presumably, the above coincidence may be due to Dirac's conjecture in the canonical theory, which claims that every first class constraint should generate gauge transformations. In the subsequent sections we will encounter the situations where a naive counting leads to mismatch of degrees of freedom in the lagrangian form.

3. $N = 2$ model

3.1 Classical action

The simplest example of the multi-local particle is the two particle system with some bilinear *interactions*. We call it bilocal particle (Hori, 1992). Let us denote the coordinates of the two particles as $x^\mu_a, (a = 1, 2; \mu = 0, 1, 2, .., D-1)$, which are functions of internal time, τ. We introduce the einbeins, $g_a, (a = 1, 2)$, for the sake of the reparametrization invariance along the world lines, which are auxiliary variables and their equations of motion make the trajectories of the particles put on the light-cones. The proposed action of the bilocal particle is written as

$$I = \int d\tau L, \qquad L = \frac{\dot{x}_1^2}{2g_1} + \frac{\dot{x}_2^2}{2g_2} + \kappa(\dot{x}_1 x_2 - \dot{x}_2 x_1), \tag{7}$$

where κ is a constant with dimension of mass squared. (Here and hereafter we suppress the spacetime indices μ, if no confusions occur.) The first two terms in the action are separately invariant under

$$\delta x_a = \epsilon_a \dot{x}_a, \qquad \delta g_a = \frac{d}{d\tau}(\epsilon_a g_a), \qquad (a = 1, 2) \tag{8}$$

where $\epsilon_a, (a = 1, 2)$ are infinitesimal parameters depending on τ. This is the well known reparametrization gauge invariance of the relativistic particle. Apparently the third term in the action would violate this invariance with independent ϵ_1 and ϵ_2, but we found larger invariance under the following transformations (Hori, 1992),

$$\delta x_1 = \epsilon_1 \dot{x}_1 + \frac{\epsilon_0}{g_2} \dot{x}_2, \qquad \delta x_2 = \epsilon_2 \dot{x}_2 + \frac{\epsilon_0}{g_1} \dot{x}_1, \tag{9}$$

$$\delta g_1 = \frac{d}{d\tau}(\epsilon_1 g_1) + 4\kappa \epsilon_0 g_1, \qquad \delta g_2 = \frac{d}{d\tau}(\epsilon_2 g_2) - 4\kappa \epsilon_0 g_2, \tag{10}$$

where the infinitesimal parameters $\epsilon_1, \epsilon_2, \epsilon_0$ are functions of τ, two of which are arbitrary, while another is subjected to the constraint

$$\dot{\epsilon}_0 + 2\kappa g_1 g_2 (\epsilon_2 - \epsilon_1) = 0. \tag{11}$$

In fact the variation of the lagrangian under (9) and (10) is

$$\delta L = \frac{d}{d\tau}\left[\epsilon_0 \left(\frac{\dot{x}_1 \dot{x}_2}{g_1 g_2} + \kappa \left(\frac{\dot{x}_2 x_2}{g_2} - \frac{\dot{x}_1 x_1}{g_1} \right) \right) + \sum_{a,b=1}^{2} \epsilon_a \left(\kappa \epsilon_{ab} x_a \dot{x}_b + \delta_{ab} \frac{\dot{x}_a \dot{x}_b}{g_b} \right) \right]$$

$$+ \left[\dot{\epsilon}_0 + 2\kappa g_1 g_2 (\epsilon_2 - \epsilon_1) \right] \frac{\dot{x}_1 \dot{x}_2}{g_1 g_2}, \qquad (\epsilon_{12} = -\epsilon_{21} = 1, \; \epsilon_{11} = \epsilon_{22} = 0). \tag{12}$$

We can regard ϵ_1 and ϵ_2 as the independent gauge parameters, and ϵ_0 as dependent one determined by (11) except its *constant mode*, a relic of the global invariance of the action. The existence of the above unexpected gauge invariance is the origin of some curious properties of our model such that the $SL(2, \mathbb{R})$ gauge symmetry in the canonical theory and the existence of the critical dimension as is shown later.

In the case $\kappa \neq 0$, the first integral to the equations of motion derived by the action (7) is

$$\frac{\dot{x}_a}{g_a} + 2\kappa \sum_b \epsilon_{ab}(x_b - c_b) = 0, \qquad (a = 1, 2), \tag{13}$$

where $c_a, (a = 1, 2)$ are constants. Since the variations of g_a give $\dot{x}_a^2 = 0$, we have

$$(x_a - c_a)^2 = 0, \qquad (a = 1, 2). \tag{14}$$

Thus the two particles are put on the light-cone with tops of arbitrary spacetime points, and moving with velocity of light.

For the sake of the reparametrization invariance we can fix the gauge as

$$x_1^0(\tau) = x_2^0(\tau) = \tau, \tag{15}$$

then from (13) we have

$$\frac{1}{g_1} = -2\kappa \left(\tau - c_2^0 \right), \qquad \frac{1}{g_2} = 2\kappa \left(\tau - c_1^0 \right). \tag{16}$$

Substituting them back into (13), and dividing by κ, we obtain

$$(\tau - c_2^0)\dot{x}_1^i = (x_2 - c_2)^i, \qquad (\tau - c_1^0)\dot{x}_2^i = (x_1 - c_1)^i. \qquad (i = 1, 2, .., D-1). \qquad (17)$$

Note that eqs.(17) do *not* depend on κ. (The case $\kappa = 0$ should be treated separately, since in that case we must set $\dot{x}_a/g_a = c_a'$ instead of (13) with other constants c_a', leading to $\dot{x}_a = c_a'$ in the above gauge. We assume $\kappa \neq 0$ henceforth.) Differentiating (17) with respect to τ we have

$$(\tau - c_1^0)(\tau - c_2^0)\ddot{x}_a^i + (\tau - c_a^0)\dot{x}_a^i - (x_a - c_a)^i = 0, \qquad (i = 1, 2, .., D-1; a = 1, 2). \qquad (18)$$

These are the ordinary linear differential equations of second rank with regular singularities, and are solved by Frobenius's method. The general solutions are written as

$$x_a^i(\tau) = c_a^i + (\tau - c_a^0)v_a^i + f_a(\tau)w_a^i, \qquad (i = 1, 2, .., D-1; a = 1, 2), \qquad (19)$$

where v_a^i, w_a^i, $(a = 1, 2)$ are arbitrary constants, and $f_a(\tau)$ are the solutions to the equations $(\tau - c_2^0)\dot{f}_1 = f_2$ and $(\tau - c_1^0)\dot{f}_2 = f_1$, with vanishing asymptotic values, the concrete form of which are written as

$$f_a(\tau) = c_1^0 - c_2^0 + (\tau - c_a^0)\ln\left|\frac{\tau - c_1^0}{\tau - c_2^0}\right|, \qquad (a = 1, 2). \qquad (20)$$

Substituting (19) once again into (17), we see $v_1^i = v_2^i \equiv v^i$, $w_1^i = w_2^i \equiv w^i$. Furthermore using (14), we see $v^2 = 1$, $v \cdot w = w^2 = 0$. Since we assume the Euclidean signature for the spacial part of the metric, w^i must vanish. Then we obtain

$$x_a^i(\tau) = c_a^i + (\tau - c_a^0)v^i, \qquad v^2 = 1, \qquad (i = 1, 2, .., D-1; a = 1, 2). \qquad (21)$$

Thus we see that the relative coordinates $x_1^i - x_2^i$ do not depend on τ, and each particle moves with velocity of light. In other words the bilocal particle is the two end points of a *rigid stick* with arbitrary length, which moves with velocity of light. Since this result is independent of κ, the system does *not* transfered to that of two free particles in the limit $\kappa \to 0$.

In the gauge choice (15), the einbeins are determined by (16) for arbitrary τ. The independent parameters determining the initial condition are $x_a^i(0), (a = 1, 2; i = 1, 2, .., D-1)$ and $v^i, (i = 1, 2, .., D-2)$. The number of them, $3D - 4$, should be the number of the physical degrees of freedom. On the other hand the number of constant parameters corresponding to the τ-independent symmetry of the action, including ϵ_0 as well as Lorenz and translations, which survives after gauge fixing is $\frac{1}{2}(D-1)(D+4)+1$. Among them the number of freedom which fixes v^i is $\frac{1}{2}(D-1)(D-2)$. The net freedom to move the initial condition counts $3D - 2$. The discrepancy, $(3D-2) - (3D-4) = 2$, suggests existence of extra gauge degrees of freedom in the case $\kappa \neq 0$, which is not explicit in the lagrangian formulation.

3.2 Canonical theory

The canonical conjugate variables corresponding to x_1 and x_2 are $p_1 = \dot{x}_1/g_1 + \kappa x_2$, and $p_2 = \dot{x}_2/g_2 - \kappa x_1$, respectively, while those corresponding to g_a, denote π_a, are subjected to primary constraint $\pi_a \sim 0$. The total hamiltonian is

$$H_T = \frac{1}{2}(g_1\chi_1 + g_2\chi_{-1}) + v_1\pi_1 + v_2\pi_2, \tag{22}$$

where

$$\chi_1 = \frac{1}{2}(p_1 - \kappa x_2)^2, \qquad \chi_{-1} = \frac{1}{2}(p_2 + \kappa x_2)^2, \tag{23}$$

and v_1, v_2 are the Dirac variables which are unphysical.

Preservation of the primary constraints $\pi_a \sim 0$ along time development gives the secondary constraints $\chi_{\pm 1} \sim 0$, while the preservation of them gives tertiary constraints

$$\chi_0 = \frac{1}{2}(p_1 - \kappa x_2)(p_2 + \kappa x_2) \sim 0. \tag{24}$$

There are no other constraints in our model. These constraint functions satisfy the $SL(2, \mathbb{R})$ algebra on account of the Poisson brackets:

$$\{\chi_n, \chi_m\} = -2\kappa(n - m)\chi_{n+m}, \qquad (n, m = 0, \pm 1). \tag{25}$$

The whole first class constraints of our model are $\chi_0 \sim \chi_{\pm 1} \sim \pi_1 \sim \pi_2 \sim 0$. The coefficients of the first class constraints in the hamiltonian are all unphysical variables, and the values of which can be arbitrarily fixed for the sake of the gauge freedom. Two hamiltonians with different coefficients are called gauge equivalent.

According to Dirac (Dirac, 1950)(Dirac, 1964), one may say that two points in the phase space are physically equivalent if there exists another point in the phase space which develops to the two points through equations of motion determined by respective guage equivalent hamiltonians. Transformations from a point in the phase space to the physically equivalent point are called gauge transformation. Dirac conjectured (Dirac, 1950) that all first class constraints generate gauge transformations. (On the validity of Dirac's conjecture it has been argued by some authors, see, *e.g.*, (Sugano & Kamo, 1982), (Frenkel, 1982).)

Now let us examine whether the first class constraints of our model, $\chi_0, \chi_{\pm 1}, \pi_1, \pi_2$, generate the gauge transformations. Consider the transformations of a canonical variable q, generated by the constraint functions $\chi_{\pm 1}, \chi_0$ and π_1, π_2;

$$\delta q = \{q, Q\}, \qquad Q = \sum_{a=0,\pm 1} \epsilon_a \chi_a + \eta_1\pi_1 + \eta_2\pi_2, \tag{26}$$

where transformation parameters $\epsilon, \eta_{1,2}$ are time dependent with $\epsilon_a(0) = \eta_{1,2}(0) = 0$. If time development of q is generated by the total hamiltonian, *i.e.*, $\dot{q} = \{q, H_T\}$, then it turns out, using the Jacobi identity, that $q' = q + \{q, Q\}$ develops as

$$\dot{q}' = \{q, H_T'\}\big|_{q=q'} + O(\epsilon^2), \tag{27}$$

$$H_T' = H_T + \tilde{Q} + \dot{\eta}_1\pi_1 + \dot{\eta}_2\pi_2, \qquad \tilde{Q} = \sum_{a=0,\pm 1} \dot{\epsilon}_a\chi_a + \{Q, H_T\}. \tag{28}$$

We get the point $q'(\tau)$ from the initial point $q(0)$ through the "hamiltonian" H'_T. The point $q(\tau)$ is also developed from the same initial point but through the hamiltonian H_T. Thus if H'_T and H_T are gauge equivalent, then the two points $q'(t)$ and $q(t)$ are physically equivalent in Dirac's sense. But this is *not* the case, since H_T does not contain the tertiary constraint χ_0 but H'_T does. Even if we set $\epsilon_0 = 0$ the situation does not change, so we see that not only χ_0 but $\chi_{\pm 1}$ do not generate gauge transformations. This indicates the breakdown of Dirac's conjecture in our model.

The above fact, however, does not contradict with the gauge invariance in the lagrangian formulation. If we restrict ourselves to the transformation parameters so that $\tilde{Q} = 0$, then we see H_T is gauge equivalent to H'_T. Thus $\chi_0, \chi_{\pm 1}$ generate the gauge transformations with the restricted parameters. These transformations coincide with those of the lagrangian form, (9),(10), as is shown bellow. Rewriting the parameters ϵ_a in (26) as ϵ'_a, the condition $\tilde{Q} = 0$ gives

$$\eta_1 = \dot{\epsilon}'_1 + 2\kappa\epsilon'_0 g_1, \qquad \eta_2 = \dot{\epsilon}'_{-1} - 2\kappa\epsilon'_0 g_1, \qquad \dot{\epsilon}'_0 + 4\kappa(\epsilon'_2 g_1 - \epsilon'_{-1} g_2) = 0. \tag{29}$$

Therefore we have

$$\delta x_1 = \epsilon'_1(p_1 - \kappa x_2) - \frac{1}{2}\epsilon'_0(p_2 + \kappa x_1), \qquad \delta x_2 = \epsilon'_{-1}(p_2 + \kappa x_1) - \frac{1}{2}\epsilon'_0(p_1 - \kappa x_2), \tag{30}$$
$$\delta g_1 = \dot{\epsilon}'_1 + 2\kappa\epsilon'_0 g_1, \qquad \delta g_1 = \dot{\epsilon}'_1 - 2\kappa\epsilon'_0 g_2. \tag{31}$$

Substituting the definition of momenta, $p_1 = \dot{x}_1/g_1 + \kappa x_2$, $p_2 = \dot{x}_2/g_2 - \kappa x_1$, into above equations, we have

$$\delta x_1 = \epsilon'_1 \frac{\dot{x}_1}{g_1} + \frac{1}{2}\epsilon'_0 \frac{\dot{x}_2}{g_2}, \qquad \delta x_2 = \epsilon'_1 \frac{\dot{x}_2}{g_2} + \frac{1}{2}\epsilon'_0 \frac{\dot{x}_1}{g_1}. \tag{32}$$

Finally redefining the parameter as $\epsilon'_1 = g_1\epsilon_1$, $\epsilon'_{-1} = g_2\epsilon_2$, $\epsilon'_0 = 2\epsilon_0$, we get (9),(10). The last condition in (29) is the same as (11) if the redefined parameters are used.

The definition of the physical equivalence in the phase space owing to Dirac and the concept of gauge transformations based on it may be cumbersome at least in the present model. In ref.(Hori, 2009) we proposed another definition of physical equivalence, which seems natural both in the lagrangian and the canonical theories, and in accordance with Dirac's conjecture. The basic observation is that every conserved quantities have the same values along the gauge invariant orbits of canonical variables. Therefore any two physically equivalent points in Dirac's sense have the same values of all conserved quantities. Our claim is that the concept of the physical equivalence should be relaxed so that the reverse proposition holds. That is, we define that if all of the conserved variables at two points in the physical phase space (lied in the constrained subspace) coincide then the two points are called physically equivalent.

In order to determine the conserved quantities in our model let us examine the global symmetries. Since the deviations of the lagrangian under the global translations $\delta x_a = E_a$ is $\delta L = \kappa \frac{d}{d\tau}(x_1 E_2 - x_2 E_1)$, the corresponding conserved charges are

$$\tilde{p}^\mu_1 = \frac{\dot{x}^\mu_1}{g_1} + 2\kappa x^\mu_2, \qquad \tilde{p}^\mu_2 = \frac{\dot{x}^\mu_2}{g_2} - 2\kappa x^\mu_1, \tag{33}$$

which, in terms of the canonical variables, are written as

$$\tilde{p}_1^\mu = p_1^\mu + \kappa x_2^\mu, \qquad \tilde{p}_2^\mu = p_2^\mu - \kappa x_1^\mu. \tag{34}$$

Similarly the conserved charges corresponding to the Lorenz invariance are

$$M_{\mu\nu} = x_{1[\mu}p_{1\nu]} + x_{2[\mu}p_{2\nu]}. \tag{35}$$

All of the Poisson brackets between the charges $\tilde{p}_a^\mu$, $M_{\mu\nu}$ and the constraints $\chi_0, \chi_{\pm 1}$ vanish. There is another global symmetry of the lagrangian, which is seen by setting ϵ_0 to a constant and $\epsilon_1 = \epsilon_2 = 0$ in (9) and (10). The corresponding conserved charge turnes out to be χ_0 which is vanishing in the physically admissible orbits. In our model the maximal set of conserved quantities are $\tilde{p}_a^\mu, M_{\mu\nu}$ and χ_0. Since these variables are invariant (up to the first class constraints) under the transformations generated by all of the first class constraints, Dirac's conjecture holds.

In the canonical theory the gauge transformations are generated by five constraints $\chi_{0,\pm 1}, \pi_{1,2}$. If one fixes the gauge by five subsidiary conditions, then the equations of motion determine unique solutions. These ten conditions eliminate ten variables among $4(D+1)$ canonical variables x_a, p_a, g_a, π_a, and the remaining $2(2D-3)$ canonical variables, *i.e.*, $2D-3$ canonical pairs become the physical variables.

3.3 Quantization

In this subsection we present the quantum theory, assuming that our model is a constrained hamiltonian system with gauge symmetries generated by $\chi_{0,\pm 1}$. This point of view is consistent with the reduction of the classical degrees of freedom mentioned in **3.1**.

Let us represent the dynamical variables as linear operators on the space of differentiable and square integrable functions of $x_{1,2}$. The momentum observables of the two particles are defined by

$$\tilde{p}_1 = -i\partial_1 + \kappa x_2, \qquad \tilde{p}_2 = -i\partial_2 - \kappa x_1. \tag{36}$$

They satisfy the commutation relation

$$[\tilde{p}_1^\mu, \tilde{p}_2^\nu] = 2\kappa i \eta^{\mu\nu}. \tag{37}$$

That is, the momenta of the two particles do not have simultaneous eigenvalues. This is the reason why we call our system a bilocal particle instead of two particles. This is the fundamental uncertainty relation of the model.

The classical constraint functions are replaced by the following operators:

$$L_1 = \frac{i}{4\kappa}(-i\partial_1 - \kappa x_2)^2, \tag{38}$$

$$L_{-1} = \frac{i}{4\kappa}(-i\partial_2 + \kappa x_1)^2, \tag{39}$$

$$L_0 = \frac{i}{4\kappa}(-i\partial_1 - \kappa x_2)(-i\partial_2 + \kappa x_1) - \alpha, \tag{40}$$

where $\partial_a = \partial / \partial x_a$, and the constant α represents the ambiguity due to the operator ordering. The above operators constitute a basis of $SL(2, \mathbb{R})$ with central term as

$$[L_n, L_m] = (n - m) \left(L_{n+m} + \left(\alpha - \frac{D}{4} \right) \delta_{n+m} \right), \qquad (n, m = 0, \pm 1), \tag{41}$$

where D is the dimension of spacetime.

According to the gauge algebra (41) the BRST charge is defined by

$$Q = \sum_{n=0,\pm 1} c_n L_n - \frac{1}{2} \sum_{n,m=0,\pm 1} (n - m) c_n c_m \frac{\partial}{\partial c_{n+m}}, \tag{42}$$

where $c_a, (a = 0, \pm 1)$ are the BRST ghost variables. The square of the BRS charge is

$$Q^2 = 2 \left(\alpha - \frac{D}{4} \right) c_1 c_{-1}. \tag{43}$$

As in the ordinary gauge theory we require the nilpotency of Q so that the ordering ambiguity is fixes as $\alpha = D/4$, which also eliminates the central term in (41).

In ref.(Hori, 1996) we have calculated the BRST cohomology classes in the bilocal model in order to get the physical Hilbert space. We found there that there exists non-trivial physical states only in the dimensions $D = 2$ or $D = 4$. In the case $D = 2$ there exists vector states, while in the case $D = 4$ only scalar states are permitted. However, the analysis is very complicated and it seems difficult to obtain simple scheme for calculations of quantum phenomena.

The reason for the difficulty is in the fact that one can not define such an inner product in the Hilbert space that L_1 is hermitian conjugate to L_{-1}. To obtain physical states represented by functions of spacetime coordinates we are forced to solve the over determined system $L_{\pm 1}|\text{pys}\rangle = L_0|\text{pys}\rangle = 0$, which has no solution.

A field theory, however, has been constructed (Hori, 1993) by using the Chern-Simons action whose exterior derivative is replaced by the BRST operator as Witten has done (Witten, 1986) in a string field theory. But the formulation is formal and a concrete calculation of physical processes has not been achieved due to lack of connections to the first quantized theory.

This situation has been partially overcome by a modification of the model, where two particles in the bilocal model are replaced by the real and the imaginary parts of one complex particle (Hori, 2009). The model is illustrated in the next section.

4. Complex particle

4.1 Action and invariance

The improved version of the $N = 2$ model is defined as follows. Let us consider the spacetime with complex coordinates $z^\mu, (\mu = 0, 1, 2, ..D - 1)$, and a particle moving in the spacetime, the complex coordinates of which are functions of the internal time τ. The einvein g is also complex valued function of τ. The proposed action is

$$I_C = \int d\tau\, L_C, \qquad L_C = \frac{\dot{z}^2}{2g} + i\kappa \dot{z}\bar{z} + c.c. \tag{44}$$

The action describes dynamics of two real coordinates corresponding to the real and the imaginary parts of $z = x + ia$. We call the object defined above as *complex particle* (Hori, 2009).

The action is invariant under the transformations

$$\delta z = \epsilon \dot{z} + \frac{\epsilon_0}{\bar{g}}\dot{z}, \qquad \delta g = \frac{d}{d\tau}(\epsilon g) + 4i\kappa\epsilon_0 g, \tag{45}$$

where ϵ and ϵ_0 depend on τ. While ϵ has arbitrary complex value, ϵ_0 is real and subjected to the constraint,

$$\dot{\epsilon}_0 - i\kappa g\bar{g}(\epsilon - \bar{\epsilon}) = 0. \tag{46}$$

Classical solutions, the constraint structure and so force are analyzed in the similar way as those of the bilocal model. Thus we recapitulate the results. In the gauge choice

$$g^{-1} = 2\kappa|\tau - \tau_0|, \tag{47}$$

it turns out that in the kinetic terms of the action x(real part) and a(imaginary part) have correct and wrong signs, respectively. Thus x's are physical variables, while a's are ghosts. The solution for $z = x + ia$ to the equations of motion is

$$x(\tau) = x_0 + (\tau - \tau_0)k + \frac{e}{\tau - \tau_0}, \tag{48}$$

$$a(\tau) = a_0 + s(\tau - \tau_0)\left((\tau - \tau_0)k + \frac{e}{\tau - \tau_0}\right), \tag{49}$$

where $s(\tau)$ is the step function, and k and e are D-dimensional light-like vectors with real valued components, which are mutually orthogonal.

The canonical momenta of z and $\bar{z}$ are

$$p = \frac{\dot{z}}{g} + i\kappa\bar{z},$$

$$\bar{p} = \frac{\dot{\bar{z}}}{\bar{g}} - i\kappa z, \tag{50}$$

while those of g and $\bar{g}$, denoting π and $\bar{\pi}$, respectively, vanish. Note that the momenta (50) are *not* conserved quantities with respect to τ, and the conserved momenta, denoted $\tilde{p}$ and $\tilde{\bar{p}}$, are

$$\tilde{p} = p + i\kappa\bar{z},$$

$$\tilde{\bar{p}} = \bar{p} - i\kappa z, \tag{51}$$

while the generators of Lorentz transformations defined by $M_{\mu\nu} = z_{[\mu}p_{\nu]} + \bar{z}_{[\mu}\bar{p}_{\nu]}$ are conserved. The deviation of p's and $\bar{p}$'s (and their c.c.) comes from the fact that under global translations the action is invariant but the lagrangian varies by total derivatives. For an arbitrary variation δz (and $\delta \bar{z}$), the identity

$$\int d\tau \left[[\text{EL}] + \frac{d}{d\tau} \left(\bar{p}\delta z + \tilde{p}\delta \bar{z} \right) \right] = 0 \tag{52}$$

holds, where

$$[\text{EL}] = \left(\frac{\partial L}{\partial z} - \frac{d}{d\tau} \frac{\partial L}{\partial \dot{z}} \right) \delta z + \text{c.c.} \tag{53}$$

vanishes if the Euler-Lagrange equations are satisfied. Since in the translations, δz and $\delta \bar{z}$ are constants, $\bar{p}$ and $\tilde{p}$ are conserved. From the invariance under the Lorenz transformations we get $M_{\mu\nu}$ as conserved quantities.

Now the total Hamiltonian generating τ development is

$$H_T = g\chi_1 + \bar{g}\chi_{-1} + v\pi + \bar{v}\bar{\pi}, \tag{54}$$

where

$$\chi_1 = \frac{1}{2}(p - i\kappa\bar{z})^2,$$

$$\chi_{-1} = \frac{1}{2}(\bar{p} + i\kappa z)^2, \tag{55}$$

and v and $\bar{v}$ are the Dirac variables corresponding to the primary constraints, $\pi \sim \bar{\pi} \sim 0$. The preservation of the primary constraints requires the secondary constraints, $\chi_1 \sim \chi_{-1} \sim 0$, and the preservation of them requires the tertiary constraint

$$\chi_0 = \frac{1}{2}(p - i\kappa\bar{z})(\bar{p} + i\kappa z) \sim 0. \tag{56}$$

These constraint functions form a $SL(2, \mathbb{R})$ algebra with regard to Poisson brackets:

$$\{\chi_n, \chi_m\} = -2i\kappa(n - m)\chi_{n+m}, \qquad (n, m = 0, \pm 1), \tag{57}$$

and generate gauge transformations as argued in the bilocal model.

4.2 1st quantization

Now let us proceed to the quantum theory. We represent the canonical variables as operators on the Hilbert space of differentiable and square integrable functions of z and $\bar{z}$. The state vectors are functions in the Hilbert space. The inner product of two states ϕ_1, ϕ_2 is defined by

$$\langle \phi_1 | \phi_2 \rangle = \int d^D z d^D \bar{z} \, \phi_1^*(z, \bar{z}) \phi_2(z, \bar{z}). \tag{58}$$

A dynamical variable q is replaced by the differential operator $-i\partial/\partial q$. The classical constraint functions are replaced by

$$L_1 = \frac{1}{4\kappa}(-i\partial - i\kappa\bar{z})^2, \tag{59}$$

$$L_{-1} = \frac{1}{4\kappa}(-i\bar{\partial} + i\kappa z)^2, \tag{60}$$

$$L_0 = \frac{1}{4\kappa}(-i\bar{\partial} + i\kappa z)(-i\partial - i\kappa\bar{z}) + \alpha, \tag{61}$$

where $\partial = \partial/\partial z$, and the constant α represents the ambiguity due to the operator ordering. $L_{0,\pm1}$ satisfy the algebra,

$$[L_n, L_m] = (n - m)\left(L_{n+m} - \left(\alpha - \frac{D}{4}\right)\delta_{n+m}\right), \qquad (n, m = 0, \pm1). \tag{62}$$

The expression for the BRST operator is the same as eq.(42), and the requirement of the nilpotency of it is guaranteed by $\alpha = D/4$.

In the classical theory the constraints, $\chi_n = 0, (n = 0, \pm1)$, are imposed for guaranteeing the equivalence of the lagrangian and the hamiltonian formulations [2]. These constraints define the physical subspace of whole phase space. In the quantum theory we cannot regard them neither as operator equations nor as the equations to physical states, $L_{0,\pm1}|\text{phys}\rangle = 0$, since they have no solution. Hence the conditions are relaxed so that a product of the constraint operators has vanishing matrix elements between any physical states, $|\varphi\rangle$ and $|\phi\rangle$:

$$\langle\varphi|L_{n_1}\cdots L_{n_N}|\phi\rangle = 0. \tag{63}$$

This is realized by requiring

$$L_1|\phi\rangle = L_0|\phi\rangle = 0, \tag{64}$$

for physical state $|\phi\rangle$, since we have $\langle\phi|L_{-1} = 0$ by virtue of the Hermiticity, $L_1^\dagger = L_{-1}$, the property lacking in the original bilocal model. The above conditions for physical states are analogous to those of string model, and seems most natural ones.

In order that our model is physically meaningful there should exist the eigenstates of momentum. As is shown shortly this requirement gives rise to restriction on the space time dimension. The conserved quantities derived by the invariance under the space time translations are

$$\tilde{p} = -i\partial + i\kappa\bar{z},$$
$$\bar{\tilde{p}} = -i\bar{\partial} - i\kappa z. \tag{65}$$

Thus the momentum should be combinations of these quantities. From the reality of eigenvalues, it should have the form $P = \beta\tilde{p} + \bar{\beta}\bar{\tilde{p}}$, with arbitrary complex constant β. Any

---

[2] Strictly speaking, only the primary constraints are involved for the equivalence, and the secondary and tertiary constraints are imposed on the *initial conditions* so that one stays on the subspace defined by the primary constraints in later τ.

two of the eigenstates of P would be taken as independent momentum eigenstates. However, we regard one of them as the physical momentum state, since these two operators are not mutually commuting and have not simultaneous eigenvalues. Any choice of β is physically equivalent because it changes by global rotations. Here we choose the momentum of the real part of z as the physical momentum, which corresponds to $\beta = 1$ (see eq.(52)).

Now let us solve the following equations:

$$L_1|k\rangle = L_0|k\rangle = 0, \tag{66}$$

$$P|k\rangle = k|k\rangle. \tag{67}$$

If one puts

$$|k\rangle = e^{-\kappa z\bar{z}}f(z,\bar{z}), \tag{68}$$

the condition $L_1|k\rangle = 0$ reduces to $\partial\bar{\partial}f = 0$, i.e., $f(z,\bar{z})$ is an harmonic function with respect to z. The eigenvalue equation $P|k\rangle = k|k\rangle$ reduces to

$$(\partial + \bar{\partial} - 2\kappa\bar{z} - ik)f(z,\bar{z}) = 0. \tag{69}$$

This equation is of the form with separate variables, and has the solution of the form $g_1(z)g_2(\bar{z})$. The solution is written as

$$f(z,\bar{z}) = e^{ik_1 z + i(k-k_1)\bar{z} + \kappa\bar{z}^2}, \tag{70}$$

with arbitrary separation constant k_1. Multiplying arbitrary function $a(k_1)$ to (70), and integrating over k_1, we obtain the general solution to (69) as

$$f(z,\bar{z}) = e^{ik\bar{z} + \kappa\bar{z}^2}g(y), \qquad y = i(\bar{z} - z), \tag{71}$$

where $g(y)$ is an arbitrary differentiable function of real arguments y's, which can be Fourier expanded. Since $f(z,\bar{z})$ is harmonic with respect to z, $g(y)$ must be an harmonic function.

Finally, the condition $L_0|k\rangle = 0$ reduces to

$$(\bar{\partial}\partial - 2\kappa z\partial - 4\kappa\alpha)f(z,\bar{z}) = 0. \tag{72}$$

Substituting (71) into this, we get

$$\left[\left(y^\mu - \frac{k^\mu}{2\kappa}\right)\frac{\partial}{\partial y^\mu} + 2\alpha\right]g(y) = 0. \tag{73}$$

If we put

$$g(y) = \left[\left(y - \frac{k}{2\kappa}\right)^2\right]^{-\alpha} h\left(y - \frac{k}{2\kappa}\right), \tag{74}$$

eq.(73) and the harmonicity of g are reduced to

$$u^\mu \frac{\partial}{\partial u^\mu} h(u) = 0, \tag{75}$$

$$\left(\Box_u - \frac{K}{u^2}\right) h(u) = 0, \qquad K = 2\alpha(D - 2(\alpha+1)) = \frac{1}{4}D(D-4), \tag{76}$$

where $\Box_u = \partial^2/\partial u^\mu \partial u_\mu$ and u's are D-dimensional real coordinates. These equations are solved by the pseudo-harmonic analysis in D-dimensions. Transforming to the pseudo-polar coordinates [3], $(r, \theta_1, \theta_2, ..., \theta_{D-1})$, we see that from eq.(75) h does not depend on r. Since the d'Alembertian $\Box_u$ is written as $r\partial/\partial r + (1/u^2)\Delta$, where Δ is the Laplace-Beltrami operator on $S^{1,D-2}$, we see from (76) that $\Delta h = Kh$. It is well known in the theory of spherical functions (Takeuchi, 1975) that if the Laplace-Beltrami operator on S^{D-1} have single valued bounded eigenfunctions, then the eigenvalues must be of the form $K = -\ell(\ell + D - 2)$ with non negative integer ℓ. Hence if we take the Eucledian signature for the metric we get $D = 4 - 2\ell$, i.e., $D = 2$ or $D = 4$. In the present case, however, the signature of the metric is Minkowskian, and the base space is $S^{1,D-2}$ which is non-compact. The theory of pseudo-spherical functions on non-compact space (Raczka et al., 1966) (Limi'c et al., 1966) (Limi'c et al., 1967) (Strichartz, 1973) shows variety of series of eigenvalues, including continuous as well as discrete ones. An explicit form of the eigenfunctions are recently obtained for $D = 3$ (Kowalski et al., 2011). The real eigenvalues of single valued eigenfunctions on the non-compact base space are of the same form as those of the compact space except some supplementary continuous series. Here we restrict ourselves to the former cases.

We have assumed here that the eigenfunctions are single valued. If one permits double valued eigenfunctions a half integer value of K should be taken into account. The double values might come from rotations around y^0 axis. Since physical meaning of the rotations around the time axis is not clear, we simply do not consider the effects.

The eigenfunctions are expressed by Gegenbauer's polynomials for general ℓ, but are constants for $\ell = 0$. In the case $D = 2$, eqs.(75) and (76) are directly solved [4], and we get $h(u) = (u^0 \pm u^1)(u^2)^{-1/2}$.

The physical eigenstates of the momentum in four and two dimensions is written as

$$|k\rangle \propto \frac{e^{ik\bar{z} + \kappa\bar{z}(\bar{z}-z)}}{\left(z - \bar{z} - \frac{ik}{2\kappa}\right)^2} \times \begin{cases} 1 & \text{(for } D = 4) \\ z^0 - \bar{z}^0 - \frac{ik^0}{2\kappa} \pm \left(z^1 - \bar{z}^1 - \frac{ik^1}{2\kappa}\right) & \text{(for } D = 2). \end{cases} \tag{77}$$

There are *spurious* states defined by $L^n_{-1}|k\rangle, (n = 1, 2, ..)$, which are orthogonal to all physical states and have zero norm. In the string theory there are many spurious states which are physical and have zero norm, especially in the critical dimension. Existence of these states in the string theory suggests some underlying gauge invariance, since they must be decoupled from physical S-matrix. In the present model, however, spurious states are all unphysical by

---

[3] According to the metric $\eta_{\mu\nu} = (-1, 1, 1, .., 1)$, the pseudo-polar coordinates are defined by $y^0 = r\sinh\theta_1$, $y^1 = r\cosh\theta_1\sin\theta_2 \cdots \sin\theta_{D-1}$, $y^2 = r\cosh\theta_1\sin\theta_2 \cdots \cos\theta_{D-1}$, ..., $y^{D-1} = r\cosh\theta_1\cos\theta_2$.
[4] The solution $D = 2$ was overlooked in ref.(Hori, 2009).

virtue of the constraint algebra without central term, and do not enter in physical S-matrix from the outset.

4.3 Toward a field theory

A field theory based on the complex particle might have some gauge symmetries in the field theoretic sense, which may have some connections with the $SL(2, \mathbb{R})$ in the first quantized theory. The most likely candidate for the action of the field theory may be the Chern-Simons form written, for example, as (Hori, 1993)

$$\mathcal{I} = \int d^3c d^D z d^D \bar{z} \, V(z) \left(A_i \star Q A^i - \frac{g}{3} \epsilon_{ijk} A^i \star A^j \star A^k \right),$$ (78)

where Q is the BRST charge and $\star$ is some associative binary operator like a convolution. $V(z)$ is a possible measure factor. The fields A_i, $(i = 1, 2, 3)$ are fermionic and may be written as $A^i = \sum_n c_n \Psi_n^i$ with ghost variables c_n.

The nomenclature of "Chern-Siomons" comes from the Chern-Simons gauge theory on three-manifold, which has been investigated in connection with knot theory. The formal resemblance of our model to the C-S gauge theory is that the wedge product corresponds to the operator $\star$, which we call star product, and the exterior derivatives correspond to Q, which are both nilpotent. The star product satisfies

$$A \star B(x) = (-1)^{F(A)F(B)} B \star A(x),$$ (79)

where $F(A) = 1$ for fermionic A and $F(A) = 0$ otherwise.

Now the action is invariant under the *gauge* transformations:

$$\delta A^i = Q \Lambda^i + g \epsilon^{ijk} \Lambda_j \star A_k,$$ (80)

where Λ_i is arbitrary bosonic parameters depending on z's and c's. A necessary condition for the invariance is the Leibniz rule for Q expressed as

$$Q(A \star B) = Q(A) \star B + (-1)^{F(A)F(B)} A \star Q(B),$$ (81)

for arbitrary fields A and B. Then the action is invariant if the integral of *total derivative* vanishes:

$$\int d^3 c d^D z d^D \bar{z} \, V(z) \, QA = 0.$$ (82)

Expanding the fields Ψ_n^i in powers of the imaginary parts of z, the coefficients may represent physical fields. After integrations over the imaginary parts of z's and ghost variables, the action is expressed as integral of these fields over the real parts of z's, which has some gauge invariance.

Since there is no guideline for defining the star product apart from the condition (81), let us examine the Leibniz rule in the following simple representation. Consider the representation of $sl(2, \mathbb{R})$, on the space of functions of a single variable x, defined by

$$\lambda_{-1} = x - x_0, \qquad \lambda_0 = (x - x_0)\frac{d}{dx} + a, \qquad \lambda_1 = (x - x_0)\frac{d^2}{dx^2} + 2a\frac{d}{dx}, \tag{83}$$

where x_0 is a constant, and a is a constant which appears due to the ordering ambiguity. λ's satisfy the algebra,

$$[\lambda_n, \lambda_m] = (n - m)\lambda_{n+m}, \qquad (n, m = 0, \pm 1). \tag{84}$$

The natural choice for the (wedge) product of two functions, which permits the Leibniz rule, may be the convolution defined by

$$A \wedge B(x) = \int_{x_0}^{x} dx' \, A(x + x_0 - x')B(x'). \tag{85}$$

The limits in the integration in the definition of the product is so chosen as it is (anti-)commuting:

$$A \wedge B(x) = (-1)^{F(A)F(B)} B \wedge A(x). \tag{86}$$

Also we see from (85),

$$A \wedge B(x_0) = 0. \tag{87}$$

This suggests that the representation space, S, should be restricted to the functions which vanish at $z = z_0$:

$$S = \{ A \mid A \in C^2, A(x_0) = 0 \}. \tag{88}$$

Now let us examine the Leibniz rule for the *exterior derivative* defined by

$$d = \sum_{n=0, \pm 1} c_n \lambda_n - \frac{1}{2} \sum_{n,m=0, \pm 1} (n - m)c_n c_m \frac{\partial}{\partial c_{n+m}}. \tag{89}$$

As in the ordinary exterior derivative, d is nilpotent. After straightforward calculations we obtain

$$d(A \wedge B) - (dA \wedge B + A \wedge dB) = c_1 \left[2(1 - a)A \wedge B' + (2a - 1)A_0 B + AB_0 \right]$$
$$+ (1 - a)c_0 A \wedge B, \tag{90}$$

where $A_0 = A(x_0)$, $B_0 = B(x_0)$, $B' = dB/dx$. Thus we find that if and only if $a = 1$ *and* $A, B \in S$ then d behaves like a derivative operator.

Next let us examine the eigenstate expansions. The basis functions $u_k = (z - z_0)^k, (k = 0, 1, 2, ..)$ satisfy

$$\lambda_0 u_0 = a u_0, \qquad \lambda_1 u_0 = 0, \qquad u_k = \lambda_{-1}^k u_0. \qquad (k = 1, 2, ..) \tag{91}$$

A function in S is expanded as

$$A(x) = \sum_{k=1}^{\infty} \frac{u_k}{k!} A_k. \tag{92}$$

Note that $u_0 = 1$, representing the 'ground state', does not belong to S. The wedge product of A and B in S is expanded as

$$A \wedge B(x) = \sum_{k=1}^{\infty} \frac{u_k}{k!} \sum_{m=0}^{k-1} A_{k-m-1} B_m. \tag{93}$$

The summation in (93) over m is, in fact, carried out from 1 to $k-2$ due to $A_0 = B_0 = 0$. The k-th component of the wedge product is thus

$$(A \wedge B)_k = \sum_{m=1}^{k-2} A_{k-m-1} B_m. \tag{94}$$

Finally, let examine whether an integration of dA vanishes for any A. Since the ghost derivative parts in dA are of the form $c_1 c_{-1} \partial / \partial c_0$, $c_0 (c_1 \partial / \partial c_1 - c_{-1} \partial / \partial c_{-1})$, they vanish after integrations by parts. Thus it is sufficient to check only that $\int dx\, V(x) \lambda_n A = 0$, $(n = 0, \pm 1)$, with some measure factor V. This leads to $V(x) = \delta(x - x_0)$ and $a = 0$. Therefore it is impossible in the present representations to satisfy all the requirements.

Now go back to the complex particle model. Let us define the basis functions v_k as follows:

$$L_0 v_0 = \alpha v_0, \qquad L_1 v_0 = 0, \qquad v_k = L_{-1}^k v_0. \qquad (k = 1, 2, ..) \tag{95}$$

where

$$v_0 = e^{(ip + \kappa(\bar{z} - z))\bar{z}}, \qquad \alpha = \frac{D}{4}. \tag{96}$$

v_0 is the eigenstate of the momentum with eigenvalue p, but not a physical state, since $L_0 v_0 \neq 0$. The basis $v_k(p, z, \bar{z})$, $(k = 0, 1, 2, ...)$ may span a dense subset of functions which are differentiable and square integrable. We consider fields which are expanded as

$$A = \sum_{k=0}^{\infty} \frac{v_k}{k!} A_k, \tag{97}$$

where A_k are functions of the ghost variables and not depend on z's . The each component of a field A is denoted as A_k. In analogy with (94) let us define the star product as

$$(A \star B)_k = \sum_{m=0}^{k+\delta} \left(A_{k-m+\beta} B_{m+\gamma} + A_{m+\gamma} B_{k-m+\beta} \right), \tag{98}$$

where integer constants, β, γ and δ, are introduced so that the Leibniz rule might be satisfied. The star product satisfies the (anti-)symmetry, $A \star B = (-1)^{F(A)F(B)} B \star A$, and the associativity, $(A \star B) \star C = A \star (B \star C)$.

The Leibniz rule can be examined merely using the commutation relations (62) with $\alpha = D/4$, and it is sufficient to check for the first term in the BRST charge, since the second terms are in the form of derivatives. For an operator O writing as

$$\text{Leib}[O; A, B] = (OA) \star B + A \star (OB) - O(A \star B), \tag{99}$$

we get

$$\text{Leib}[L_0; A, B]_k = (\beta + \gamma + \alpha) \sum_{m=0}^{k+\delta} (A_{k-m+\beta} B_{m+\gamma} + A_{k-m+\gamma} B_{m+\beta}), \tag{100}$$

$$\text{Leib}[L_{-1}; A, B]_k = (\beta + \gamma + 1) \sum_{m=0}^{k+\delta} (A_{k-m+\beta} B_{m+\gamma-1} + A_{m+\gamma-1} B_{k-m+\beta})$$
$$-(1 + \beta)(A_{k+\beta} B_{\gamma-1} + A_{\gamma-1} B_{k+\beta}), \tag{101}$$

$$\text{Leib}[L_1; A, B]_k = (\beta + \gamma + 2\alpha - 1) \sum_{m=0}^{k+\delta} (A_{k-m+\beta} B_{m+\gamma+1} + A_{m+\gamma+1} B_{k-m+\beta})$$
$$+(A_{k+\beta+1} B_{\gamma} + A_{\gamma} B_{k+\beta+1})$$
$$-(\beta + 2\alpha - 1 - \delta)(A_{-\alpha+\beta} B_{k+\alpha+\gamma+1} + A_{k+\alpha+\gamma+1} B_{-\alpha+\beta}). \tag{102}$$

The bulk parts (the summations) of these quantities vanish if we put

$$\beta + \gamma + \alpha = 0, \tag{103}$$
$$\beta + \gamma + 1 = 0, \tag{104}$$
$$\beta + \gamma + 2\alpha - 1 = 0, \tag{105}$$

which are equivalent to

$$(D/4 =)\alpha = 1, \qquad \beta + \gamma = -1. \tag{106}$$

The marginal parts (single terms) vanish if

$$\beta = -1, \qquad \gamma = 0, \qquad \delta = 0, \tag{107}$$

and $A_0 = B_0 = 0$. Thus we see that the Leibniz rule for the BRST charge is valid only if $D = 4$ and restricting the function space to

$$S = \{A | A_0 = 0\}. \tag{108}$$

The star product should be

$$(A \star B)_k = \sum_{m=1}^{k} (A_{k-m-1} B_m + A_m B_{k-m-1}). \tag{109}$$

Note $(A \star B)_k = 0$ for $k \leq 2$. The reason for restriction to $D = 4$ seems a technical one in building a field theory, while the restriction to $D = 2$ or $D = 4$ in the first quantized theory is intrinsic in the model.

Finally let us examine the vanishing of the integral of the *total derivatives* of the form QA. As in the simple representation (83), it is sufficient to check $\int V L_n A = 0, (n = 0, \pm 1)$. Integrating by parts, the conditions become $K_n V = 0$, $(n = 0, \pm 1)$, where

$$K_1 = \frac{1}{4\kappa}(i\partial - i\kappa\bar{z})^2, \tag{110}$$

$$K_{-1} = \frac{1}{4\kappa}(i\bar{\partial} + i\kappa z)^2, \tag{111}$$

$$K_0 = \frac{1}{4\kappa}(i\partial - i\kappa\bar{z})(i\bar{\partial} + i\kappa z) + \alpha, \qquad \alpha = \frac{D}{4}. \tag{112}$$

Now let us find the explicit form of $V(z, \bar{z})$. Putting

$$V(z, \bar{z}) = e^{i\kappa\bar{z}y}G(y), \qquad y = i(\bar{z} - z), \tag{113}$$

we find after straightforward calculations that the conditions are

$$K_1 V(z) = -\frac{1}{4\kappa}e^{i\kappa\bar{z}y}\Box_y G(y) = 0, \tag{114}$$

$$K_{-1} V(z) = e^{i\kappa\bar{z}y}\left(y\partial_y + \frac{1}{2}(y^2 + D)\right)G(y) = 0, \tag{115}$$

$$K_0 V(z) = -\frac{1}{2}e^{i\kappa\bar{z}y}\left(y\partial_y + D - 2\alpha\right)G(y) = 0. \tag{116}$$

From the last two equations we see $y^2 G(y) = 0$, so we find $G(y) \propto (y^2)$. Thus the solution must be

$$V(z, \bar{z}) = e^{i\kappa\bar{z}y}\delta(y^2). \tag{117}$$

Substituting this back into (114)-(116), we get

$$K_1 V(z) = -\frac{1}{2\kappa}e^{i\kappa\bar{z}y}\left[(D - 4)\delta'(y^2) + 2(2\delta'(y^2) + y^2\delta''(y^2))\right] \tag{118}$$

$$K_{-1} V(z) = \frac{1}{2}e^{i\kappa\bar{z}y}\left[(D - 4)\delta(y^2) + y^2\delta(y^2) + 4(\delta(y^2) + y^2\delta'(y^2))\right] \tag{119}$$

$$K_0 V(z) = -\frac{1}{4}e^{i\kappa\bar{z}y}\left[(D - 2 - 2\alpha)\delta(y^2) + 4(\delta(y^2) + y^2\delta'(y^2))\right]. \tag{120}$$

From the identity $x\delta(x) = 0$, we see $\delta(x) + x\delta'(x) = 0$ and $2\delta'(x) + x\delta''(x) = 0$. Hence we see that $K_n V = 0$, $(n = 0, \pm 1)$ if and only if $D = 4$. Once again $D = 4$ makes us happy!

The action of the field theory is an integral over z and z, where the imaginary part of z's are restricted on the light-cone.

5. Extension to $N \geq 3$

5.1 Actions and gauge invariance

The action of the N-extended multi-local particle is defined by

$$I_N = \int d\tau\, L_N, \qquad L_N = \sum_{a=1}^{N}\frac{1}{2g_a}\dot{x}_a^2 + \sum_{a,b}\kappa_{ab}\dot{x}_a x_b, \tag{121}$$

where x_a are the (real) coordinates of the N particles and each g_a is the einbein of the a-th world line which is parametrized by τ, and dots denote the derivatives with respect to internal time τ. The difference of this N particle system from the ordinary free particles comes from the second term in eq.(121), where κ_{ab} is an arbitrary anti-symmetric constant matrix.

The action has the hidden local symmetry generated by

$$\delta x_a = \epsilon_a \dot{x}_a + \sum_b s_{ab} \frac{\dot{x}_b}{g_b}, \tag{122}$$

$$\delta g_a = \frac{d}{d\tau}(\epsilon_a g_a) + k_a g_a, \tag{123}$$

where ϵ_a, s_{ab} and k_a are infinitesimal local parameters constrained by $s_{ab} = s_{ba}$ and

$$\dot{s}_{ab} + 2\kappa_{ab} g_a g_b (\epsilon_b - \epsilon_a) + 2\sum_c (\kappa_{ac} g_a s_{cb} + \kappa_{bc} g_b s_{ca}) = g_a k_a \delta_{ab}, \tag{124}$$

In fact the variation of the lagrangian is

$$\delta L_N = \frac{d}{d\tau}\left[\sum_a \frac{\epsilon_a \dot{x}_a^2}{2g_a} + \sum_{a,b}(\kappa_{ab}\epsilon_a x_b \dot{x}_a + \frac{s_{ab}\dot{x}_a\dot{x}_b}{2g_a g_b}) + \sum_{a,b,c}\frac{\kappa_{ab}s_{ac}x_b\dot{x}_c}{g_c}\right]$$

$$+\frac{1}{2}\sum_{ab}(s_{ab} - s_{ba})\frac{\dot{x}_a}{g_a}\frac{d}{d\tau}\left(\frac{\dot{x}_b}{g_b}\right)$$

$$+\sum_{a,b}\left[\frac{1}{2}\dot{s}_{ab} + 2\kappa_{ab}g_a g_b \epsilon_b + g_a\left(2\sum_c \kappa_{ac}s_{cb} - \frac{1}{2}k_a\delta_{ab}\right)\right]\frac{\dot{x}_a\dot{x}_b}{g_a g_b}. \tag{125}$$

In order to fix the model we set the non-vanishing components of the anti-symmetric parameter κ_{ab} as in the following two cases:

(i) Closed N-particle ($N \geq 3$):

$$\kappa_{aa+1} = \kappa, \quad (a = 1, ..., N-1), \quad \kappa_{N1} = -\kappa, \quad \kappa_{ab} = -\kappa_{ba}, \tag{126}$$

(ii) Open N-particle ($N \geq 2$):

$$\kappa_{aa+1} = \kappa, \quad (a = 1, ..., N-1), \quad \kappa_{ab} = -\kappa_{ba}, \tag{127}$$

and other κ's are set to zero, where κ is the coupling constant. The closed N-particle system is characterized by the anti-symmetric matrix κ_{ab}, each row (or column) of which has two non vanishing elements, while in the open N-particle system this is valid except for the first (or N-th) row (or column) corresponding to the two ends of the N particles. The bilocal particle is the open 2-particle. In what follows we restrict ourselves to the open N-particle, since the constraint structures in the canonical theories of the closed N-particle are rather complicated compared with the open ones.

Now the number of the gauge degrees of freedom can be counted in the similar way as in the bilocal particle, where the degrees of freedom of the initial condition are counted in suitable gauge condition. But this procedure is rather cumbersome in the lagrangian formalism compared with the hamiltonian one.

However, a shortcut derivation of the physical degrees of freedom in the lagrangian formalism is possible. The result coincides precisely with the hamiltonian one, if Dirac's conjecture holds. The reasoning is as follows. The number of the unphysical, *i.e.*, the gauge degrees of freedom is the number of the independent parameters and their time derivatives appeared in the transformation rules, where a parameter and all of its (higher order) time derivative(s) are formally regarded as *independent*. (For a skeptical reader we recommend to check the above rule in the case of the Yang-Mills or the local Lorentz symmetries.)

The counting argument in the open N-particle system is as follows. The independent transformation parameters are extracted by solving the constraint, eq.(124). For $a = b$ in eq.(124), we obtain k_a in terms of $\dot{s}_{ab}$ and s_{ab}. For $b = a + 1$, we see $\epsilon_a (a = 2, ..., N)$ are expressed in terms of $\epsilon_1, \dot{s}_{aa+1}, (a = 1, ..., N-1)$ and s_{ab}. Next, for $b \geq a + 2$, we see $\dot{s}_{ab}$ are expressed by s_{ab}. Thus we have the independent parameters, $s_{ab}(\frac{N(N+1)}{2}), \dot{s}_{aa}(N), \dot{s}_{aa+1}(N - 1)$ and $\epsilon_1(1)$, where the numbers of each independent parameter are written in the parentheses.

Substituting the above parameters into eqs.(122) and (123), we get the extra independent parameters, $\ddot{s}_{aa+1}(N - 1)$ and $\dot{\epsilon}_1(1)$. Thus we have the total of $\frac{1}{2}N(N + 1) + 3N$ independent parameters in eqs.(122) and (123). However, a short manipulation shows that s_{aa} and $\dot{s}_{aa}$ actually do not appear or be absorbed into ϵ_a by shifting $\epsilon_a \to \epsilon_a + \frac{s_{aa}}{g_a}$. Hence, finally, we see the number of the gauge degrees of freedom is $\frac{1}{2}N(N + 1) + N$. Among them N degrees of freedom are used for fixing g_a, and the remaining $\frac{1}{2}N(N + 1)$ are of our interest. $\frac{1}{2}N(N + 1)$ constraints and the same number of gauge fixing conditions eliminate a part of the canonical variables, $x_a, p_a, (a = 1, 2, .., N)$, leaving $\frac{1}{2}N(2D - N - 1)$ canonical pairs as physical. Hence if $N \leq 2(D - 1)$ there are at least one physical degrees of freedom.

In the next subsection we show that the number of the first class constraints in the hamiltonian formalism coincides precisely with the above number. This is in accordance with Dirac's conjecture, *i.e.*, all of the first class constraints generate the gauge symmetry of the system.

5.2 Canonical theory

The algebraic structure of the symmetry is clarified in the canonical formalism. Introducing the momenta p_a and π_a conjugate to x_a and g_a, respectively, and defining

$$V_{ab} = \frac{1}{2} p_a^{(-)} p_b^{(-)}, \tag{128}$$

$$p_a^{(-)} = p_a - \sum_b \kappa_{ab} x_b, \tag{129}$$

we can express the total hamiltonian as

$$H_T^{(N)} = \sum_a g_a V_{aa} + \sum_a \Lambda_a \pi_a, \tag{130}$$

where Λ's are Dirac variables which can be set to arbitrary functions of canonical variables. The Poisson brackets of V's are given by

$$\{V_{ab}, V_{cd}\} = \kappa_{c(a}V_{b)d} + \kappa_{d(a}V_{b)c}. \tag{131}$$

Now let us derive the constraints for the canonical variables in the open N-particle system. The primary constraints are $\pi_a \sim 0$, since the lagrangian does not contain $\dot{g}$'s. The stability of the primary constraints along the time development requires the secondary constraints $V_{aa} \sim 0$. The stability of the latter, in turn, requires $V_{aa+1} \sim 0, (a = 1, ..., N - 1)$. In general, the stability of $V_{aa+k} \sim 0$ requires $V_{aa+k+1} \sim 0$. After all we have $\frac{1}{2}N(N + 1)$ secondary and tertiary constraints, V_{ab}, which close under the Poison brackets, and form the first class constraints.

V_{ab} generate the gauge symmetry which has the form of eqs.(122) and (123) in the lagrangian formalism, and transform the hamiltonian, eq.(130), into the same form but with different coefficients of V_{aa}. This ambiguity of the coefficients is a reflection of the gauge invariance and is removed by the gauge fixing.

5.3 Quantization

In order to quantize the system we replace p_a^{μ}'s by $-i\partial_{\mu a}$, and we denote the quantum operators obtained by this replacement by writing hats on these quantities. The generators of the gauge transformations are defined by

$$\hat{V}_{ab} = \frac{1}{4}\left(\hat{p}_a^{(-)}\hat{p}_b^{(-)} + \hat{p}_b^{(-)}\hat{p}_a^{(-)}\right). \tag{132}$$

The gauge algebra is expressed as
$$[\hat{V}_{ab}, \hat{V}_{cd}] = i\kappa_{c(a}\hat{V}_{b)d} + i\kappa_{d(a}\hat{V}_{b)c}. \tag{133}$$

The ambiguities from the operator ordering are fixed by requiring the nilpotency of the BRST operator as in the $N = 2$ theories, then the central terms in the gauge algebra also vanish.

The generators for the kinematic symmetry are as follows:

translations : $\qquad \hat{p}_a^{(+)} = \hat{p}_a + \sum_b \kappa_{ab}x_b, \quad (a = 1, ..., N) \tag{134}$

Lorentz tfm. : $\qquad M_{\mu\nu} = \sum_{a=1}^{N} \hat{p}_{a[\mu}x_{a\nu]} \tag{135}$

where $\hat{p}_{a\mu} \equiv -i\partial_{a\mu}$. $\hat{p}_{+a}^{(+)}$ generate the translation of a-th particle. They form the following algebra with a central term:

$$[M_{\mu\nu}, M_{\lambda\rho}] = i\eta_{\rho[\mu}M_{\nu]\lambda} - i\eta_{\lambda[\mu}M_{\nu]\rho}, \tag{136}$$

$$[M_{\mu\nu}, \hat{p}_{a\lambda}^{(+)}] = -i\eta_{\lambda[\mu}\hat{p}_{\nu]a}^{(+)}, \tag{137}$$

$$[\hat{p}_{a\mu}^{(+)}, \hat{p}_{b\nu}^{(+)}] = 2i\kappa_{ab}. \tag{138}$$

The algebra defined above contains the Poincaré algebra as a subalgebra. The crucial point is the uncertainty relation (138). The momentum of each particle does not have a certain value irrespective to the momentum of the neighboring particles. Another important feature is the commutativity of the kinematic generators and those of gauge generators:

$$[\hat{V}_{ab}, \hat{p}_{\mu a}^{(+)}] = [\hat{V}_{ab}, M_{\mu\nu}] = 0. \tag{139}$$

These relations assure the consistency of the gauge structure and the kinematic properties of the model.

The first quantizations of the N-extended models may be achieved in the similar way as that of the bilocal models. Chern-Simons type actions may be used in field theories. It is interesting to know whether the critical dimensions exist also in the N-extended models. However, there might be similar difficulties as in the bilocal model, and they may be overcome by improving them to those like complex particle as is done in the bilocal model. We leave these problems to future studies.

6. Summary

In the present paper we have analyzed the multi-local particle models especially emphasizing on the complex particle. At first sight the guage degrees of freedom of the multi-local particle are less than those of the canonical theory, which may lead to breakdown of Dirac's conjecture. The concept of physical equivalence is argued to be modified so that the guage transformations are extend to whole algebra, recovering Dirac's conjecture.

The constraint structure of the model of the complex particle is suited for the ordinary quantization scheme as opposed to the original bilocal model, due to the Hermiticity property of $L_{\pm 1}$. In the first quantization we see that, requiring the existence of the momentum eigenstates which satisfy the physical state conditions, the dimension of the spacetime is restricted to be two or four. The most natural action of the field theory might be of the form of Chern-Simons one, where the exterior derivative is replaced by the BRST charge. It is rather unexpected that the action has gauge invariance only in the four dimensions. This fact is caused by the Leibniz rule of the BRST charge and the vanishing of the total derivative, *i.e.*, $\int QA = 0$, which are satisfied only in four dimensions.

Although the complex particle model is favorable in many respects than the original bilocal model, the latter is more intuitive in that the classical solution is interpreted as a rigid stick. As far as we know the bilocal model is the first example of relativistically admissible rigid stick.

We extend the bilocal model to $N \geq 3$ particle system, and obtain large classes of actions. The larger the guage algebra, the less physical degrees of freedom. The models categorized into two classes, *i.e.*, open and closed types. In open N-particle system it turns out that the number of the constraints and the corresponding gauge symmetry is $\frac{1}{2}N(N+1)$. Consequently the physical degrees of freedom survives only if $N \leq 2(D-1)$.

The models proposed here have not been aimed so far to phenomenological applications but to the analysis in their theoretical aspects such as the gauge invariance or critical dimensions.

Of course we do not intend to claim that the present model is the theory of the nature and for that reason the dimension of our spacetime should be four. However, it is interesting that there exist simple models other than the string, which have critical dimensions. We hope that the future investigations along with the direction described here may open a new perspective in the area of quantum gravity where some non-locality of a fundamental object should play the central role in the Planck scale.

7. References

Dirac, P. (1950). Generalized hamiltonian dynamics, *Can. J. Math.* 2.: 129.

Dirac, P. (1964). *Lectures on Quantum Mechanics*, Dover Pub., Mineola, New York.

Frenkel, A. (1982). Comment on cawley's counterexample to a conjecture of dirac, *Phys. Rev.* D21: 2986.

Hori, T. (1992). Hidden symmetry of relativistic particles, *J. Phys. Soc. Jap* 61.: 744.

Hori, T. (1993). Bilocal field theory in four dimensions, *Phys. Rev.* D48.: R444.

Hori, T. (1996). Brs cohomology of a bilocal model, *Prog. Theor. Phys.* 95.: 803.

Hori, T. (2009). Relativistic particle in complex spacetime, *Prog. Theor. Phys.* 122.(2.): 323–337.

Kowalski, K., Rembieli'nski, J. & Szcze'sniak, A. (2011). Pseudospherical functions on a hyperboloid of one sheet, (arXiv:1104.3715v1).

Limi'c, N., Niederle, J. & Raczka, R. (1966). Continuous degenerate representation of noncompact rotation grouts. ii, *J. Math. Phys.* 7: 2026.

Limi'c, N., Niederle, J. & Raczka, R. (1967). Eigenfunction expansions associated with the second order invariant operator on hyperboloids and cones. iii, *J. Math. Phys.* 8: 1079.

Raczka, R., Limi'c, N. & Niederle, J. (1966). Discrete degenerate representation of noncompact rotation grouts. i, *J. Math. Phys.* 7: 1861.

Strichartz, R. S. (1973). Harmonic analysis on hyperboloids, 12: 341.

Sugano, R. & Kamo, H. (1982). Poincare-cartan invariant form and dynamical systems with constraints, *Prog. Theor. Phys.* 67: 1966.

Takeuchi, M. (1975). *Modern Spherical Functions*, Iwanami Shoten, Tokyo, Japan.

Witten, E. (1986). Noncommutative geometry and string field theory, *Nucl.Phys.* B268: 253.

2

Quantum Theory as Universal Theory of Structures – Essentially from Cosmos to Consciousness

T. Görnitz

FB Physik, J. W. Goethe-Universität Frankfurt/Main
Germany

1. Introduction

Quantum theory is the most successful physical theory ever. About one third of the gross national product in the developed countries results from its applications. These applications range from nuclear power to most of the high-tech tools for computing, laser, solar cells and so on. No limit for its range of validity has been found up to now.

Quantum theory has a clear mathematical structure, so a physics student can learn it in a short time. However, very often quantum theory is considered as being "crazy" or "not understandable". Such a stance appears reasonable as long as quantum theory is seen primarily as a theory of small particles and the forces between them. However, if quantum theory is understood more deeply, namely as a general theory of structures, not only the range is widely expanded, but it also becomes more comprehensible (T. Görnitz, 1999).

Quantum structures can be material, like atoms, electrons and so on. They can also be energetic, like photons, and finally they can be mere structures, such as quantum bits. Keeping this in mind it becomes apprehensible that quantum theory has two not easily reconcilable aspects: on the one hand, it possesses a clear mathematical structure, on the other hand, it accounts for *well-known experiences of everyday life*: e.g. a whole is often more than the sum of its parts, and not only the facts but also the possibilities can be effective.

If henadic and future structures (i.e. structures which are related to unity [Greek "hen"] and to future) become important in scientific analysis, then the viewable facts in real life differ from the calculated results of models of classical physics, which suppose elementary distinctions between matter and motion, material and force, localization and extension, fullness and emptiness and which describe any process as a succession of facts. From quantum theory one can learn two elementary insights:

1. *Not only facts but also possibilities can influence the way in which material objects behave.*
2. *The elementary distinctions made in classical models are often useful, but not fundamental.*

Quantum theory shows that there are equivalences between the concepts of matter and motion, material and force, localization and extension, fullness and emptiness, and so on, and these equivalences can be reduced to one fundamental equivalence: the equivalence of matter, energy and abstract quantum information.

2. How to understand the laws of nature?

Mankind searched for laws of nature to be braced for future events and to react on them. A rule and even more a law is only reasonable for a multitude of equal events. For a singular and therefore unique event the idea of a rule is meaningless because of the lack of a recurrence. The required equality for applying rules or laws will be achieved by ignoring differences between distinct events. Therefore, as a matter of principle, laws of nature are always approximations, eventually very good approximations at the present time. If a law of nature is expressed in the form of a mathematical structure - which is always the case in physics - then this structure may conceal the approximate character of the law. This can lead to confusion about the interpretation of some laws and their correlations. One should keep this in mind when interpretational questions of quantum effects are to be deliberated.

3. What is the central structure for an understanding of quantum theory?

To understand the central structure of quantum theory one has to inspect how composite systems are formed.

In classical physics the composition of a many-body system is made in an additive way. The state space of the composed system is the direct sum of the state spaces of the single particle systems. This results in a "Lego world view" of smallest building blocks – of one or another kind of "atoms". In this view the world has to be decomposed into ultimately elementary objects – which never change - and the forces between such objects. This picture about the structure of reality was generally accepted for more than two millennia.

Composite systems in quantum physics are constructed in a fundamentally different way. The state space of a composed quantum system is the tensor product of the state spaces of the single particle systems. *To explain quantum theory we have to start with this structural difference.* However, "tensor product" is a very technical concept. Is it possible to relate this concept to something familiar in every day live?

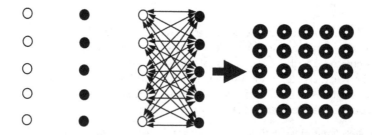

Fig. 1. The additive composition of two objects in classical physics, the states of the parts are outlined by white and black circles. The relational composition in quantum physics, marked by arrows, create the new states of the composed object. They are neither black nor white.

Let us recall that "relations" create a product structure. One can say that the new states of a composed object are the relational structures between the states of its parts. Therefore *quantum theory can be characterized as the physics of relations;* it can be seen as a clear mathematical implementation of a familiar experience of life: A whole is often more than the sum of its parts.

Up to now this central aspect of quantum theory is often misunderstood. In physics, and also in the philosophy of sciences, one speaks, for example, of "particles", i.e. electrons and so on, in a weakly bound system. This is certainly useful from a the practical point of view, but does not apply to the basic issue. In principle, two interacting electrons are "one object with charge -2" – and not two independently existing particles.

Relational structures create networks, in the essence they are plurivalent. In such a network many different connections between two outcomes are possible. This leads us to a further characterization of quantum theory, namely, *quantum theory as "the physics of possibilities".*

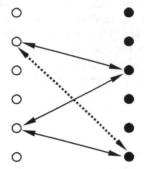

Fig. 2. Relations are not unique, they constitute possibilities.

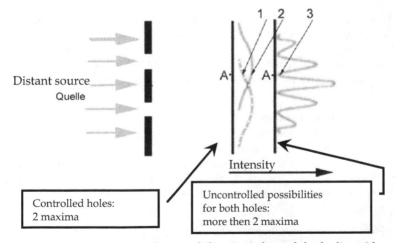

Fig. 3. When quantum particles have the possibility to go through both slits without controlling their passage, then one will find more then two maxima on a screen behind the slits. Position "A" can be reached if one of the holes is open, but no longer if both are open and not controlled.

In our daily life we are influenced not only by facts of the past but also by future possibilities, which we anticipate, wish for, or are afraid of. Quantum theory deals with possibilities only. We have to understand that also possibilities can have an impact – not only facts.

If in a double slit experiment quantum particles have the possibility of going through both slits without controlling their passage, will find more than two intensity maxima on a screen behind the slits. If the passage through the slits is controlled, which means that the passage through one of the slits becomes a fact, then only two maxima will result. This is comparable to experience of our everyday life: control restricts possibilities and thereby influences human behavior.

4. The indissoluble relation between classical physics and quantum physics: The dynamic layering-process

There are some popular but insufficient ideas about the distinctions of quantum physics and classical physics. One misleading distinction concerns the scope of application to microphysics and macrophysics, respectively. It is true that in microphysics only quantum theory is applicable; nevertheless there are also many macroscopic quantum phenomena. Another topic is the difference between continuous and discontinuous effects, the former being attributed to classical physics. However, it should be recalled that many operators in quantum physics have continuous spectra. Popular but false is also the distinction between a "fuzzy" quantum theory and a "sharp" classical physics. It ignores that quantum theory provides for the most accurate description of nature we ever had. Classical physics nourishes the illusion of exactness. Its mathematical structure is based on the assumption of "arbitrarily smooth changes" of any variable. While this is a precondition for calculus, it is by no means always afforded by nature. At very high precision the quantum structure will become important anyway, as *quantum physics is the physics of preciseness*.

Often there is no need for the precision of quantum theory. At first sight most of the processes in nature appear to be smooth. However, upon closer inspection, all actions are quantized, they appear in discrete "numbers" or "steps". One may say that, strictly speaking, all changes are quantum jumps. So a quantum jump is the smallest non-zero change in nature - which may explain why this concept is so attractive in politics and economics.

Since the early days of quantum mechanics Bohr has insisted that classical physics is a precondition for speaking about quantum results. It is impossible to ignore that for humans; there are not only possibilities but also facts. For an adequate description of nature we need both parts of physics, classical and quantum physics. Its connection can be described as a "dynamic layering-process". The classical limit transforms a quantum theoretical description into a classical one, the process of quantization converts classical physics into quantum physics.

It seems evident that quantum theory is the foundation of classical physics. The existence of all the objects handled so successfully by classical physics can only be understood adopting quantum theory. It may be recalled that the existence of atoms, having opposite charges inside, is forbidden by classical electrodynamics. On the other hand, classical physics is a precondition for the appearance of quantum properties. The quantum properties of a system become visible only if its entanglement with the environment is cut off. Such a cut can be modeled mathematically only in classical physics.

The laws of classical physics ignore the relational aspects of nature. While thus being inferior to the quantum laws, they are potentially much easier to apply. For large objects, the relational aspects are very small, so that often there is no need to employ a quantum description.

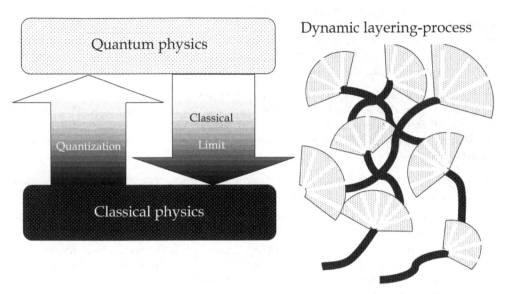

Fig. 4. Dynamic layering process between classical and quantum physics.

5. The meaning of quantization

Concerning quantization many concepts have been proposed (a good overview can be found in Ali & Engliš, 2005), and one may wonder whether here a simple fundamental structure can possibly be established.

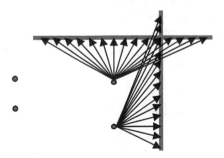

Fig. 5. Quantization of a bit: From the two states {0,1} of a bit to a two-dimensional complex state space of a qubit.

Ignoring for the moment the canonical quantization, a general structure can be inferred from the quantization of a bit with its two states {0,1} to a qubit: here quantization is obtained by constructing all complex-valued functions on the set of two points, resulting in a two-dimensional complex space C^2.

In a related way, path-integral quantization can be interpreted as constructing all functions on the set of the classical pathways. At first glance, second quantization seems to be different. However, the construction of a state of the quantum field in terms of states of quantum particles is analogous to the construction of an analytical function in terms of powers of the variable. The analytical functions are dense, e.g., in the set of continuous or measurable functions, and even distributions conceived as limits of analytical functions. So, in a certain sense, the analytical functions represent "all functions", and a quantum field can be interpreted as "the functions on the set of quantum particles".

In conclusion, we may say: *Quantization is the transition from the manifold of the facts to the possibilities over the facts* – where the possibilities are given in the form of functions on the manifold of facts. It even seems possible to state: "*Quantization is (actually) second quantization.*" In this sense Einstein's invention of photons was the first demonstration of quantization.

How can the canonical quantization of classical mechanics, characterized by a bisection ("polarization") of variables, be interpreted in the present context?

The explanation can be given as follows: Mechanics is the classical limit of quantum mechanics in that all operators commute. In quantum mechanics the position operator acts multiplicatively on wave functions over position space, while the momentum operator acts multiplicatively on wave functions over momentum space. A duplication replacing $\Phi(x)$ or $\Psi(p)$ by $R[x,p]$ allows for the commutation of position and momentum. Accordingly, in classical mechanics positions and momenta are the fundamental variables; the polarization reverses this duplication.

The essence of quantization can summarized in the sentence: Quantization is the transition from the facts to the relational network of possibilities associated with the facts - mathematically represented by a linear space of functions defined on the set of the facts.

A further characterization is as follows: The quantization of a system is the transition from a nonlinear description in a low-dimensional space, where the system may have many or infinitely many degrees of freedom (e.g. classical mechanics or electromagnetic fields), to a linear description of many or infinitely many systems with few degrees of freedom in an infinite-dimensional space (e.g. quantum bits, photons, or other field quanta). This reminds of the exponential map and its conversion of products, being nonlinear, into sums, which are linear.

6. Quantum theory relativizes distinctions

Quantum theory is consistent with everyday experience indeed, but in non-living nature quantum effects become essential only at a high precision scale. At high precision, though, effects may appear that are not so evident from the everyday experience in the world around us.

Already in school the so-called wave-particle-duality is a subject. According to quantum theory, one and the same quantum object can act, depending on the circumstances, more like a wave or more like a particle, that is, as a more extended or more localized object. As we have discussed, quantum states can be understood as extended functions on facts, and it is thus an essential non-local theory. We may say: *Quantum physics is the physics of non-locality.* A strict distinction between locality and non-locality is relativized by quantum theory.

Quantum theory demonstrates that transformations between matter and motion or between force and material are possible. Of course, Einstein's famous formula $E = mc^2$ was found in special relativity, but in any related experiment antimatter is involved. This genuine quantum concept shows that the transformation between matter and motion is an effect of quantum theory.

Motion is often declared as a *property* of matter, but *quantum theory shows that matter and motion are equivalent.* This happens always in the large accelerators, but it is also related to the central philosophical aspect of second quantization: The distinction between object and attribute depends on the context. One and the same quantum particle is an object in quantum mechanics and is an attribute of a quantum field. Therefore we can state that *quantum theory discloses an equivalence between objects and attributes.* That quantum theory has relativized the distinctions between objects, structures and attributes (or tropes as some philosophers say) is also of philosophical relevance.

Matter is visible and inert, forces are invisible and not impenetrable. From the quantum point of view, however, the distinction of force and matter reduces to the difference between quanta of integer or half-integer spin. In the large accelerators, those quanta are transformed among each other. So *quantum theory unveils an equivalence between forces and matter.*

The model of the Dirac sea shows up that even *the distinction between emptiness and plenitude is relativized by quantum theory.*

7. The quantum theoretical equivalence of matter, energy and quantum information

In addition to what was discussed above, quantum theory allows for a completely new perspective on the three entities matter, energy and quantum information. Already since 1955 C. F. v. Weizsäcker has speculated on the possibility of founding physics on quantum information. His "Ur-Theory" grows up from the intention "Physics is an extension of logics" (Weizsäcker, 1958, p. 357). As the basis for the envisaged reduction he has proposed quantized binary alternatives referred to as "Ur-Alternativen" or urs. Werner Heisenberg wrote about Weizsäcker's concept "... that the realization of this program requires thinking at such a high degree of abstraction that up to now – at least in physics – has never happened." For him, Heisenberg, "it would be too difficult", but v. Weizsäcker and his coworkers should definitely carry on. (Heisenberg, 1969, p. 332) For a long time, however, v. Weizsäcker's project was hardly appreciated, and one may wonder about the lack of recognition.

One reason may be that the concept was far too abstract. Moreover, there were almost no relation to experimental evidence. At that time, the quantities v. Weizsäcker proposed were beyond the imagination of the physicists. That one proton is made up of 10^{40} qubits is a hard sell in physics even today. Another serious problem was that v. Weizsäcker's models were inconsistent with general relativity at that time.

As essential step forward, it proved necessary to go beyond the urs. v. Weizsäcker (1982, p. 172) proposes „An »absolute« value of information is meaningless". But this is a contradiction to his claim (1971, p. 361): "Matter is information". Matter has an absolute value, as zero grams of matter is a clearly defined quantity. Therefore, with regard to *an equivalence of matter and information, the latter must have an absolute value* as well. So there was the need to extend the concept of »information« to one which is "absolute". At that absolute level, one must do without reference to an "emitter" or "receiver", and – even more important - dispense with the concepts of meaning or knowledge, at least initially. This is the basic precondition for establishing the equivalence of matter and information.

Here it proved necessary to make a connection to modern theoretical und empirical structures of physics, especially Bekenstein's and Hawking's entropy of black holes, and a rational cosmology. Physics is more than an »extension of logics«, and, in physics information differs from destination, or meaning, or knowledge. Meaning always has a subjective aspect too, so meaning cannot be a basis for science and objectivity.

If quantum information is to become the basis for science it must be conceived as absolute quantum information, free of meaning. It is denominated as "Protyposis" to avoid the connotation of information and meaning. Protyposis enables a fundamentally new understanding of matter which can seen as "formed", "condensed" or "designed" abstract quantum information. Absolute quantum information provides a base for a new understanding of the world ranging from matter to consciousness. Protyposis adds to E=mc², that is, the equivalence of matter and energy, a further formula (Görnitz, T 1988², Görnitz, T., Görnitz, B. 2008) :

$$N = m\, c^2\, t_{cosmos}\, 6\pi/\hbar \tag{1}$$

A mass m or an energy mc^2 is equivalent to a number N of qubits. The proportionality factor contains t_{cosmos}, the age of the universe. Today a proton is 10^{41} qubits. A hypothetical black hole with the mass of the universe would have an entropy of order 10^{123}. If a particle is added, the entropy of the black hole increases proportionally to the mass-energy of the particle. If a single proton is added to the cosmic mass black hole, the entropy will rise by 10^{41} bits. These 10^{41} qubits "are" the proton, and only very few of those qubits will appear as meaningful information. All the others are declared as mass or energy. The cosmic mass black hole has an extension corresponding to the curvature radius of the universe. If the hypothetical proton disappears behind the horizon, any information on the proton is lost, and thus the unknown information, that is, the entropy, becomes maximal.

8. Relativistic particles from quantum bits

For a precise definition of a particle one has to employ Minkowski space. Here, a relativistic particle is then represented by an irreducible representation of the Poincaré group. Such a representation can be constructed from quantum information by Parabose creation and destruction operators for qubits and anti-qubits (urs and anti-urs) with state labels running from 1 to 4.

Let be $|\Omega\rangle$ the vacuum for qubits, p the order of Parabose statistics and r,s,t $\in\{1,2,3,4\}$. The commutation relations for Parabose are

$$[\hat{a}_t,\{\hat{a}_r^+,\hat{a}_s^+\}] = -2\delta_{rt}\hat{a}_s^+ - 2\delta_{st}\hat{a}_r^+ \quad [\hat{a}_t^+,\{\hat{a}_r,\hat{a}_s\}] = -2\delta_{rt}\hat{a}_s - 2\delta_{st}\hat{a}_r \quad [\hat{a}_t^+,\{\hat{a}_r^+,\hat{a}_s\}] = -2\delta_{st}\hat{a}_r^+$$

$$[\hat{a}_s,\{\hat{a}_r,\hat{a}_s\}] = 0 \qquad [a_s^+,\{a_r^+,a_s^+\}] = 0 \qquad \hat{a}_s\hat{a}_r^+|\Omega\rangle = \delta_{rs}p|\Omega\rangle \tag{2}$$

If we make the following abbreviations:

$$\{\hat{a}_r^+,\hat{a}_s^+\} = 2f[r,s] \qquad \{\hat{a}_r,\hat{a}_s\} = 2w[r,s] \qquad \{\hat{a}_r^+,\hat{a}_s\} = 2d[r,s] \quad , \tag{3}$$

then the operators for the Poincaré-group get the form:

Boosts

$$M_{10} = i\left(\; w[1,4] - f[4,1] + w[2,3] - f[3,2]\;\right)/2$$
$$M_{20} = \left(\; w[1,4] + f[4,1] - w[2,3] - f[3,2]\;\right)/2$$
$$M_{30} = i\left(\; w[1,3] - f[3,1] - w[2,4] + f[4,2]\;\right)/2$$

Rotations

$$M_{21} = \left(\; d[1,1] - d[2,2] - d[3,3] + d[4,4]\;\right)/2$$
$$M_{31} = i\left(\; d[2,1] - d[1,2] - d[3,4] + d[4,3]\;\right)/2 \tag{4}$$
$$M_{32} = \left(\; d[2,1] + d[1,2] - d[3,4] - d[4,3]\;\right)/2$$

Translations

$$P_1 = \left(-\,w[2,3]\; -f[3,2]\; -w[1,4]\; -f[4,1]\; -d[1,2]\; -d[2,1] - d[4,3]\; -d[3,4]\right)/2$$
$$P_2 = i\left(-w[2,3]\; +f[3,2]\; +w[1,4]\; -f[4,1]\; -d[1,2]\; +d[2,1] - d[4,3]\; +d[3,4]\right)/2$$
$$P_3 = \left(-w[1,3]\; -f[3,1]\; +w[2,4]\; +f[4,2]\; -d[1,1]\; +d[2,2] - d[3,3]\; +d[4,4]\right)/2$$
$$P_0 = \left(-w[1,3]\; -f[3,1]\; -w[2,4]\; -f[4,2]\; -d[1,1]\; -d[2,2] - d[3,3]\; -d[4,4]\right)/2$$

The vacuum of Minkowski space $|0\rangle$ is an eigenstate of the Poincaré group with vanishing mass, energy, momentum and spin. The Minkowski vacuum can be constructed (Görnitz, T., Graudenz, Weizsäcker, 1992) from the vacuum of the qubits $|\Omega\rangle$. In conventional notation (with $\hat{a}_i^+$) it looks like:

$$|0\rangle = \sum_{n_1=0}^{\infty}\sum_{n_2=0}^{\infty}\frac{(-1)^{n_1+n_2}}{n_1!\,n_2!}\left(\frac{\hat{a}_1^+\hat{a}_3^+ + \hat{a}_3^+\hat{a}_1^+}{2}\right)^{n_1}\left(\frac{\hat{a}_2^+\hat{a}_4^+ + \hat{a}_4^+\hat{a}_2^+}{2}\right)^{n_2}|\Omega\rangle \tag{5}$$

With respect to the Minkowski-vacuum a massless boson with helicity $-\sigma$ in z-direction and momentum m can be constructed as follows:

$$\Phi(m,\sigma) = \sum_{n_1=0}^{\infty}\frac{(-m)^{n_1}(2\sigma+p-1)!}{(p-1+n_1+2\sigma)!n_1!}\left(\frac{\hat{a}_1^+\hat{a}_3^+ + \hat{a}_3^+\hat{a}_1^+}{2}\right)^{n_1}\left(\hat{a}_1^+\hat{a}_1^+\right)^{\sigma}|0\rangle \tag{6}$$

For a photon is $|\sigma| = 1$.

A massive spinless boson at rest, constructed on the Minkowski-vacuum $|0\rangle$, with rest mass m = P0 $\neq$ 0, momentum P1=P2=P3= 0, and Parabose-order p >1 is given by:

$$\Phi(m) = \sum_{n_1=0}^{\infty} \sum_{n_2=0}^{\infty} \sum_{n_3=0}^{\infty} \frac{(-1)^{(n_1+n_2+n_3)} m^{(2\cdot n_1+n_2+n_3)} (p-2+n_1+n_2+n_3)!}{(p-1+2\cdot n_1+n_2+n_3)!(p-2+n_1+n_3)!(p-2+n_1+n_2)!} \bullet$$

$$\bullet \frac{1}{n_1!\cdot n_2!\cdot n_3!} \left(\frac{\hat{a}_4^+\hat{a}_2^+ + \hat{a}_2^+\hat{a}_4^+}{2} \right)^{n_3} \left(\frac{\hat{a}_4^+\hat{a}_1^+ + \hat{a}_1^+\hat{a}_4^+}{2} \right)^{n_1} \left(\frac{\hat{a}_3^+\hat{a}_2^+ + \hat{a}_2^+\hat{a}_3^+}{2} \right)^{n_1} \left(\frac{\hat{a}_3^+\hat{a}_1^+ + \hat{a}_1^+\hat{a}_3^+}{2} \right)^{n_2} |0\rangle \tag{7}$$

Another example (Görnitz, T., Görnitz B., 2002) is a massive fermion at rest with spin 1/2, constructed on the Minkowski-vacuum $|0\rangle$, with mass m = $P_0 \neq 0$, $spin_z= -\frac{1}{2}$, momentum $P_1=P_2=P_3= 0$ and Parabose-order p >1:

$$\Phi(m) = \sum_{n_1=0}^{\infty} \sum_{n_2=0}^{\infty} \sum_{n_3=0}^{\infty} \frac{(-1)^{(n_1+n_2+n_3)} m^{(2\cdot n_1+n_2+n_3)} (p-1+n_1+n_2+n_3)!}{(p+2\cdot n_1+n_2+n_3)!(p-2+n_1+n_3)!(p-1+n_1+n_2)!} \bullet$$

$$\left[\hat{a}_1^+ + m\hat{a}_2^+ \left(\frac{\hat{a}_4^+\hat{a}_1^+ + \hat{a}_1^+\hat{a}_4^+}{2} \right) \right] \frac{\left(\frac{\hat{a}_4^+\hat{a}_2^+ + \hat{a}_2^+\hat{a}_4^+}{2} \right)^{n_3} \left(\frac{\hat{a}_4^+\hat{a}_1^+ + \hat{a}_1^+\hat{a}_4^+}{2} \right)^{n_1} \left(\frac{\hat{a}_3^+\hat{a}_2^+ + \hat{a}_2^+\hat{a}_3^+}{2} \right)^{n_1} \left(\frac{\hat{a}_3^+\hat{a}_1^+ + \hat{a}_1^+\hat{a}_3^+}{2} \right)^{n_2}}{n_1!\cdot n_2!\cdot n_3!\cdot (p+1+2\cdot n_1+n_2+n_3)\cdot (p-1+n_1+n_3)} |0\rangle \tag{8}$$

Indeed, matter can be seen as a special form of abstract quantum information.

The possibility to create relativistic particles via quantum bits is essential for many aspects in the scientific description of nature. The change of the state of such an object amounts to a transformation associated with an element of the Poincaré group. In this operation the number and structure of its qubits will be changed. Interactions between matter – i.e. fermions – is effected by exchanging bosons. On the basis of the theory outlined here, this always appears as an exchange of qubits.

For everyday purposes, one may state:

- *Matter is inactive, it resists change.*
- *Energy can move matter.*
- *Information can trigger energy.*

9. The relationships between quantum information, particles, living beings and consciousness

Einstein's equivalence, E=mc², does not imply that the distinction between matter (having a restmass) and energy is always dispensable. Pure energy, e.g. massless photons, behave differently than massive particles. However, particles with mass can emit and absorb such photons. On the other hand, photons of sufficiently high energy can be transformed into particles with rest mass.

Analogous relations apply to quantum information. If it should be localized then the mathematical structure implies that this information must have an material or at least an energetic carrier. However, a qubit needs not to be fixed to such a carrier. Here it should be recalled that the carriers themselves are special forms of the protyposis, in the same way as material objects can be understood as being special forms of energy.

A further aspect is even more interesting: there is an analogy to the conversion of the energy of motion of a massive body into massless photons. The photons can separate from the mass and travel apart, but then they are no longer localized in space, only in time, as the structure of a rest mass in Minkowski space any longer applies. Of course, the equivalence E=mc² is not affected by this.

Qubits too may change a carrier or separate from it. In the latter case, they are no longer localizable in space and time, because neither the structure of an energy nor of a rest mass in Minkowski space applies any longer. While qubits can form particles with and without mass – and therefore fields – there is no reason to assume that they always have to form particles. Obviously, quantum information does not necessarily become manifest in the form of particles or fields.

The autonomous existence of the protyposis, of absolute quantum information, can solve some of present problems in science. The dark energy in cosmology, for example, could be a non-particle form of protyposis, as will be addressed below.

In connection with a living body qubits can become meaningful. For animals, meaning not only depends on the "incoming" information, but also on the respective situation and the particular living conditions of the animal.

Meaningful information can change its carriers, for instance from a sound wave to electrical nerve impulses, and allows for control unstable systems, such as living beings. Obviously, this finding will greatly influence the scientific understanding of life and mind. (see Görnitz, T., Görnitz, B., 2002, 2006, 2008). Mind is neither matter nor energy, rather it is protyposis in the shape of meaningful quantum information.

For a scientific understanding of the mind a dualistic conception would be in contradiction to all science. Concerning the materialistic alternative, mind is clearly no matter, and a reduction of mind to small material particles will not succeed. Protyposis, more specifically, the equivalence of protyposis, matter, and energy, and its eventual manifestation as meaningful information, offers a solution to this problem.

Life is characterized as control and timing, enabled by quantum information. Only unstable systems can be controlled. Living systems are unstable because they are far from the thermodynamical equilibrium. In the self-regulation of organisms – extending even to consciousness in the later stages of the biological evolution – quantum effects can become operational at the macroscopic level.

Consciousness is quantum information carried by a living brain, it is quantum information that experiences and knows itself.

This is not an analogy, but rather a physical characterization. It means that it is no longer required to perceive the interaction between quantum information in the shape of matter and quantum information in the shape of consciousness as a phenomenon beyond the field of science.

The scientific description of consciousness opens also the way to extend the Copenhagen interpretation of quantum theory. Usually it is stated that a measuring process has happened when an observer has notified a result. But as long as the observer and his conscious mind are not subjects of physics, a theory of measurement seems to be beyond physics as well. Here, the central role of quantum information will become important. In an abstract way, a measurement can be seen as the transition from a quantum state comprising all its possibilities to a conclusive fact. Generating a fact by measuring, results in the loss of information on all other potential states. The quantum eraser experiments show that a virtual measurement can produce a real fact only if the information on the quantum possibilities is lost.

However, before addressing the intended extension of the Copenhagen interpretation, we take a closer look to the structure of the cosmic space, as there is an essential connection.

10. Quantum information and the introduction of the cosmic position space

The Minkowski-space is a very good approximation in the domain of our laboratories and our environment. However, while the Minkowski-space is essential for an exact description of particles, the real position space in cosmology is different from this idealization. Since Einstein we know that the physical space can be curved.

The idea of understanding position space as a consequence of the symmetry of quantum bits was first proposed by von Weizsäcker (1971, p. 361; 1982, p. 172). He and Drieschner (1979) showed how qubits can explain that the space of our physical experience is three-dimensional. This was the first attempt to establish the dimensionality of physical space from first principles. (As an aside, to argue that space has in reality 10 or even 26 dimensions is not really convincing.) However, their models were not consistent with general relativity. This problem can be overcome by group theoretical considerations.

Any decision that can scientifically be decided, can be reduced to quantum bits. The states of a quantum bit are represented and transformed into each other by its symmetry group. The symmetry group for a quantum bit is spanned by the groups SU(2), U(1), and the complex conjugation. The essential part of the quantum bit symmetry group is the SU(2), a three-parameter compact group. Any number of quantum bits can be represented in the Hilbert space of measurable functions on the SU(2), which as its largest homogeneous space is an S^3. This Hilbert space is the carrier space for the regular representation of the SU(2) that contains every irreducible representation of this group. The three-dimensional S^3 space is identified with the three-dimensional position space.

Using group theoretical arguments, a relation can by established between the total number of qubits and the curvature radius R of the S^3 space (Görnitz, T., 1988). A spin-1/2-representation of the SU(2) group, the representation of a single qubit, is formed from functions on the S^3 space having a wave length of the order of R. If the tensor product of N of such spin-1/2-representations – i.e. of N qubits - is decomposed into irreducible representations, then representations associated with much shorter wavelengths can be found. The multiplicities of such representations increase with decreasing wavelengths. They are high up to functions with a wavelength in the order $R/\sqrt{N}$. Here the multiplicities reach a maximum. Because of an exponential decrease of the multiplicities, the shorter wavelengths seem not to be of physical relevance. An N-dependent metric on this S^3 is established by introducing a length related to this maximum:

$$\lambda_0 = R/\sqrt{N} \tag{9}$$

as the length-unit is introduced.

If the S^3 space is identified with the position space, a cosmological model results using three physically plausible assumptions (Görnitz, T. 1988[2]). They correspond to the basic assumptions in the three fundamental theories of physics , i.e. special relativity, quantum theory, thermodynamics:

1. There exists a universal and distinguished velocity.
2. The energy of a quantum system is inversely proportional to its characteristic wave length.
3. The first law of thermodynamics is valid.

The first assumption introduces the velocity of light, c. The second assumption is the familiar Planck relation, $E=h\nu=hc/\lambda$, while the third allows us to define a cosmological pressure p according to $dU+pdV=0$.

The result is a compact Friedman-Robertson-Walker-Space-Time. Measured in units of the fundamental length λ_0, the cosmic radius R grows with the velocity of light. Therefore in this model the horizon problem as well as the flatness problem are absent. The horizon problem is related to the fact that the background radiation from opposed directions in space is entirely identical, whereas according to most cosmological models those regions could never have been in causal contact. As a remedy, inflation was invented. However, the *ad hoc* assumptions necessary here violate an important energy condition (Hawking, Ellis, 1973). This suggests that another solution of the horizon problem should be sought. More recently, the inflation concept has been criticized on other grounds, too (Steinhart, 2011).

By the group theoretical argument, the number of qubits increases quadratically with the age of the universe, i.e., with the cosmic radius R. The energy attributed to a single qubit is inversely proportional to R. Therefore, the total energy U rises with R and the energy density decreases with $1/R^2$. According to the first law of thermodynamics, the resulting state equation for the cosmic substrate, the protyposis, follows as $\mu=-p/3$.

From the metric of this model the Einstein-tensor $G_i{}^k$ can be computed, and, using the relations between energy density and pressure, the energy-momentum-tensor $T_i{}^k$ is obtained. Both tensors appear as being proportional to each other.

If it is demanded that this proportionality between $G_i{}^k$ and $T_i{}^k$ is conserved also for local variations of the energy, Einstein's equations of general relativity emerge as a consequence of the abstract quantum information. If a smallest physical meaningful length – the Planck length – according to $\lambda_{Pl} = \lambda_0 \sqrt{3}/2$ is introduced, one obtains with $\kappa= 8\pi G/c^4$:

$$G_i{}^k = \kappa \; T_i{}^k \qquad\qquad (10)$$

In this cosmological model the dark energy can be interpreted as protyposis, i.e., as absolute quantum information, that is homogeneous and isotropic, and not organized in the form of quantum particles.

11. The measuring process – reinterpreting the Copenhagen interpretation

After the clarification of the relation between abstract quantum theory and space, we will turn to the strangest concept in quantum physics, namely the measuring process. Here the strongest discomfort results because the unitary time evolution of a quantum system appears to be interrupted.

A comprehensive review on the different attempts to solve this problem has been given by Genovese (2010). In this review most of the modern attempts are addressed, but fundamental earlier work, e.g. by Heisenberg or v. Weizsäcker, is not cited. I think, "the transition from a microscopic probabilistic world to a macroscopic deterministic world described by classical mechanics (macro-objectification)" is less of a problem, and also one should not say that the Copenhagen interpretation "is weak from a conceptual point of view since it does not permit to identify the border between quantum and classical worlds. How many particles should a body have for being macroscopic?"(Genovese, 2010).

Rather the problem seems to stem from the conceptual fixation of physics on the more than 2000 years old notion of "atoms" of one or another kind as basic structures. Quantum theory opens the possibility to recognize that more abstract structures – quantum information – should be viewed as the fundamental entities. This will open a new perspective for the measuring process as well.

As already mentioned, the measuring process is often seen as the most controversial aspect of quantum theory. The "normal" time evolution in quantum theory is a unitary process. In the Schrödinger picture the time-evolution of the wave function is given by

$$\Phi_{Schr}(t) = \hat{U}(t)\Phi_{Schr}(0) = e^{-\frac{i}{\hbar}\hat{H}t}\Phi_{Schr}(0) \tag{11}$$

where $\hat{H}$ is the Hamiltonian of the system under consideration. In an equivalent way, referred to as Heisenberg picture, the time- dependence can be shifted to the physical operators:

$$\hat{A}_{Heis}(t) = \hat{U}(t)^{\dagger}\hat{A}_{Heis}(0)\hat{U}(t) = e^{\frac{i}{\hbar}\hat{H}t}\hat{A}_{Heis}(0)e^{-\frac{i}{\hbar}\hat{H}t} \tag{12}$$

The time evolution according to the Schrödinger equation of the system conserves the absolute value of the scalar products and does not change the total probability. In the measuring process, by contrast, the so-called "collapse of the wave function" is no longer unitary. There has been much discussion about this disruption in the description of the regular time evolution and whether that disruption can possibly be avoided. Let me very briefly recall the essential aspects.

If a quantum system is in a state Φ, then in the measuring process every state Ψ can be found if

$$< \Phi \mid \Psi> \neq 0 \tag{13}$$

The probability ω to find Ψ if the system is in the state Φ is given by

$$\omega = \mid < \Phi \mid \Psi> \mid^2 \tag{14}$$

The so-called many-worlds-interpretation of QM assumes that there is no break in the unitary time evolution. It is postulated that any possible Ψ will be a real outcome of the measurement, but for every Ψ there is a separate universe in which this outcome is realized as a factum. For most people this interpretation is not acceptable because of the fantastical ontological overload thereby introduced. However, with a simple "one-word-dictionary" (Görnitz, T., Weizsäcker 1987) it can be translated into the normal world view: just replace "many worlds" by "many possibilities".

The Copenhagen interpretation introduces an observer who is responsible for stating that a result has been found. If the observer can verify that the process has occurred, the result can be seen as a factum. Given that any description of reality needs a person to do the description, the introduction of an observer into the description of nature does not seem to be a serious constraint. However, the problem here is that this construct does not allow one to include the observer himself in the scientific description, which ultimately has to be based on physics, that is, quantum theory. In a conversation, reported to me by v. Weizsäcker about the necessity of a "cut" between quantum and classical physics and the movability of that cut, Heisenberg argued that the cut cannot be moved into the mind of the observer. According to v. Weizsäcker, his friend Werner Heisenberg said: "In such a case no physics would remain ".

This point of view is easily understandable, because at that time the range of physics was limited to material and energetical objects, but did not yet extend to quantum information. Since then much experimental and theoretical work on quantum information has emerged, indicating the mind can be understood as a very special form of quantum information. Therefore it is no longer warranted to keep the mind totally outside the realm of physics. (see Görnitz, T., Görnitz, B., 2002, 2008) However, for a scientific description of the observer and even of his mind, the original Copenhagen interpretation has to be extended.

What is the role of the observer. Let the quantum system be in a state Φ. After the measuring process the observer has to realize that all the possible states Ψ did not turn into real facts except for the final Ψ_f, which is associated with the actual outcome of the measurement.

But left with Ψ_f , the information about the former state Φ is no longer available. The only remaining information on Φ is that Φ is not orthogonal to Ψ_f. Obviously, this only very vague and imprecise information as, in general, infinitely many states will be not orthogonal to Ψ_f.

If the observer comes to know the result of the measurement associated with Ψ_f, he will use this new wave function for the future description of the system. It is useful to describe this change of the wave function as a result of the change in the observer's knowledge. (Görnitz, T., v. Weizsäcker 1987, see also: Görnitz, T., Lyre 2006). The measurement provides new knowledge and the observer can take that into account.

Now the intriguing question is how does a fact come about in physics if there is no observer to constitute it?

In a pure quantum description no facts can arise (at least when one does not resort to infinite many degrees of freedom, as done in the algebraical description by Primas (1981)). While classical physics does describe facts, the classical description does not make any difference between past and future facts, all events being determined in the same way. More specifically, in classical physic the "real character" of time with its difference between past and future does not appear. Neither in classical nor in quantum physics the irreversibility of a factum is a consequence of the respective mathematical structure. Both theories have a reversible structure, and the irreversibility encountered, for example, in thermodynamics is attributed to the describers imperfect knowledge of the microscopic configurations of the system. However, imperfect knowledge cannot be the cause of a physical occurrence.

Information plays the central role in the measurement process. However, as long as information is only understood as being "knowledge", a human observer has to be supposed. My proposal is to expand the role of information, which will allow us to explain how facts arise in the scientific description of nature, even without supposing an observer in the first place.

Apparently, only in the measurement, i.e. in the transition from the quantum to classical description, physics does discriminate *before* the event and *after* the event. This means that the problem of how events occur must be solved. In this regard, I think the experiments with the quantum eraser (Scully et al., 1991, Zajonc et al., 1991, Herzog et al., 1995) may shed some light on the role of the observer.

The first such experiments were "double slit" experiments. If it is possible — at least in principle — to get the "which-way" information about the slits, there will be no interference; otherwise, interference patterns should appear.

As the quantum eraser shows, it is not a disturbance by the observer which causes the measuring process; rather, the crucial factor here is the loss of information concerning the original state Φ. As to Scully and Walther (1998) state: "It is simply knowing (or having the ability to know even if we choose not to look at the Welcher-Weg detector) which eliminates the pattern. This has been verified experimentally. Hence one is led to ask: what would happen if we put a Welcher-Weg detector in place (so we lose interference even if we don't look at the detector) and then erase the which-way information after the particles have passed through? Would such a "quantum eraser" process restore the interference fringes? The answer is yes and this has also been verified experimentally."

To make clear what happens, it is useful to describe the process somewhat differently.

The authors say "… and then erase the which-way information." However, the information is not taken out of the system and then destroyed. On the contrary, the which-way information can leave the system only potentially. If indeed the information had left the system, a factum concerning the "Weg" would have been produced and interference would be absent. However, this was not the case; the information was returned into the system and, in fact, not able to leave it. Both "ways" remained possible, no path became factual, and the interference appeared.

Thus, the essential aspect of a measurement is whether some information on the state of the system under consideration is lost.

As long as there is the possibility that the information can come back into the system, no real factum has been created and no measurement has occured. Only if it is guaranteed that the information (or at least a part of it) has left the system for good, a factum has been created and a measuring result can be established.

Each state of a quantum system is "co-existing" with all other states not orthogonal to it. In this huge manifold of states there are eigenstates of the measuring interaction. In the measuring process associated with the respective measuring interaction one of the eigenstates becomes factual when the information on all the other states has been lost.

That the measuring process relies on a loss of information seems contradictory, but, in fact, it is not: in the measurement a huge amount of quantum possibilities is reduced to the distinct classical information on a fact. The information about the measuring result is factual and, that is, classical information which can be replicated. Such classical information can be repeatedly taken out from the system; the measuring result can be read out repeatedly.

How can this loss of information modelled?

An essential step towards a better understanding was accomplished by the theory of quantum decoherence, originating with the work by Zeh (1970). Decoherence can explain how, in an approximate way, a quantum object acquires classical properties. There is a recent and detailed book (Schlosshauer, 2007) to which the reader is referred to for technical details. The central point is that for the composite system of the object, the measuring device, and the environment, the non-diagonal elements in the partial density matrix of the object become exponentially small in a short time. Therefore, the object density matrix rapidly assumes a form reflecting the situation of a classical probability for an unknown fact within an ensemble of possible facts.

It is occasionally stated, however misleadingly, that decoherence solves the measuring problem. This is not the case. Zeh wrote (1996, S. 23) that an environmental induced decoherence alone does not solve the measuring problem; this applies even more to a microscopic environment, where decoherence does not necessarily lead to an irreversible change. Joos (1990) wrote: "that the derivation of classical properties from quantum mechanics remains insufficient in one essential aspect. The ambiguity of the quantum mechanical dynamics (unitary Schrödinger dynamics versus indeterministic collapse) remains unsolved. The use of local density matrices presupposes implicitly the measuring axiom, i.e., the collapse." And he proceeds in his text as follows: "certain objects for a local observer appear classical (so defining what a classical object is), but the central question remains unsolved; why in the non-local quantum world local observers exist at all?"

To answer this question one has to take into account, besides quantum information, the cosmological aspects of quantum theory.

Decoherence in a system is caused by the interaction with a macroscopic device, which in turn is embedded in an even larger environment. As a consequence, information flows out from the quantum object. As long as this information is restricted to a finite volume, there is no fundamental obstacle preventing the information from coming back into the quantum object. However, any environment is ultimately coupled to the cosmic space, being presently nearly empty and dark. This cosmic boundary condition is the reason that in the end, perhaps with some intermediate steps, the information can escape without any realistic chance for ever coming back.

As long as a quantum system is completely isolated, so that not even information can escape, it will remain in its quantum state, comprising all its respective possibilities. Only if a system is no longer isolated, allowing information to escape, which usually will be effected by outgoing photons, a factum can arise.

Of course, mathematically one is confronted with a limiting procedure. To prove rigorously that no information will ever come back, an infinite time limit has to be considered. Alternatively, an environment is needed with actually infinite many degrees of freedom in order to have superselection rules in a strong sense. (Primas, 1983).

As far as an observer of the process is concerned, he may decide that for all practical purposes the appearance of a factum can be acknowledged. In view of the cosmological conditions, dispelling any expectations that the information will come back, one does not have to assume that the creation of a factum depends on the perception of the observer. On the other hand, there may be still a role for the observer, namely, to assure the end of the limiting process. Loosely speaking, one may say that the observer has to "guarantee" that the information on all the other possible states Ψ does not come back and turn the measurement into an illusion.

Usually, outgoing information will be carried mostly by photons. Because the cosmos is empty, dark and expanding with high velocity there is no chance that an outgoing photon will be replaced by an equal incoming photon. As is well known, the cosmos was not always as it is today. In the beginning, the cosmos was dense and hot. Going back to the earliest stages of the universe, it was ever more likely that an outgoing photon was matched by an equal incoming one. This means that the idea of creating facts becomes more and more obsolete when approaching the singular origin. Concomitantly, the structure of time with the difference between past, present and future looses gradually its significance, to disappear completely in the neighbourhood of the singularity. Obviously, without the familiar structure of time the conceptuality of empiricism has no meaning, nor has the concept of empirical science.

The generalization of "knowledge" to "information," more exactly to "quantum information," opens the possibility to extend the Copenhagen interpretation in such a way that the observer and his consciousness can be included in the scientific description. (see Görnitz, T., Görnitz, B., 2002, 2008).

The intention here is not only to describe quantum processes in the brain but also quantum processes of the mind.

That biological processes, like vision, i.e., the absorption of photons in the retina, or the absorption of photons in a plant and the subsequent transport of the electrons in the process of photosynthesis, must be described as quantum processes is already well-known. (Engel et al., 2007).

This suggests that in the brain, perhaps the most complicated organ in biology, quantum processes will play a crucial role, even when decoherence happens just as much. However, the effect of decoherence decreases for decreasing mass of the quantum objects. Decoherence is weakest for massless objects, such as photons. Photons are essential as carriers of information in the brain, as the supporter of thoughts. As a hint at the role of photons in the brain, one may see the forensic fact that the personality of a patient has passed away if photons can no longer be found in the EEG.

12. Conclusions

As all science is approximation, a good description of nature incorporates its decomposition into objects and forces between them and their factual description – done by classical physics – as well as taking into account the possibilities and the aspect of wholeness as done by quantum theory. The dynamical layering-process describes the interrelations between the classical and the quantum approaches.

Quantum theory, being the most successful fundamental physical theory, can be understood as the physics of relations, possibilities, and nonlocality. At the core of quantum theory is the equivalence of locality and nonlocality, matter and force, wholeness and emptiness, and, last but not least, the equivalence of matter, energy, and quantum information.

These fundamental equivalences, based on absolute quantum information, allow for a foundation of the cosmological concepts as well as the inclusion of consciousness in the scientific description of nature.

13. Acknowledgments

I thank Jochen Schirmer and Ludwig Kuckuck for helpful advices.

14. References

Ali, S T., Engliš, M. (2005). Rev. in Math. Physics, 17, 391

Drieschner, M. (1979). Voraussage – Wahrscheinlichkeit – Objekt, Springer, Berlin

Engel, G. S., Calhoun, T. R. Read, E. L., Ahn, T.-K., Manclal, T., Cheng, Y.-C. , Blankenship, R. E. and Fleming, G. R. (2007). Nature 446, 782

Genovese, M., (2010). Adv. Sci. Lett. 3, 249–258 ., C. F. v.

Görnitz, T., Weizsäcker, C. F. v. (1987) Intern. Journ. Theoret. Phys. 26, 921;

Görnitz T., (1988). Intern. J. of Theoret. Phys. 27 527

Görnitz T (1988^2) Intern. J. of Theoret. Phys. 27 659

Görnitz T., Graudenz D., Weizsäcker, C. F. v. (1992). Intern. J. Theoret. Phys. 31, 1929

Görnitz T., (1999). Quanten sind anders - Die verborgene Einheit der Welt, Spektrum Akadem. Verl., Heidelberg

Görnitz T., Görnitz, B. (2002). Der kreative Kosmos - Geist und Materie aus Information. Spektrum Akadem. Verl., Heidelberg

Görnitz, T., Görnitz, B. (2008). Die Evolution des Geistigen, Vandenhoeck & Ruprecht, Göttingen,

Hawking, S. W., Ellis, G. F. R. (1973). The Large Scale Structure of the Universe, University Press, Cambridge

Heisenberg, W. (1969). Der Teil und das Ganze, Piper, München

Herzog, Th. J., Kwiat, P. G., Weinfurter, H. and Zeilinger, A. (1995). Phys. Rev. Lett. 75, 3034

Joos, E., (1990). Philosophia Naturalis 27, 31

Primas, H., (1981). Chemistry, Quantum Mechanics and Reductionism, Springer, Berlin

Schlosshauer, M., (2007) Decoherence and the Quantum-to-Classical Transition, Berlin; (see also Schlosshauer, M., Rev. Mod. Phys. 76, 1267 (2004); or arXiv:quant-ph/0312059v4.

Scully, M. O., Englert, B.-G. and Walther, H. (1991). Nature 351, 112

Scully, M. O. and Walther, H. (1998). Found. Phys. 28, 399

Steinhart, P. J. (2011). Kosmische Inflation auf dem Prüfstand, Spektrum d. Wissenschaft, Nr. 8 40-48

Weizsäcker, C. F. v. (1958). Weltbild der Physik, Hirzel, Stuttgart

Weizsäcker C. F. v. (1971). Einheit der Natur, Hanser, München

Weizsäcker, C. F. v. (1982). Aufbau der Physik, Hanser, München, Engl.: Görnitz, T., Lyre, H., (Eds.) (2006). C. F. v. Weizsäcker – Structure of Physics, Springer, Berlin

Zajonc, A. G., Wang, L. J., Zou,X. Y. and Mandel, L. (1991). Nature 353, 507

Zeh, H. D., (1970). *Found. Phys.* 1, 69; for a pedagogical introduction see: C. Kiefer, E. Joos, (1998). arXiv:quant-ph/9803052v1 19 Mar

Zeh, H. D. (1996), in Editors D. Giulini, E. Joos, C. Kiefer, J. Kupsch, I.-O. Stamatescu, and H. D. Zeh, *Decoherence and Appearance of a Classical World in Quantum Theory*, Springer, Berlin, Heidelberg

3

Effects on Quantum Physics of the Local Availability of Mathematics and Space Time Dependent Scaling Factors for Number Systems

Paul Benioff

Physics division, Argonne National Laboratory, Argonne, IL
USA

1. Introduction

The relation of mathematics to physics and its influence on physics have been a topic of much interest for some time. A sampling of the literature in this area includes Wigner's paper, "The Unreasonable Effectiveness of Mathematics in the Natural Sciences" (Wigner, 1960) and many others (Bendaniel, 1999; Bernal et al, 2008; Davies, 1990; Hut et al, 2006; Jannes, 2009; Omnès, 2011; Tegmark, 2008; Welch, 2009). The approach taken by this author is to work towards a comprehensive theory of physics and mathematics together (Benioff, 2005; 2002). Such a theory, if it exists, should treat physics and mathematics as a coherent whole and not as two separate but closely related entities.

In this paper an approach is taken which may represent definite steps toward such a coherent theory. Two ideas form the base of this approach: The local availability of mathematics and the freedom to choose scaling factors for number systems. Local availability of mathematics is based on the idea that all mathematics that an observer, O_x, at space time point, x, can, in principle, know or be aware of, is available locally at point x. Biology comes in to the extent that this locally available knowledge must reside in an observers brain. Details of how this is done, biologically, are left to others to determine.

This leads to the association of a mathematical universe, $\bigvee_x$, to each point x. $\bigvee_x$ contains all the mathematics that O_x can know or be aware of. For example, $\bigvee_x$ contains the various types of numbers: the natural numbers, $\bar{N}_x$, the integers, $\bar{I}_x$, the rational numbers, $\overline{Ra}_x$, the real numbers, $\bar{R}_x$, and the complex numbers, $\bar{C}_x$. It also contains vector spaces, $\bar{V}_x$, such as Hilbert spaces, $\bar{H}_x$, operator algebras, $\overline{Op}_x$, and many other structures.

The universes are all equivalent in that any mathematical system present in one universe is present in another. It follows that $\bigvee_y$ contains systems for the different types of numbers as $\bar{N}_y, \bar{I}_y, \overline{Ra}_y, \bar{R}_y, \bar{C}_y$. It also contains $\bar{H}_y, \overline{Op}_y$, etc. Universe equivalence means here that for any system type, S, $\bar{S}_y$ is the same system in $\bigvee_y$ as $\bar{S}_x$ is in $\bigvee_x$.

For the purposes of this work, it is useful to have a specific definition of mathematical systems. Here the mathematical logical definition of a system of a given type as a structure (Barwise, 1977; Keisler, 1977) is used. A structure consists of a base set, a few basic operations, none or a few basic relations, and a few constants. The structure must satisfy a set of axioms appropriate for the type of system being considered. For example,

$$\bar{N}_x = \{N_x, +_x, \times_x, <_x, 0_x, 1_x\} \tag{1}$$

satisfies a set of axioms for the natural numbers as the nonnegative elements of a discrete ordered commutative ring with identity (Kaye, 1991),

$$\bar{R}_x = \{R_x, +_x, -_x, \times_x, \div_x, <_x, 0_x, 1_x\} \tag{2}$$

is a real number structure that satisfies the axioms for a complete ordered field (Randolph, 1968), and

$$\bar{C}_x = \{C_x, +_x, -_x, \times_x, \div_x, 0_x, 1_x\} \tag{3}$$

is a complex number structure that satisfies the axioms for an algebraically closed field of characteristic 0 (Shoenfield, 1967).

$$\bar{H}_x = \{H_x, +_x, -_x, \cdot_x, \langle -, - \rangle_x, \psi_x\} \tag{4}$$

is a structure that satisfies the axioms for a Hilbert space (Kadison, 1983). Here ψ_x is a state variable in $\bar{H}_x$. There are no constants in $\bar{H}_x$. The subscript, x, indicates that these structures are contained in V_x.

The other idea introduced here is the use of scaling factors for structures for the different number types. These scale structures are based on the observation (Benioffa, 2011; Benioffb, 2011; Benioffc, 2011) that it is possible to define, for each number type, structures in which number values are scaled relative to those in the structures shown above. The scaling of number values must be compensated for by scaling of the basic operations and constants in such a way that the scaled structure satisfies the relevant set of axioms if and only if the original structure does.

Scaling of number structures introduces scaling into other mathematical systems that are based on numbers as scalars for the system. Hilbert spaces are examples as they are based on the complex numbers as scalars.

The fact that number structures can be scaled allows one to introduce scaling factors that depend on space time or space and time. If $y = x + \hat{\mu}dx$ is a neighbor point of x, then the real scaling factor from x to y is defined by

$$r_{y,x} = e^{\vec{A}(x) \cdot \hat{\mu}dx}. \tag{5}$$

Here $\vec{A}$ is a real valued gauge field that determines the amount of scaling, and $\hat{\mu}$ and dx are, respectively, a unit vector and length of the vector from x to y. Also $\cdot$ denotes the scalar product. For y distant from x, $r_{y,x}$ is obtained by a suitable path integral from x to y.

Space time scaling of numbers would seem to be problematic since it appears to imply that comparison of theoretical and experimental numbers obtained at different space time points have to be scaled to be compared. This is not the case. As will be seen, number scaling plays no role in such comparisons. More generally, it plays no role in what might be called, "the commerce of mathematics and physics".

Space time dependent number scaling is limited to expressions in theoretical physics that require the mathematical comparison of mathematical entities at different space time points.

Effects on Quantum Physics of the Local Availability of
Mathematics and Space Time Dependent Scaling Factors for Number Systems

49

Typical examples are space time derivatives or integrals. Local availability of mathematics makes such a comparison problematic. If f is a space time functional that takes values in some structure $\bar{S}$, then "mathematics is local" requires that for each point, y, $f(y)$ is an element of $\bar{S}_y$. In this case space time integrals or derivatives of f make no sense as they require addition or subtraction of values of f in different structures. Addition and subtraction are defined only within structures, not between structures.

This problem is solved by choosing some point x, such as an observers location, and transforming each $\bar{S}_y$ into a local representation of $\bar{S}_y$ on $\bar{S}_x$. Two methods are available for doing this: parallel transformations for which the local representation of $\bar{S}_y$ on $\bar{S}_x$ is $\bar{S}_x$ itself, and correspondence transformations. These give a local, scaled representation of $\bar{S}_y$ on $\bar{S}_x$ in that each element of $\bar{S}_y$ corresponds to the same element of $\bar{S}_x$, multiplied by the factor $r_{y,x}$.

The rest of this paper explains, in more detail, these ideas and some consequences for physics. The next section describes representations of number types that differ by scaling factors. Sections 3 and 4 describe space time fields of complex and real number structures and the representation of $r_{y,x}$ in terms of a gauge field, as in Eq. 5. This is followed by a discussion of the local availability of mathematics and the assignment of separate mathematical universes to each space time point. Section 6 describes correspondence and parallel transforms. It is shown that $\vec{A}$ plays no role in the commerce of mathematics and physics. This involves the comparison and movement of the outcomes of theoretical predictions and experiments and the general use of numbers.

Section 7 applies these ideas to quantum theory, both with and without the presence of $\vec{A}$. Parallel and correspondence transformations are used to describe the wave packet representation of a quantum system. It is seen that there is a wave packet description that closely follows what what is actually done in measuring the position distribution and position expectation value. The coherence is unchanged in such a description.

The next to last section uses "mathematics is local" and the scaling of numbers to insert $\vec{A}$ into gauge theories. The discussion is brief as it has already been covered elsewhere (Benioffa, 2011; Benioffc, 2011). $\vec{A}$ appears in the Lagrangians as a boson for which a mass term is not forbidden. The last section concludes the paper.

The origin of this work is based on aspects of mathematical locality that are already used in gauge theories (Montvay & Münster, 1994; Yang & Mills, 1954) and their use in the standard model (Novaes, 2000). In these theories, an n dimensional vector space, $\vec{V}_x$, is associated with each point, x, in space time. A matter field $\psi(x)$ takes values in $\vec{V}_x$. Ordinary derivatives are replaced by covariant derivatives, $D_{\mu,x}$, because of the problem of comparing values of $\psi(x)$ with $\psi(y)$ and to introduce the freedom of choice of bases. These derivatives use elements of the gauge group, $U(n)$, and their representations in terms of generators of the Lie algebra, $u(n)$, to introduce gauge bosons into the theories.

2. Representations of different number types

Here the mathematical logical definition Barwise (1977); Keisler (1977) of mathematical systems as structures is used. A structure consists of a base set, basic operations, relations, and constants that satisfy a set of axioms relevant to the system being considered. As each type of number is a mathematical system, this description leads to structure representations of each number type.

The types of numbers usually considered are the natural numbers, $\bar{N}$, the integers, $\bar{I}$, the rational numbers, $\overline{Ra}$, the real numbers, $\bar{R}$, and the complex numbers, $\bar{C}$. Structures for the real and complex numbers can be defined by

$$\bar{R} = \{R, +, -, \times, \div, <, 0, 1\}$$
$$\bar{C} = \{C, +, -, \times, \div, 0, 1\}. \tag{6}$$

A letter with an over line, such as $\bar{R}$, denotes a structure. A letter without an over line, as R in the definition of $\bar{R}$, denotes the base set of a structure.

The main point of this section is to show, for each type of number, the existence of many structures that differ from one another by scale factors. To see how this works it is useful to consider a simple case for the natural numbers, $0, 1, 2, \cdots$. Let $\bar{N}$ be represented by

$$\bar{N} = \{N, +, \times, <, 0, 1\} \tag{7}$$

where $\bar{N}$ satisfies the axioms of arithmetic (Kaye, 1991).

The structure $\bar{N}$ is a representation of the fact that $0, 1, 2 \cdots$ with appropriate basic operations and relations are natural numbers. However, subsets of $0, 1, 2, \cdots$, along with appropriate definitions of the basic operations, relations, and constants are also natural number structures.

As an example, consider the even numbers, $0, 2, 4, \cdots$ in $\bar{N}$ Let $\bar{N}_2$ be a structure for these numbers where

$$\bar{N}_2 = \{N_2, +_2, \times_2, <_2, 0_2, 1_2\} \tag{8}$$

Here N_2 consists of the elements of N with even number values in $\bar{N}$. The structure $\bar{N}_2$ shows that the elements of N that have value $2n$ in $\bar{N}$ have value n in $\bar{N}_2$. Thus the element that has value 2 in $\bar{N}$ has value 1 in $\bar{N}_2$, etc. The subscript 2 on the constants, basic operations, and relations in $\bar{N}_2$ denotes the relation of these structure elements to those in $\bar{N}$.

The definition of $\bar{N}_2$ floats in the sense that the specific relations of the basic operations, relation, and constants to those in $\bar{N}$ must be specified. These are chosen so that $\bar{N}_2$ satisfies the axioms of arithmetic if and only if $\bar{N}$ does. A suitable choice that satisfies this requirement is another representation of $\bar{N}_2$ defined by

$$\bar{N}_2^2 = \{N_2, +, \frac{\times}{2}, <, 0, 2\}. \tag{9}$$

This structure is called the representation of $\bar{N}_2$ on $\bar{N}$.

$\bar{N}_2^2$ shows explicitly the relations between the basic operations, relations, and number values in $\bar{N}_2$ and and those in $\bar{N}$. For example, $1_2 \leftrightarrow 2, +_2 \leftrightarrow +, \times_2 \leftrightarrow \times/2, <_2 \leftrightarrow <$. These relations are such that $\bar{N}_2^2$, and thereby $\bar{N}_2$, satisfies the axioms of arithmetic if and only if $\bar{N}$ does.

$\bar{N}_2^2$ also shows the presence of 2 as a scaling factor. Elements of the base set N_2 that have value n in $\bar{N}_2$ have value $2n$ in $\bar{N}$. Note that, by themselves, the elements of the base set have no intrinsic number values. The values are determined by the axiomatic properties of the basic operations, relations, and constants in the structure containing them.

This description of scaled representations applies to the other types of numbers as well. For real numbers let r be a positive real number in $\bar{R}$, Eq. 6. Let

$$\bar{R}_r = \{R, +_r, -_r, \times_r, \div_r, <_r, 0_r, 1_r\} \tag{10}$$

Effects on Quantum Physics of the Local Availability of
Mathematics and Space Time Dependent Scaling Factors for Number Systems

51

be another real number structure. Define the representation of $\bar{R}_r$ on $\bar{R}$ by the structure,

$$\bar{R}_r^r = \{R, +, -, \frac{\times}{r}, r\div, <, 0, r\}. \tag{11}$$

$\bar{R}_r^r$ shows that number values in $\bar{R}_r$ are related to those in $\bar{R}$ by a scaling factor r.

$\bar{R}_r^r$ gives the definitions of the basic operations, relation, and constants in $\bar{R}_r$ in terms of those in $\bar{R}$. These definitions must satisfy the requirement that $\bar{R}_r$ satisfies the real number axioms if and only if $\bar{R}_r^r$ does if and only if $\bar{R}$ does.[1]

Note that the base set R is the same for all three structures. Also the elements of R do not have intrinsic number values independent of the structure containing R. They attain number values only inside a structure where the values depend on the structure containing R.

The relationships between number values in $\bar{R}_r^r$, $\bar{R}_r$, and $\bar{R}$ can be represented by a new term, correspondence. One says that the number value a_r in $\bar{R}_r$ *corresponds* to the number value ra in $\bar{R}$. This is different from the notion of *sameness*. In $\bar{R}$, ra is different from the value a. However, a is the *same* value in $\bar{R}$ as a_r is in $\bar{R}_r$ as ra is in $\bar{R}_r^r$. The distinctions between the concepts of correspondence and sameness does not arise in the usual treatments of numbers. The reason is that sameness and correspondence coincide when $r = 1$.

For complex numbers, the structures, in addition to $\bar{C}$, Eq. 6, are

$$\bar{C}_r = \{C, +_r, -_r, \times_r, \div_r, 0_r, 1_r\}, \tag{12}$$

and the representation of $\bar{C}_r$ on $\bar{C}$ as

$$\bar{C}_r^r = \{C, +, -, \frac{\times}{r}, r\div, 0, r\}. \tag{13}$$

Here r is a real number value in $\bar{C}$. a is the same number value in $\bar{C}$ as a_r is in $\bar{C}_r$. Otherwise the description is similar to that for the natural and real numbers. More details on these and other number type representations are given in (Benioffb, 2011).

3. Fields of mathematical structures

As was noted in the introduction, the local availability of mathematics results in the assignment of separate structures, $\bar{S}_x$, to each point, x, of space time. Here S denotes a type of mathematical structure. The discussion is limited to the main system types of concern. These are the real numbers, the complex numbers, and Hilbert spaces. Hilbert spaces are included here because the freedom of choice of scaling factors for number types affects Hilbert spaces as they are based on complex numbers as scalars.

3.1 Complex numbers

Parallel transformations between $\bar{C}_x$ and $\bar{C}_y$ for two points, x, y, define the notion of *same* number values between the structures. Let $F_{y,x}$ be an isomorphism from $\bar{C}_x$ onto $\bar{C}_y$. With

$$\bar{C}_x = \{C_x, +_x, -_x, \times_x, \div_x, 0_x, 1_x\}$$
$$\bar{C}_y = \{C_y, +_y, -_y, \times_y, \div_y, 0_y, 1_y\}, \tag{14}$$

---

[1] The relations between the structures are also valid for negative values of r, provided $<$ in Eq. 11 is replaced by $>$, and appropriate changes are made in the axioms to reflect this replacement.

and

$$F_{y,x}\bar{C}_x = \bar{C}_y, \tag{15}$$

$F_{y,x}$ defines the notion of same number value and same operation in $\bar{C}_y$ as that in $\bar{C}_x$. This is expressed by

$$a_y = F_{y,x}a_x$$
$$Op_y = F_{y,x}Op_x. \tag{16}$$

Here a_y is the same (or $F_{y,x}$-same) number value in $\bar{C}_y$ as a_x is in $\bar{C}_x$. Op_y is the same operation in $\bar{C}_y$ as Op_x is in $\bar{C}_x$. Op denotes any one of the operations, $+, -, \times, \div$.

Note that $F_{y,x}$ is independent of paths between x and y. This follows from the requirement that for a path P from x to y and a path Q from y to z,

$$F_{z,x}^{Q*P} = F_{z,y}^{Q}F_{y,x}^{P}. \tag{17}$$

Here $Q * P$ is the concatenation of Q to P. If $z = x$ then the path is cyclic and the final structure is identical to the initial one. This gives the result that

$$F_{x,x}^{Q*P} = 1. \tag{18}$$

This shows that $F_{y,x}^{P}$ is path independent so that a path label is not needed. Note that

$$F_{y,x} = F_{x,y}^{-1}. \tag{19}$$

The subscript order in $F_{y,x}$ gives the path direction, from x to y.

At this point the freedom to choose complex number structures at each space time point is introduced. This is an extension, to number structures, of the freedom to choose basis sets in vector spaces as is used in gauge theories (Montvay & Münster, 1994; Yang & Mills, 1954) This can be accounted for by factoring $F_{y,x}$ into a product of two isomorphisms as in

$$F_{y,x} = W_r^y W_x^r. \tag{20}$$

Here $y = x + \hat{v}dx$ is taken to be a neighbor point of x.

The action of W_r^y and W_y^r is given by

$$\bar{C}_y = W_r^y \bar{C}_x^r = W_r^y W_x^r \bar{C}_x = F_{y,x}\bar{C}_x. \tag{21}$$

Here $r_{y,x}$ is a real number in $\bar{C}_x$ that is associated with the link from x to y. As was the case for $F_{y,x}$ the order of the subscripts determines the direction of the link. Thus $r_{x,y}$ is a number in $\bar{C}_y$ for the same link but in the opposite direction and

$$(r_{x,y})_x r_{y,x} = 1. \tag{22}$$

Here $(r_{x,y})_x$ is the same number value in $\bar{C}_x$ as $r_{x,y}$ is in $\bar{C}_y$. In the following, the subscripts y, x are often suppressed on $r_{y,x}$ to simplify the notation.

The structure $\bar{C}_x^r$ is defined to be the representation of $\bar{C}_y$ on $\bar{C}_x$. As is the case for $\bar{C}_x^r$, Eq. 13, the number values and operations in $\bar{C}_x^r$ are defined in terms of the corresponding number values and operations in $\bar{C}_x$:

Effects on Quantum Physics of the Local Availability of
Mathematics and Space Time Dependent Scaling Factors for Number Systems

53

$$\bar{C}_x^r = \{C_x, +_x, -_x, \frac{\times_x}{r}, r \div_x, 0_x, r\}. \tag{23}$$

The multiplication and division by r, shown in $\times_x/r, r\div_x$, are operations in $\bar{C}_x$. Note that the number value r in $\bar{C}_x$ is the multiplicative identity in $\bar{C}_x^r$. Also $\bar{C}_x^r$ has the same base set, C_x, as does $\bar{C}_x$.

The corresponding definition of W_x^r is given by

$$W_x^r(a_x) = ra_x, \quad W_x^r(\pm_x) = \pm_x$$

$$W_x^r(\times_x) = \frac{\times_x}{r} \quad W_x^r(\div_x) = r \div_x. \tag{24}$$

W_x^r is an isomorphism in that

$$W_x^r(a_x O_x b_x) = W_x^r(a_x)W_x^r(O_x)W_x^r(b_x). \tag{25}$$

Here a_x and b_x are number values in $\bar{C}_x$ and O_x denotes the basic operations in $\bar{C}_x$.

W_r^y has a similar definition as it is an isomorphism from $\bar{C}_x^r$ to $\bar{C}_y$. Since the definition is similar it will not be given here.

$\bar{C}_x^r$ can also be represented in a form similar to that of Eq. 12 as

$$\bar{C}_{r,x} = \{C_x, \pm_{r,x}, \times_{r,x}, \div_{r,x}, 0_{r,x}, 1_{r,x}\}. \tag{26}$$

This structure can be described as the representation of $\bar{C}_y$ at x. The relation between the number values and operations in $\bar{C}_{r,x}$ and those in $\bar{C}_x$ is provided by $\bar{C}_x^r$ which defines the number values and operations of $\bar{C}_{r,x}$ in terms of those in $\bar{C}_x$. In this sense both $\bar{C}_{r,x}$ and $\bar{C}_x^r$ are different representations of the same structure. From now on $\bar{C}_{r,x}$ and $\bar{C}_x^r$ will be referred to as the representation of $\bar{C}_y$ at x and on $\bar{C}_x$ respectively.

The relations between the basic operations and constants of $\bar{C}_x^r$ and those of $\bar{C}_x$ lead to an interesting property. Let $f_x^r(a_x^r)$ be any analytic function on $\bar{C}_x^r$. It follows that

$$f_x^r(a_x^r) = b_x^r \Leftrightarrow rf_x(a_x) = rb_x \Leftrightarrow f_x(a_x) = b_x. \tag{27}$$

Here f_x is the same function on $\bar{C}_x$ as f_x^r is on $\bar{C}_x^r$. Also a_x and b_x are the same number values in $\bar{C}_x$ as a_x^r and b_x^r are in $\bar{C}_x^r$.

This result follows from the observation that any term $(a_x^r)^n/(b_x^r)^m$ in $\bar{C}_x^r$ satisfies the relation

$$\frac{(a_x^r)^n}{(b_x^r)^m}x = r\frac{(a_x)^n}{(b_x)^m}x. \tag{28}$$

The n factors and n-1 multiplications in the numerator contribute a factor of r. This is canceled by a factor of r in the denominator. The one r factor arises from the relation of division in $\bar{C}_x^r$ to that in $\bar{C}_x$.

Eq. 27 follows from the fact that Eq. 28 holds for each term in any convergent power series. As a result it holds for the power series itself.

3.2 Real numbers

Since the treatment for real numbers is similar to that for complex numbers, it will be summarized here. The representations of $\bar{R}_y$ at x and on $\bar{R}_x$ are given by Eqs. 10 and 11 as

$$\bar{R}_{r,x} = \{R_x, \pm_{r,x}, \times_{r,x}, \div_{r,x}, <_{r,x}, 0_{r,x}, 1_{r,x}\}$$
$$\bar{R}_x^r = \{R_x, \pm_x, \tfrac{\times_x}{r}, r \div x, <_x, 0_x, r_x\}. \tag{29}$$

Here $r = r_{x,y}$ is a positive real number.

The definition of parallel transforms for complex number structures applies here also. Let $F_{y,x}$ transform $\bar{R}_x$ to $\bar{R}_y$. $F_{y,x}$ defines the notion of same real number in that $a_y = F_{y,x}(a_x)$ is the same real umber in $\bar{R}_y$ as a_x is in $\bar{R}_x$. As was shown in Eqs. 20 and 21, $F_{y,x}$ can be factored into two operators as in $F_{y,x} = W_r^y W_x^r$ where

$$\bar{R}_y = W_r^y \bar{R}_x^r = W_r^y W_x^r \bar{R}_x = F_{y,x} \bar{R}_x. \tag{30}$$

W_x^r defines the scaled representation of $\bar{R}_y$ on $\bar{R}_x$. It is given explicitly by Eq. 24. W_r^y maps the scaled representation onto $\bar{R}_y$. Eqs. 27 and 28 also hold for the relations between any real valued analytic function on $\bar{R}_x^r$ and its correspondent on $\bar{R}_x$ in that $f_x^r(a_x^r) = r f_x(a_x)$.

3.3 Hilbert spaces

As noted in the introduction, Hilbert space structures have the form shown in Eq. 4 as

$$\bar{H}_x = \{H_x, +_x, -_x, \cdot_x, \langle -, - \rangle_x, \psi_x\}. \tag{31}$$

Complex numbers are included implicitly in that Hilbert spaces are closed under multiplication of vectors by complex numbers. Also scalar products are bilinear maps with complex values.

As was the case for numbers, parallel transformation of $\bar{H}_y$ to x maps $\bar{H}_y$ onto $\bar{H}_x$. If scaling of the numbers is included, then the local representation of $\bar{H}_y, \bar{C}_y$ on $\bar{H}_x, \bar{C}_x$ is given by $\bar{H}_x^r, \bar{C}_x^r$. The structure, $\bar{C}_x^r$, is shown in Eq. 23. $\bar{H}_x^r$ is given by

$$\bar{H}_x^r = \{H_x, \pm_x, \frac{\cdot_x}{r}, \frac{\langle -, - \rangle_x}{r}, r\psi_x\} \tag{32}$$

This equation gives explicitly the relations of operations and vectors of the local representation of $\bar{H}_y$ to those in $\bar{H}_x$. The relations are defined by the requirement that $\bar{H}_x^r$ satisfy the Hilbert space axioms (Kadison, 1983) if and only if $\bar{H}_x$ does.[2] Here $r\psi_x$ is the same vector in $\bar{H}_x^r$ as ψ_x is in $\bar{H}_x$.

---

[2] Support for the inclusion of r as a vector multiplier, as in $\bar{H}_x^r$, Eq. 32, is based on the equivalence between finite dimensional vector spaces and products of complex number fields (Kadison, 1983). If $\bar{H}_y$ and $\bar{H}_x$ are n dimensional spaces, then $\bar{H}_y \simeq \bar{C}_y^n$ and $\bar{H}_x \simeq \bar{C}_x^n$.

These equivalences extend to the local representation of $\bar{H}_y$ on $\bar{H}_x$. As the local representation of $\bar{C}_y$ on $\bar{C}_x, \bar{C}_x^r$ is the scalar field base for the local Hilbert space representation. It follows that $\bar{H}_x^r$ is equivalent to $(\bar{C}_x^r)^n$. A vector in $(\bar{C}_x^r)^n$ corresponds to an n-tuple, $\{a_{x,j}^r : j = 1, \cdots, n\}$ of number values in $\bar{C}_x^r$. Use of the fact that the value $a_{x,j}^r$ in $\bar{C}_x^r$, corresponds to the number value $ra_{j,x}$ in $\bar{C}_x$ shows that the n-tuple in $(\bar{C}_x^r)^n$ corresponds to the n-tuple, $r\{a_{j,x} : j = 1, \cdots, n\}$ in $\bar{C}_x^n$. These equivalences should extend to the case where $\bar{H}_y$ and $\bar{H}_x$ are separable, which is the case here.

Effects on Quantum Physics of the Local Availability of
Mathematics and Space Time Dependent Scaling Factors for Number Systems

55

The description of $\bar{H}_x^r$ given here is suitable for use in section 7 where wave packets for quantum systems are discussed. For gauge theories, the Hilbert spaces contain vectors for the internal variables of matter fields. In this case one has to include a gauge field to account for the freedom to choose bases (Montvay & Münster, 1994; Yang & Mills, 1954). The local representation of $\bar{H}_y$ on $\bar{H}_x$ is then given by (Benioffa, 2011; Benioffc, 2011)

$$\bar{H}_x^{r,V} = \{H_x, \pm_x, \frac{\cdot x}{r}, \frac{\langle -, -\rangle_x}{r}, rV\psi_x\}. \tag{33}$$

If the $\bar{H}_x$ are n dimensional, then V is an element of the gauge group, $U(n)$.

4. Gauge fields

As was noted in the introduction, for $y = x + \hat{v}dx$, $r_{y,x}$ can be represented as the exponential of a vector field:

$$r_{y,x} = e^{\vec{A}(x)\cdot\hat{v}dx} = e^{A_v(x)dx^v} \tag{34}$$

(sum over repeated indices implied). $\vec{A}(x)$ is also referred to as a gauge field as it gives the relations between neighboring complex number structures at different space time points. To first order in small quantities,

$$r_{y,x} = 1 + \vec{A}(x)\cdot\hat{v}dx. \tag{35}$$

The use of $r_{y,x}$ makes clear the fact that the setup described here is a generalization of the usual one. To see this, set $\vec{A}(x) = 0$ everywhere. Then $r_{y,x} = 1$ for all y, x and the local representations of $\bar{C}_y$ and $\bar{R}_y$ on $\bar{C}_x$ and $\bar{R}_x$ are $\bar{C}_x$ and $\bar{R}_x$. Since the $\bar{C}_x$ and $\bar{R}_x$ are then independent of x, one can replace $\bar{C}_x$ and $\bar{R}_x$ with just one complex and real number structure, $\bar{C}$ and $\bar{R}$.

4.1 Scale factors for distant points

The description of $r_{y,x}$ can be extended to points y distant from x. Let P be a path from x to y parameterized by a real number, s, such that $P(0) = x$ and $P(1) = y$. Let $r_{y,x}^P$ be the scale factor associated with the path P. If a_y is a number value in $\bar{C}_y$, then a_y corresponds to the number value, $r_{y,x}^P a_x$, in $\bar{C}_x$ where $a_x = F_{x,y}a_y$ is the same number value in $\bar{C}_x$ as a_y is in $\bar{C}_y$.

One would like to express $r_{y,x}^P$ as an exponential of a line integral along P of the field $\vec{A}(x)$. However this is problematic because the integral corresponds to a sum over s of complex number values in $\bar{C}_{P(s)}$. Such a sum is not defined because addition is defined only within a number structure. It is not defined between different structures.

This can be remedied by referring all terms in the sum to one number structure such as $\bar{C}_x$. To see how this works, consider a two step path from x to $y = x + \hat{v}_1\Delta_x$ and from y to $z = y + \hat{v}_2\Delta_y$. Δ_y is the same number in $\bar{C}_y$ as Δ_x is in $\bar{C}_x$.

Let a_z be a number value in $\bar{C}_z$. a_z corresponds to the number value $r_{z,y} \times_y a_y$ in $\bar{C}_y$. Here $a_y = F_{y,z}a_z$ is the same number value in $\bar{C}_y$ as a_z is in $\bar{C}_z$. In $\bar{C}_x$, $r_{z,y} \times_y a_y$ corresponds to the number value given by

$$r_{y,x} \times_x (r_{z,y})_x(\frac{\times_x}{r_{y,x}})(r_{y,x} \times_x a_x) = (r_{z,y})_x r_{y,x} a_x. \tag{36}$$

Here $(r_{z,y})_x = F_{x,y}(r_{z,y})$ is the same number in $\bar{C}_x$ as $r_{z,y}$ is in $\bar{C}_y$ and $\times_x/r_{y,x}$ is the representation of $\times_y$ on $\bar{C}_x$. The $\bar{C}_x$ multiplications are implied in the righthand term and $a_x = F_{x,y}a_y$.

The factor $(r_{z,y})_x r_{y,x}$ can be expressed in terms of the field $\vec{A}$. It is

$$(r_{z,y})_x r_{y,x} = e^{(\vec{A}(y))_x \cdot \hat{v}_2 \Delta_x + \vec{A}(x) \cdot \hat{v}_1 \Delta_x}. \tag{37}$$

Δ_y is replaced here by its same value Δ_x in $\bar{C}_x$.

Let P be an n step path where $P(0) = x_0 = x, P(j) = x_j, P(n-1) = x_{n-1} = y$ and $x_{j+1} = x_j + \hat{v}_j \Delta_{x_j}$. Then $r^P_{y,x} a_x$ is given by

$$r^P_{y,x} = \prod_{j=0}^{n-1} (r_{x_{j+1},x_j})_x = \exp\left(\sum_{j=0}^{n-1} [(\vec{A}(x_j)) \cdot \hat{v}_j \Delta_{x_j}]_x\right). \tag{38}$$

The subscript x denotes the fact that all terms in the product and in the exponential, are values in $\bar{C}_x$. For example $(r_{x_{j+1},x_j})_x = F_{x,x_j} r_{x_{j+1},x_j}$ is the same value in $\bar{C}_x$ as r_{x_{j+1},x_j} is in $\bar{C}_{x_j}$. An ordering of terms in the product of Eq, 38 is not needed because the different r factors commute with one another.

This can be extended to a line integral along P. The result is (Benioffa, 2011)

$$r^P_{y,x} = \exp\left\{ \int_0^1 (\vec{A}(P(s)))_x \cdot \left(\frac{dP(s)}{ds}\right)_x ds \right\} = \exp\left\{ \int_P (\vec{A}(\vec{z}))_x d\vec{z} \right\}. \tag{39}$$

The subscript x on the factors in the integral mean that the terms are all evaluated in $\bar{C}_x$.

It is unknown if the field $\vec{A}$ and thereby $r^P_{y,x}$ is or is not independent of the path P from x to y. If $\vec{A}$ is not integrable, then the path dependence introduces complications. In particular it means that for y distant from x, there is no path independent way to describe the local representation of $\bar{C}_y$ on $\bar{C}_x$. The local representation would have to include a path variable as in $\bar{C}_x^{r^P_{y,x}}$.

In this work, this complication will be avoided by assuming that $\vec{A}$ is integrable. Then $r^P_{y,z} = r_{y,x}$, independent of P.

Let P be a path from x to y and Q be another path from y to x. Then integrability gives

$$r^{Q*P}_{x,x} = (r^Q_{x,y})_x r^P_{y,x} = 1_x \tag{40}$$

Here $Q * P$ is the concatenation of Q to P. This result gives

$$(r^Q_{x,y})_x = (r^P_{y,x})^{-1}. \tag{41}$$

5. Local availability of mathematics

The local availability of mathematics means that for an observer, O_x, at point x, the mathematics that O_x can use, or is aware of, is locally available at x. Since mathematical systems are represented by structures, (Barwise, 1977; Keisler, 1977), one can use V_x to denote the collection of all these structures. V_x includes real and complex number and Hilbert space

Effects on Quantum Physics of the Local Availability of
Mathematics and Space Time Dependent Scaling Factors for Number Systems

57

structures $\bar{R}_x, \bar{C}_x, \bar{H}_x$, structures for operator algebras as well as many other structure types. All the mathematics that O_x uses to make physical predictions and physical theories use the systems in $\bigvee_x$. Similarly, all the mathematics available to an observer O_y at point y is contained in $\bigvee_y$.

An important requirement is that the mathematics available, in principle at least, to an observer must be independent of the observers location. This means that $\bigvee_y$ must be equivalent to $\bigvee_x$. For each system structure in $\bigvee_y$, there must be a corresponding structure in $\bigvee_x$. Conversely, for each system structure in $\bigvee_x$ there must be a corresponding structure in $\bigvee_y$. Furthermore the corresponding structures in $\bigvee_y$ and $\bigvee_x$ must be related by parallel transforms that map one structure to another. These parallel transforms define what is meant by the *same* structure and the same structure elements and operations in $\bigvee_x$ as in $\bigvee_y$, and conversely.

This use of parallel transforms is an extension to other types of mathematical systems, of the definitions and use of parallel transforms, Section 3, to relate complex and real number structures at different space time points. It is based on the description of each type of mathematical system as structures, each consisting of a base set, basic operations, relations and constants, that satisfy a set of axioms relevant to the system type.(Barwise, 1977; Keisler, 1977).

The association of an observer to a point, as in O_x, is an idealization, mainly because observers, e.g. humans, have a finite size. Because of this, an observer's location is a region and not a point. This is the case if one notes that the observer's brain is the seat of all mathematical knowledge and limits consideration to the brain. In addition, quantum mechanics introduces an inherent uncertainty to the location of any system. In spite of these caveats, the association of an observer to a point will be used here.

An important aspect of $\bigvee_x$ is that O_x must be able to use the systems in $\bigvee_x$ to describe the systems in $\bigvee_y$. This can be done by means of parallel transform maps or correspondence maps from systems in $\bigvee_y$ to those in $\bigvee_x$. Parallel transforms map elements and operations of system structures $\bar{S}_y$ to the same elements and operations of $\bar{S}_x$. In this case O_x can use the mathematics of $\bar{S}_x$ as a stand in for the mathematics of $\bar{S}_y$.

Correspondence maps take account of scaling of real and complex numbers in relating systems at y to those at x. In this case O_x describes the mathematics of $\bar{S}_y$ in terms of the local representation, $\bar{S}_x^r$, of $\bar{S}_y$ on $\bar{S}_x$. If $S = R$ or $S = C$ then O_x would describe the properties of $\bar{R}_y$ or $\bar{C}_y$ in terms of the local scaled systems $\bar{R}_x^r$ and $\bar{C}_x^r$.

The existence of correspondence maps means that for each system type, S, $\bigvee_x$ contains all the scaled systems, $\bar{S}_x^r$, for each point y, in addition to $\bar{S}_x$. (Recall $r = r_{y,x}$.) They include scaled real numbers, $\bar{R}_x^r$, complex numbers, $\bar{C}_x^r$, and scaled Hilbert spaces, $\bar{H}_x^r$, as well as many other system types.

All these scaled systems are available to an observer, O_x at x. Since they are locally available, O_x takes account of scaling by using them to make theoretical calculations that require inclusion of numbers or vectors at different space or space time points. If O_x does not use these correspondence maps and restricts use to parallel transform maps only, then the setup becomes simpler in that each $\bar{S}_x^r$ is identical to $\bar{S}_x$.

This raises the question of when correspondence maps can be used instead of parallel transform maps. This will be discussed in the next sections. Here the use of correspondence maps follows from the inclusion into physics of the freedom to choose number systems at different space time points. In this sense it extends the freedom to choose bases in vector spaces in gauge field theory (Montvay & Münster, 1994; Yang & Mills, 1954) to the underlying scalars.

6. Correspondence maps and parallel transform maps

It is proposed here that correspondence maps be used in any theoretical physics expression that requires the mathematical comparison of mathematical entities at different space time points. Typical examples are space and time derivatives or integrals of functions or functionals as maps from space time (or space and time) to elements of mathematical systems that are based on the real or complex numbers as scalars. An example is an n component complex scalar field.

At this point it is not completely clear if there are other cases in which correspondence maps should be used instead of parallel transform maps. However it is clear that there are many situations where these maps should not be used. These involve what is referred to here as the commerce of mathematics and physics.[3]

To see this suppose O_x wants to compare the numerical output of either an experiment or a theoretical computation, done at x, with the numerical output of either an experiment or computation done at y. Let b_x and d_y be the real valued numerical outcomes obtained. Use of the correspondence maps means that O_x would compare b_x with the local representation of d_y at x, that is, with the number $r_{y,x}d_x$. Here d_x is the same number in $\bar{R}_x$ as d_y is in $\bar{R}_y$.

This is contradicted by experience. There is no hint of a factor $r_{y,x}$ in comparing outcomes of repeated experiments, or comparing experimental outcomes with theoretical predictions, or in any other use of numbers in commerce. If one ignores statistical and quantum uncertainties, numerical outcomes of repeated experiments or repeated computations are the same irrespective of when and where they are done.[4]

The reason for this is a consequence of a basic fact. This is that no experiment and no computation *ever* directly yields a numerical value as an outcome. Instead the outcome of any experiment is a physical system in a physical state that is *interpreted* as a numerical value. Similarly the outcome of a computation is a physical system in a state that is *interpreted* as a numerical value.

The crucial word here is *interpreted*. If ψ_y is the output state of a measurement apparatus for an experiment at point y, and ϕ_x is the output state of a computation at point x, then the numerical values of these output states are given by $a_y = I_y(\psi_y)$ and $b_x = I_x(\phi_x)$. Here I_y and I_x are interpretive maps from the output states of the measurement system into $\bar{R}_y$ and from the computation output states into $\bar{R}_x$ respectively. The space time dependence of the maps is indicated by the x, y subscripts.

---

[3] Mathematical and physical commerce refers to the use of numerical values as outcomes of computations and experiments in science, business, and communication.

[4] A similar criticism of a suggestion by Weyl was made almost 100 years ago by Einstein (O'Raifeartaigh, 1997).

The "Naheinformationsprinzip", no information at a distance principle (Mack, 1981; Montvay & Münster, 1994), forbids direct comparison of the information in ψ_y with that in ϕ_x. This means that $I_y(\psi_y)$ and $I_x(\phi_x)$ cannot be directly compared. Instead the information contained in ψ_y and that contained in ϕ_x must be transported, by physical means, to a common point for comparison.

There are many different methods of physical information transmission. Included are optical and electronic methods as well as older slower methods. All methods involve motion of an information carrying physical system from one point to another, "information is physical" Landauer (1991). The physical system used should be such that the state of the information carrying degrees of freedom does not change during transmission from one point to another.

Figure 1 illustrates schematically, in one dimensional space and time, the nonrelativistic transmission of a theory computation output state obtained at x', u and an experimental output state obtained at y, v to a common point, x, t, for comparison. Here x, x', y are space locations and u, v, t are times.

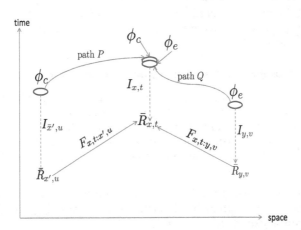

Fig. 1. A simple example of comparing theory with experiment. The ovals denote the output computation and experiment systems in states ϕ_c and ϕ_e at space and times, x', u and y, v. P and Q denote the paths followed by these systems. One has $P(u) = x'$ and $Q(v) = y$. The double oval in the center denotes the the two systems at the point, x, of path intersection where $P(t) = Q(t) = x$. The interpretation maps are denoted by $I_{P(s),s}$ and $I_{Q(s),s}$ for different times s. The real number structures $\bar{R}_{P(s),s}$ and $\bar{R}_{Q(s),s}$ are associated with each point in the paths P and Q. $F_{x,t;x',u}$ and $F_{x,t;y,v}$ are parallel transform operators that map the real number structures at the points of theory and experiment completion to the point of path intersection.

The figure, and the discussion, illustrate a general principle. All activities in the commerce of mathematics and physics consist of physical procedures and operations that generate physical output systems in states that are interpreted as numerical values. The "no information at a distance" principle forbids direct comparison of the associated number values at different points. Instead the systems or suitable information carriers must be brought to a common point where the numerical information, as number values in just one real number structure, can be locally compared.

Similar considerations apply to storage of outcomes of experiments or computations either in external systems or in the observers brain. As physical dynamic systems, observers move in space time. If P is a path taken by an observer, with $P(\tau)$ the observers location at proper time τ, the mathematics available to $O_{P(\tau)}$ is that in $\vee_{P(\tau)}$. If $\phi(P(\tau))$ denotes the state of a real number memory trace in an observers brain, then the number value represented by $\phi(P(\tau))$ is given by $I_{P(\tau)}(\phi(P(\tau)))$. This is a number value in $\bar{R}_{P(\tau)}$. At a later proper time τ', the number value represented by the memory trace is $I_{P(\tau')}(\phi(P(\tau')))$. If there is no degradation of the memory trace, then $I_{P(\tau)}(\phi(P(\tau)))$ is the same number value in $\bar{R}_{P(\tau)}$ as $I_{P(\tau')}(\phi(P(\tau')))$ is in $\bar{R}_{P(\tau')}$. Correspondence maps play no role here either.

7. Quantum theory

As might be expected, the local availability of mathematics and the freedom of choice of number scaling factors, have an effect on quantum theory. This is a consequence of the use of space time integrals and derivatives in the theory. For example, one would expect to see the gauge field $\vec{A}$ appear in quantum descriptions of physical systems. To see how this effect arises, it is useful to limit the treatment to nonrelativistic quantum mechanics on three dimensional Euclidean space, R^3.

7.1 Effect of the local availability of mathematics on quantum theory

The local availability of mathematics requires that the usual setup of just one $\bar{C}, \bar{R}, \bar{H}$ is replaced by separate number structures, $\bar{R}_x, \bar{C}_x$ and separate Hilbert spaces, $\bar{H}_x$, associated with each x in R^3.[5] It follows that mathematical operations, such as space or time derivatives or integrals, which involve nonlocal mathematical operations on numbers or vectors at different points, cannot be done. The reason is that these operations violate mathematical locality.

To preserve locality, one must use either parallel transformations or correspondence transformations. These two methods are well illustrated by considering a single particle wave packets. The usual representation has the form

$$\psi = \int \psi(y)|y\rangle dy \tag{42}$$

where the integral is over all space points in R^3.

One result of the local availability of mathematics is that, for each y, the vector, $\psi(y)|y\rangle$, is in $\bar{H}_y$, just as $\psi(y)$ is a number value in $\bar{C}_y$. It follows that the space integral over y makes no sense. It describes a suitable limit of adding vectors that belong to different Hilbert spaces. Addition is not defined between different spaces; it is defined only within one Hilbert space and complex number structure.

---

[5] The association of separate Hilbert spaces to each point x is different here from that used in gauge theory (Montvay & Münster, 1994). In gauge theory, the spaces are all finite dimensional and apply to internal states of the Fermion fields. Here the Hilbert spaces describe states of systems spread over space, e.g. as wave packets.

Effects on Quantum Physics of the Local Availability of
Mathematics and Space Time Dependent Scaling Factors for Number Systems

61

The use of parallel transformations replaces Eq. 42 by

$$\psi_x = \int_x \psi(y)_x |y_x\rangle_x dy_x. \tag{43}$$

Here $\psi(y)_x = F_{x,y}\psi(y)$ is the same number value in $\bar{C}_x$ as $\psi(y)$ is in $\bar{C}_y$, the number triple, y_x, in $|y_x\rangle_x$ is the same triple in $\bar{R}_x^3$ as y is in $\bar{R}_y^3$, and $|y_x\rangle_x$ is the same state in $\bar{H}_x$ as $|y\rangle$ is in $\bar{H}_y$. The differentials $dy_x = dy_x^1 dy_x^2 dy_x^3$ refer to $\bar{R}_x^3$. The subscript x in $\int$ indicates that the integral is based on $\bar{H}_x, \bar{C}_x$. The representations of sameness given above are shown explicitly by

$$\begin{aligned}
\psi(y)_x &= F_{x,y}\psi(y) \\
y_x &= \bar{F}_{x,y}(y) \\
|y_x\rangle_x &= |\bar{F}_{x,y}(y)\rangle_x \\
dy_x &= \bar{F}_{x,y}(dy).
\end{aligned} \tag{44}$$

Also $|\bar{F}_{x,y}(y)\rangle_x$ is the same basis vector in $\bar{H}_x$ as $|y\rangle$ is in $\bar{H}_y$.

Note that the point x on which the integral is based is arbitrary. Eq. 43 holds if the subscript x is replaced by another point z. Then the integral is based on $\bar{C}_z, \bar{H}_z$.

The use of parallel transforms is applicable to other aspects of quantum mechanics. For each y in R^3, the momentum operator, $\mathbf{p}_y$, for vectors in $\bar{H}_y$ is given by

$$\mathbf{p}_y = i_y \hbar_y \nabla_y = i_y \hbar_y \sum_{j=1}^{3} \partial_{j,y}. \tag{45}$$

Here, $i_y, \hbar_y$ are numbers in $\bar{C}_y$. The action of $\mathbf{p}_y$ on a vector ψ at point y gives for the jth component

$$\mathbf{p}_{y,j}\psi = i_y \hbar_y \partial_{j,y}\psi = \frac{\psi(y + dy^j) - \psi(y)}{dy^j}. \tag{46}$$

As was the case for the space integral, this expression makes no sense because $\psi(y + dy^j)$ is in $\bar{C}_{y+d^j y}$ and $\psi(y)$ is in $\bar{C}_y$.

This can be remedied by replacing $\partial_{j,y}$ by $\partial'_{j,y}$ where

$$\partial'_{j,y}\psi = \frac{\psi(y + dy^j)_y - \psi(y)}{dy^j}. \tag{47}$$

Here $\psi(y + dy^j)_y = F_{y,y+d^j y}\psi(y + dy^j)$. It follows from this that the expression for the momentum becomes

$$\mathbf{p}'_y\psi = \sum_{j=1}^{3} p'_{j,y}\psi = i_y \hbar_y \sum_{j=1}^{3} \partial'_{j,y}\psi. \tag{48}$$

The Hamiltonian for a single quantum system in an external potential, acting on a state, ψ_y at point y, is given by

$$H_y\psi(y) = -\frac{\hbar_y^2}{2m_y} \sum_{j=1}^{3} (\partial')_{y,j}^2 \psi(y) + V(y)\psi(y). \tag{49}$$

Here, $\hbar_y$ and m_y are Planck's constant and the particle mass. They have values in $\bar{R}_y$. The values of the external potential, $V(y)$, are also in $\bar{R}_y$.

The main difference between this and the usual expression for a Hamiltonian is the replacement of $\partial_{j,y}$ with $\partial'_{j,y}$. Otherwise, the expressions are the same.

For a single particle state, ψ, the momentum representation is $\psi = \int \psi(p)|p\rangle dp$. Here $dp = dp_1 dp_2 dp_3$. Since the amplitude $\psi(p)$ is a complex number value and no location for the value is specified, one may choose any location, x, such as that of an observer, O_x, to assign $\psi(p)$ as a number value in $\bar{C}_x$ and the integral as an element of V_x.

The relation between $\psi(p)$ and $\psi(x)$ is given by the Fourier transform. The components of the space integral in

$$\psi(p) = \int e^{i_z p_z z} \psi(z) dz \tag{50}$$

must all be mapped to a common point, x, for the integral to make sense. This gives

$$\psi(p)_x = \int_x (e^{i_z p_z z})_x \psi(z)_x dz_x \tag{51}$$

Here

$$(e^{i_z p_z z})_x = F_{x,z} e^{i_z p_z z} = e^{i_x p_x z_x} \tag{52}$$

is the same number in $\bar{C}_x$ as $e^{i_z p_z z}$ is in $\bar{C}_z$.

The treatment described can be extended to multiparticle entangled states. It is sufficient to consider two particle states. For example a two particle state $\psi_{1,2}$ where the total momentum of the two particles is 0 can be expressed by

$$\psi_{1,2} = \int \psi_1(p)\psi_2(-p)|p\rangle_1|-p\rangle_2 dp. \tag{53}$$

Use of Fourier transforms gives

$$\psi_{1,2} = \int dz_1 dz_2 (\int e^{iz_1 p} \psi_1(p)|p\rangle_1 e^{-iz_2 p} \psi_2(-p)|-p\rangle_2 dp). \tag{54}$$

The integral must be transformed to a Hilbert space with just one scalar field. For a point x with $\bar{C}_x, \bar{H}_x$, the integrand factors are parallel transformed to obtain

$$(\psi_{1,2})_x = \int_x (dz_1)_x (dz_2)_x (\int (e^{iz_1 p})_x \psi_1(p)_x|p_x\rangle_1 (e^{-iz_2 p})_x \psi_2(-p)_x|-p_x\rangle_2 dp_x). \tag{55}$$

Here

$$\begin{aligned}(e^{iz_1 p})_x &= F_{x,z_1}(e^{iz_1 p})\\(e^{-iz_2 p})_x &= F_{x,z_2}(e^{-iz_2 p})\end{aligned} \tag{56}$$

Also p_x is the same value in $\bar{C}_x$ as p is in $\bar{C}_{z_1}$ in the z_1 integral, Eq. 54, and $-p_x$ is the same value in $\bar{C}_x$ as $-p$ is in $\bar{C}_{z_2}$ in the z_2 integral.

Effects on Quantum Physics of the Local Availability of
Mathematics and Space Time Dependent Scaling Factors for Number Systems

63

7.2 Inclusion of number system scale factors

The above shows that the imposition of "mathematics is local" on quantum theory is more complex than the usual treatment with just one scalar field and one Hilbert space for all space points. Since the description with parallel transforms is equivalent to the usual one, the added complexity is not needed if one goes no further with it.

This is not the case if one extends the treatment to include space dependent scaling factors for the different $\bar{C}_x, \bar{R}_x$. For a given x, the local representations of $\bar{C}_y$ on $\bar{C}_x$ are given by scaled representations, $\bar{C}_x^{r_{y,x}}$, of $\bar{C}_y$ on $\bar{C}_x$. Also the local representation of $\bar{H}_y$ on $\bar{H}_x$ with effects of the number scaling included, is given by $\bar{H}_x^{r_{y,x}}$.

For $y = x + \hat{v}dx$, a neighbor point of x, the scaling factor, $r_{y,x}$ is given by, $r_{y,x} = e^{\vec{A}(x)\cdot\hat{v}dx}$, Eq. 34. If y is distant from x and $\vec{A}(x)$ is integrable, then, expressing $r_{y,x}$ as an integral along a straight line path from x to y gives, (Eq. 39)

$$r_{y,x} = \exp\left(\int_0^1 \sum_{i=1}^3 A_i(sx_i)_x s(x_i)_x ds\right) = \exp\left(\sum_{i=1}^3 \int_{x_i}^{y_i} A_i(z_i)_x dz_x^i\right) = \prod_{i=1}^3 r_{y,x,i} \tag{57}$$

Here $x_i = \vec{x} \cdot \hat{i}$ and $y_i = \vec{y} \cdot \hat{i}$ are the components of $\vec{x}$ and $\vec{y}$ in the direction i. The last equality assumes that the components of $\vec{A}$ commute with one another.[6] The subscript x indicates that the integrals are defined on $\bar{R}_x$.

The presence of $\vec{A}(x)$ affects the expression of a wave packet state ψ as given by Eq. 43. In this case the wave packet expansion of ψ_x is given by

$$\psi_x = \int_x r_{y,x}\psi(y)_x|y_x\rangle_x dy_x \tag{58}$$

where $r_{y,x}$ is given by Eq. 57.

This result is obtained by noting that for each y the local representation of $\bar{H}_y, \bar{C}_y$ on $\bar{H}_x, \bar{C}_x$, with scaling factor included, is $\bar{H}_x^{r_{y,x}}, \bar{C}_x^{r_{y,x}}$. The vector in $\bar{H}_x^{r_{y,x}}$ that is the same vector as $\psi(y)|y\rangle$ is in $\bar{H}_y$, is denoted by $\psi(y)_x^r \cdot_x^r |y_x^r\rangle$. Here $\psi(y)_x^r$ is the same number value in $\bar{C}_x^{r_{y,x}}$ as $\psi(y)$ is in $\bar{C}_y$, and $|y_x^r\rangle$ is the same vector in $\bar{H}_x^{r_{y,x}}$ as $|y\rangle$ is in $\bar{H}_y$ as $|y_x\rangle_x$ is in $\bar{H}_x$. This follows from the observation that y, y_x^r, and y_x are the respective same numbers in $\bar{R}_y^3, (\bar{R}_x^r)^3$, and $\bar{R}_x^3$. Scalar vector multiplication in $\bar{H}_x^{r_{y,x}}$ is shown by $\cdot_x^r$.

The corresponding state on $\bar{H}_x$ is obtained by noting that

$$\psi_x^r \cdot_x^r |y_x^r\rangle_x^r \Rightarrow r_{y,x}\psi(y)_x(\cdot_x^r)_x|r_{y,x}y_x\rangle_x = r_{y,x}\psi(y)_x \frac{\cdot_x}{r_{y,x}} r_{y,x}|y_x\rangle_x = r_{y,x} \cdot_x |y_x\rangle_x. \tag{59}$$

This is the result shown in Eq. 58. The use of $(\cdot_x^r)_x = \frac{\cdot_x}{r_{y,x}}$, Eq. 32, is based on the requirement that $\bar{H}_x^{r_{y,x}}$ satisfies the same Hilbert space axioms (Kadison, 1983) as does $\bar{H}_x$.

It must be emphasized that the usual predictions of the quantum mechanical properties of wave packets, with $\vec{A}(x) = 0$ everywhere, do a good job of prediction of experimental results. So far, quantum mechanical predictions and experiments have not shown the need for the

---

[6] Here, and in what follows, $\vec{A}(x)$ is assumed to be integrable.

presence of $\vec{A}$. This shows that the effect of the $\vec{A}$ field must be very small, either through the values of the field itself or by use of a very small coupling constant, g, of the field to numbers and vectors. This would be accounted for by replacing $\vec{A}$ in Eqs. 34 and 57 by $g\vec{A}$.

In this sense the presence of $\vec{A}$ is no different than the presence of the gravitational field. In theory, a proper description of quantum mechanics of systems should include the effects of the gravitational field. However, it can be safely neglected because the field is so small, at least far away from black holes where quantum physics is done.

Another feature of Eq. 58 is the dependence on the reference point x. This can be removed by restricting the integration volume to a region of space, excluding x, where the region contains essentially all of ψ. This is what one does in any experiment since ψ is prepared in a region that does not include the observer.

The removal of x dependence then follows from expressing Eq. 58 as a sum of two terms, one as the integral over V and the other over all space outside V:

$$\psi_x = \int_{x,V} r_{y,x}\psi(y)_x|y_x\rangle_x dy_x + \int_{x,W} r_{y,x}\psi(y)_x|y_x\rangle_x dy_x. \tag{60}$$

The subscript, W, on the second integral means that it refers to integration over all space outside V. Here x is a point in W.

Assume that V is chosen so that the integral over W can be neglected. Then

$$\psi_x \cong \int_{x,V} r_{y,x}\psi(y)_x|y_x\rangle_x dy_x. \tag{61}$$

This equation has a problem in that the correspondence transforms are extended from any point in V to a point outside V. However these transforms are restricted here to apply within space or space time integrals, and not outside the integration volume.

This can be fixed by choice of a point z on the surface of V and replacing $r_{y,x}$ by $U_{x,z}r_{y,z}$. The factor $r_{y,z}$ accounts for the correspondence transform from a point y in V to a point z on the surface of V, and $U_{x,z}$ is a unitary operator that parallel transforms the result from z to x. Then one has

$$\psi_{x,z} \cong U_{x,z}\psi_z = U_{x,z}\int_{z,V} r_{y,z}\psi(y)_z|y_z\rangle_z dy_z = \int_{x,V}(r_{y,z})_x\psi(y)_x|y_x\rangle_x dy_x. \tag{62}$$

The subscript z on $\psi_{x,z}$ indicates a possible dependence on the choice of z on the surface of V. Figure 2 illustrates the setup for two points y, u in V.

This result shows that the wave packet representation is independent of x, such as an observers location, provided it is outside of V. However, to the extent that $\vec{A}$ cannot be neglected, the representation does depend on the location of z. To see what this dependence is, let w be another point on the surface of V. Then, following Eq. 62, $\psi_{x,w}$ is given by

$$\psi_{x,w} \cong U_{x,w}\psi_w = U_{x,w}\int_{w,V} r_{y,w}\psi(y)_w|y_w\rangle_w dy_w = \int_{x,V}(r_{y,w})_x\psi(y)_x|y_x\rangle_x dy_x. \tag{63}$$

For the comparison, at x, of $\psi_{x,z}$ with $\psi_{x.w}$, it is sufficient to compare, in $\bar{H}_z$, the parallel transformation of ψ_w to z with ψ_z, in $\bar{H}_z$. The parallel transformation of ψ_w to z is given by

$$(\psi_w)_z = \int_{z,V}(r_{y,w})_z(\psi(y)_w)_z|(y_w)_z\rangle_z d(y_w)_z. \tag{64}$$

Effects on Quantum Physics of the Local Availability of
Mathematics and Space Time Dependent Scaling Factors for Number Systems

65

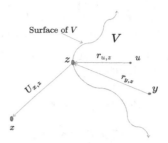

Fig. 2. Representation of scaling factors in the integrals from point z on the surface of V to points y and u. The direction implied in the order of the subscripts of r is opposite to the direction of the parallel transformations taking the integrand factors from y and u to z. $U_{x,z}$ parallel transforms ψ_z to $(\psi_z)_x = \psi_x$.

Use of the fact that parallel transforms of numbers and vectors from y to w and then to z are the same as transforms from y to z gives

$$(\psi_w)_z = \int_{z,V} (r_{y,w})_z \psi(y)_z |y_z)_z dy_z. \tag{65}$$

Since $\vec{A}$ is integrable, one can write

$$(r_{y,w})_z = (r_{y,z} r_{z,w})_z = r_{y,z}(r_{z,w})_z \tag{66}$$

to obtain

$$(\psi_w)_z = (r_{z,w})_z \int_{z,V} r_{y,z} \psi(y)_z |y_z)_z dy_z. \tag{67}$$

The subscript, z, denotes parallel transformation to z, of mathematical elements that are not in V_z. No subscript appears on $r_{y,z}$ as it is already a number value in $\bar{R}_z$.

This shows that $(\psi_w)_z$ differs from ψ_z by a factor, $(r_{z,w})_z = (r_{w,z})^{-1}$, Eq. 41. The difference is preserved on parallel transformation to x in that $(\psi_w)_x = \psi_{x,w}$ differs from $(\psi_z)_x = \psi_{x,z}$ by a factor $(r_{z,w})_x$.

If the effect of $\vec{A}$ is small, then it is useful to express $r_{y,x}$ as an expansion to first order in the exponential. For example the expression for ψ_x in Eq. 58 becomes,

$$\psi_x \cong \int_x (1 + \int_x^y \vec{A}(w)_x \cdot \hat{v} dw_x) \psi(y)_x |y_x)_x dy_x. \tag{68}$$

Here $\hat{v}$ is a unit vector along the direction from x to y. The first term of the expansion corresponds to the usual case with $\vec{A}$ equal to 0 everywhere. The x dependence arises from the second term, which gives the correction due to the presence of $\vec{A}$.

The restriction of the integration to a finite volume V, as in Eq. 62, removes the dependence on x in that ψ_x is the same vector in $\bar{H}_x$ as ψ_y is in $\bar{H}_y$ provided y is not in V. Expansion of $(r_{y,z})_x$ to first order in small terms shows that the z dependence arises from the $\vec{A}$ containing term as in Eq. 68.

The dependence on z can be appreciable because z can be any point on the surface of V. What is interesting is that this dependence can be greatly reduced by using the properties of actual measurements to minimize the effect of $\vec{A}$ on the predicted expectation value.

Consider a position measurement on a system in state ψ_x, Eq. 58. The expectation value for this measurement, calculated at x, is given by

$$\langle \psi_x | \bar{y} | \psi_x \rangle_x = \int_x r_{y,x} y_x |\psi(y)|_x^2 dy_x. \tag{69}$$

This expectation value[7] is an idealization or what one does. It does not take account of what one actually does.

Typically, position measurements are done by dividing a volume of space up into cubes and measuring the relative frequency of occurrence of the quantum system in the different cubes. A measurement consists of a large number of repetitions of this measurement on repeated preparations of the system in state ψ. Assume the cubes in space have volume Δ^3 where Δ is the length of a side. Then outcomes of the repeated experiment are "yeses" from the cube detectors whose locations are denoted by triples, j, k, l of integers. Each "yes" means that the location is somewhere in the volume of the responding detector cube located at position, $z_{j,k,l} = j\Delta, k\Delta, l\Delta$. The local availability of mathematics means that $z_{j,k,l}$ is a triple of numbers in $\bar{R}_{z_{j,k,l}}$.

The presence of parallel and correspondence transformations enables physical theory to express exactly what is done experimentally. Eq. 58 for ψ_x is replaced by an expression that limits integrals with scaling factors to the cube volumes and parallel transports these integrals to a common point x where they can be added together. The result, ψ'_x, is given by

$$\psi'_x = \sum_{j,k,l} U_{x,z} \int_{V_{j,k,l}} r_{w,z} \psi(w)_z |w_z\rangle_z dw_z. \tag{70}$$

Here the sum is over all cubes. Each cube is labeled by a point $z = z_{j,k,l}$ on the cube surface. Each integral over the volume, $V_{j,k,l} = \Delta^3$, of cube, j, k, l, is a vector in $\bar{H}_z$. Within each integral, the r factor scales the values of each integrand at point w to values at z. $U_{x,z}$ parallel transforms the integrals at different z to a common point x.

The theoretical expectation value for the experimental setup described here is given by

$$\langle \psi' | \bar{y} | \psi' \rangle_x = \sum_{j,k,l} U_{x,z} \int_{V_{j,k,l}} r_{w,z} w_z |\psi(w)|_z^2 dw_z. \tag{71}$$

The effect of the r factor is smaller here than it is in the expectation value using ψ_x. The reason is that it is limited to integrations over small volumes.

This representation of the prediction is supported by the discussion on mathematical and physical commerce. The "no information at a distance" principle requires that the information contained in the outcomes of each position measurement, as physical systems in "yes" or "no" states for each point $z_{j,k,l}$, be transmitted by physical means to x where the results of the

---

[7] All but one of the r factors appearing in the integrand are canceled by the r factors in the denominators of the multiplication operations.

Effects on Quantum Physics of the Local Availability of
Mathematics and Space Time Dependent Scaling Factors for Number Systems

67

repeated measurements are tabulated. The tabulation is all done at x. No factor involving $\vec{A}$ appears in the transmission or tabulation.

As noted, the effect of $\vec{A}$ appears only in the integrals over the volumes Δ^3. In these integrals, to first order,

$$r_{w,z} \cong 1_z + \int_z^w \vec{A}(y)_z \cdot \hat{n} dy_z. \tag{72}$$

Since the integral is limited to points within the volume Δ^3, it is clear that as $\Delta \to 0$, the integrals for each cube also approach 0.

This shows that the effect of the $\vec{A}$ field diminishes as the accuracy of the measurement increases. In the limit $\Delta = 0$, the $\vec{A}$ field disappears and one gets the usual theoretical prediction without $\vec{A}$ present. However, the Heisenberg uncertainty principle prevents the limit, $\Delta = 0$, from actually being achieved.

The presence of the $\vec{A}$ field affects other quantum mechanical properties of systems. For example, the description of the momentum operator with $\vec{A} \neq 0$, replaces Eq. 48 by

$$\mathbf{p}_{A,y}\psi = \sum_{j=1}^{3} p_{A,j,y}\psi = i_y \hbar_y \sum_{j=1}^{3} D_{j,y}\psi. \tag{73}$$

$D_{j,y}\psi$ is given by altering Eq. 47 to read

$$D_{j,y}\psi = \frac{r_{y+dy^j,y}\psi(y+dy^j)_y - \psi(y)}{dy^j}. \tag{74}$$

Here $r_{y+dy^j,z}\psi(y+dy^j)_y$ is the number value in $\bar{C}_y$ that corresponds to $\psi(y+dy^j)$ in $\bar{C}_{y+dy^j}$.

Using the fact that

$$r_{y+d^jy,y} = e^{A_j(y)d^jy} \tag{75}$$

and expansion to first order in the exponential gives,

$$D_{j,y}\psi = \partial'_{j,y}\psi + A_j(y)\psi(y). \tag{76}$$

The momentum components become

$$p_{A,j,y} = i_y \hbar_y D_{j,y} = i_y \hbar_y (\partial'_{j,y} + A_j(y)). \tag{77}$$

This expression is similar to that for the canonical momentum in the presence of an electromagnetic field. Note that $A_j(y)$ is pure real.

The expressions for the Hamiltonian for a single particle remain as shown in Eq. 49 except that ∂' is replaced by D. For example Eq. 49 becomes

$$(H_y)\psi(y) = -\frac{\hbar_y^2}{2m_y} \sum_{j=1}^{3} (D)_{j,y}^2 \psi(y) + V(y)\psi(y) \tag{78}$$

with $D_{y,j}$ given by Eq. 76.

Inclusion of scaling factors into the two particle state entangled by momentum conservation is straightforward. This is achieved by including scale factors in the two particle space integral, $\int_x (dz_1)_x (dz_2)_x$, in Eq. 55. The result is

$$(\psi_{1,2})_x = \int_x r_{z_1,x}(dz_1)_x r_{z_2,x}(dz_2)_x \Big(\int (e^{iz_1 p})_x \psi_1(p)_x |p_x\rangle_1 (e^{-iz_2 p})_x \psi_2(-p)_x| - p_x\rangle_2 dp_x \Big). \quad (79)$$

Here $r_{z_1,x}$ and $r_{z_2,x}$ are given by Eq. 57.

8. Gauge theories

One approach (Montvay & Münster, 1994) to gauge theories already makes partial use of the local availability of mathematics with the assignment of an n dimensional vector space to each x. Here the vector space is assumed to be a Hilbert space, $\bar{H}_x$, at each x. This $\bar{H}_x$ is quite different from that discussed in the previous section in that the vectors in $\bar{H}_x$ refer to the internal states of matter fields. Matter fields ψ are functionals where for each space time point x, $\psi(x)$ is a vector in $\bar{H}_x$.

The freedom of choice of a basis (Montvay & Münster, 1994; Yang & Mills, 1954) in each $\bar{H}_x$ is reflected in the factorization,

$$U_{y,x} = Y_{y,x} V_{y,x}, \quad (80)$$

of a parallel transform operator, $U_{y,x}$, (Mack, 1981) from $\bar{H}_x$ to $\bar{H}_y$ where $y = x + \hat{v}dx$ is a neighbor point of x.[8]

The unitary operator $V_{y,x}$ expresses the freedom of basis choice. As such it is an element of the gauge group $U(n)$ with a Lie algebra representation (Cheng & Li, 1984; Utiyama, 1956)

$$V_{y,x} = e^{i\Xi_\mu(x)dx^\mu} e^{i\Omega_\mu^j(x)\tau_j dx^\mu}. \quad (81)$$

Sum over repeated indices is implied. The τ_j are the generators of the Lie algebra $su(n)$ and the $\Omega_\mu^j(x)$ are the components of the n different gauge fields, $\vec{\Omega}^j(x)$. $\Xi(x)$ is the gauge field for the $U(1)$ factor of $U(n)$.

The covariant derivative of the field, ψ, is expressed by

$$D_{\mu,x}\psi = \frac{V_{\mu,x}\psi(x + dx^\mu)_x - \psi(x)}{dx^\mu}. \quad (82)$$

Here $V_{\mu,x}$ is the μ component of $V_{y,x}$. Expansion of the exponential to first order in small quantities gives

$$D_{\mu,x}\psi = \partial'_{\mu,x}\psi + i(g_1\Xi_\mu(x) + g_2\Omega_\mu^j(x)\tau_j)\psi(x). \quad (83)$$

Coupling constants, g_1 and g_2, have been added. The definition of $\partial'_{\mu,x}\psi$ is essentially the same as that given in Eq. 47. It is given by

$$\partial'_{\mu,x}\psi = \frac{\psi(x + dx^\mu)_x - \psi(x)}{dx^\mu}. \quad (84)$$

---

[8] Factorization is necessary because $U_{y,x}$ cannot be expressed as an exponential of Lie algebra elements or as a matrix of numbers. The reason is that the action of such a representation on a vector in $\bar{H}_x$ gives another vector in $\bar{H}_x$. It is not a vector in $\bar{H}_y$. Factorization remedies this in that $V_{y,x}$ is a unitary map from $\bar{H}_x$ to $\bar{H}_x$ and $Y_{y,x}$ is a unitary map from $\bar{H}_x$ to $\bar{H}_y$.

Effects on Quantum Physics of the Local Availability of
Mathematics and Space Time Dependent Scaling Factors for Number Systems

69

Here $\psi(x+dx^\mu)_x = U_{x,x+dx^\mu}\psi(x+dx^\mu)$ is the same vector in $\bar{H}_x$ as $\psi(x+dx^\mu)$ is in $\bar{H}_{x+dx^\mu}$.

The covariant derivative, Eq. 83, accounts for the local availability of mathematics and the freedom of basis choice. It does not include the effects of scaling factors for numbers. This is taken care of by replacing $V_{\mu,x}\psi(x+dx^\mu)_x$ in Eq. 82 by $r_{x+dz^\mu,x}V_{\mu,x}\psi(x+dx^\mu)_x$. This is a vector in the local representation, $\bar{H}_x^r$, Eq. 32, of $\bar{H}_y$ on $\bar{H}_x$.

Expansion of the exponentials to first order adds another term to $D_{\mu,x}$ in Eq. 83. One obtains (Benioffa, 2011; Benioffc, 2011)

$$D_{\mu,x}\psi = \partial'_{\mu,x}\psi + g_r A_\mu(x) + i(g_1\Xi_\mu(x) + g_2\Omega^j_\mu(x)\tau_j)\psi(x). \tag{85}$$

A coupling constant, g_r, for $\vec{A}(x)$ has been added. The coupling constants, and i are all number values in $\bar{C}_x$.

The physical properties of the gauge fields in $D_{\mu,x}$ are obtained by restricting the Lagrangians to only those terms that are invariant under local and global gauge transformations (Cheng & Li, 1984). For Abelian gauge theories, such as QED, $\vec{\Omega}(x)$ is absent. Invariance under local gauge transformations, $\Lambda(x)$, requires that the covariant derivative satisfy (Cheng & Li, 1984)

$$D'_{\mu,x}\Lambda(x)\psi(x) = \Lambda(x)D_{\mu,x}\psi(x). \tag{86}$$

$D'_{\mu,x}$ is obtained from $D_{\mu,x}$ by replacing $A_\mu(x)$ and $\Xi_\mu(x)$ with their primed values, $A'_\mu(x), \Xi'_\mu(x)$. The presence of the primes allows for the possible dependence of the fields on the local $U(1)$ gauge transformation, $\Lambda(x)$ where

$$\Lambda(x) = e^{i\phi(x)}. \tag{87}$$

Use of Eq. 84 and separate treatment of real and imaginary terms gives the following results:

$$\begin{aligned} A'_\mu(x) &= A_\mu(x) \\ g_1\Xi'_\mu(x) &= g_1\Xi_\mu(x) - \partial'_{\mu,x}\phi(x). \end{aligned} \tag{88}$$

This shows that the real field $\vec{A}$ is unaffected by a $U(1)$ gauge transformation. It also shows that $\Xi_\mu(x)$ transforms in the expected way as the electromagnetic field.

As is well known the properties of the $\vec{\Xi}$ field show that it is massless. The reason is that a mass term for this field is not locally gauge invariant (Cheng & Li, 1984; Montvay & Münster, 1994).

Unlike the case for the $\vec{\Xi}$ field, a mass term can be present for the real $\vec{A}$ field. This suggests that it represents a gauge boson for which mass is optional. That is, depending on what physical system $\vec{A}$ represents, if any, the presence of a mass term in Lagrangians is not forbidden.

For nonabelian gauge theories, such as $U(2)$ theories, Eq. 88 still holds. However there is an additional equation giving the transformation properties of the three vector gauge fields under local $SU(2)$ gauge transformations. These properties result in the physical representation of these fields in Lagrangians as charged vector bosons (Cheng & Li, 1984). The $\vec{A}$ and $\vec{\Xi}$ bosons are still present.

8.1 Physical properties of the $\vec{A}$ field from the gauge theory viewpoint

At this point it is not known what physical system, if any, is represented by the $\vec{A}$ field. Candidates include the inflaton field (Albrecht, 1982; Linde, 1982), the Higgs boson, the graviton, dark matter, and dark energy. One aspect that one can be pretty sure of is that the ratio of the $\vec{A}$ field - matter field coupling constant, g_r, to the fine structure constant, α, must be very small. This is a consequence of the great accuracy of the QED Lagrangian and the fact that the $\vec{A}$ field appears in covariant derivatives for all gauge theory (and other) Lagrangians.

As was noted, $\vec{\Xi}$ is the photon field. Inclusion of this field and a Yang Mills term for the dynamics of this field into the Dirac Lagrangian gives the QED Lagrangian (Cheng & Li, 1984).

9. Conclusion

This work is based on two premises: the local availability of mathematics and the existence of scaling factors for number systems. Local availability is based on the idea that the only mathematics that is directly available to an observer is that which is, or can be, in his or her head. Mathematical information that is separate from an observer, O_x, at space time point x, such as a textbook or a lecturer at point y, must be physically transmitted, e.g. by acoustic or light waves, to O_x where it becomes directly available.

This leads to a setup in which mathematical universes, V_x, are associated with each point x. If an observer moves through space time on a world line, $P(\tau)$, parameterized by the proper time τ, the mathematics directly available to $O_{P(\tau)}$ at time τ is that in $V_{P(\tau)}$.

Each V_x contains many types of mathematical systems. If $\bar{S}_x$ is in V_x, then V_y contains the same system type, $\bar{S}_y$, and conversely. Each V_x contains the different types of number systems and many other systems that are based on numbers. Included are the real and complex numbers $\bar{R}_x$ and $\bar{C}_x$.

Here the mathematical logical definition (Barwise, 1977; Keisler, 1977) of each type of system as a structure is used. A structure consists of a base set, basic operations, relations, and constants that satisfies axioms appropriate for the type of structure considered. Examples are $\bar{R}$ and $\bar{C}$, Eq. 6 for the real and complex numbers.

For each type of number structure it is possible to define many structures of the same type that differ by scaling factors (Benioffb, 2011). For each real number r, one can define structures $\bar{R}^r$, Eq. 11, and $\bar{C}^r$, Eq. 13, in which a scale factor r relates the number values in $\bar{R}^r$ and $\bar{C}^r$ to those in $\bar{R}$ and $\bar{C}$. The scaling of number values must be compensated for by scaling of the basic operations and relations in a manner such that $\bar{R}^r$ and $\bar{C}^r$ satisfy the relevant axioms for real and complex numbers if and only if $\bar{R}$ and $\bar{C}$ do.

The local availability of mathematics requires that one be able to construct local representations of $\bar{C}_y$ on $\bar{C}_x$. Two methods were described. One uses parallel transformations. These define or represent the notion of sameness between mathematical systems at different points. If $F_{x,y}$ is a parallel transform map from $\bar{S}_y$ onto $\bar{S}_x$, then for each element, w_y, in $\bar{S}_y$, $w_x = F_{x,y}(w_y)$ is the same element in $\bar{S}_x$ as w_y is in $\bar{S}_y$. In this case the local representation, $W_{x,y}\bar{S}_y$, of $\bar{S}_y$ on $\bar{S}_x$ is $\bar{S}_x$ itself.

Effects on Quantum Physics of the Local Availability of
Mathematics and Space Time Dependent Scaling Factors for Number Systems

71

The other method uses what are called correspondence maps. These combine parallel transformations with scaling. The local representation of $\bar{C}_y$ on $\bar{C}_x$ is $\bar{C}_x^r$, which is a scaling of $\bar{C}_x$ by a factor $r = r_{y,x}$. (From now on $\bar{R}$ is not explicitly mentioned as it is implicitly assumed to be part of $\bar{C}$.) The local representation of an element, a_y, of $\bar{C}_y$ corresponds to the element $r_{y,x}a_x$ in $\bar{C}_x$. Here $a_x = F_{y,x}a_y$ is the same element in $\bar{C}_x$ as a_y is in $\bar{C}_y$.

It was seen that the scaling of numbers plays no role in the general use of numbers in mathematics and physics. This includes such things as comparing outcomes of theory predictions with experimental results or in comparing outcomes of different experiments. More generally it plays no role in the use of numbers in the commerce of mathematics and physics. The reason is that theory computations and experiment outcomes obtained at different locations are never directly compared. Instead the information contained in the outcomes as physical states must be transmitted by physical systems to a common point. There the states of the physical transmittal systems are interpreted locally as numbers, and then compared.

In this work, number scaling was limited to theory calculations that involve space time derivatives or integrals. Examples of this were described in quantum theory and in gauge theories. An example discussed in some detail was the expansion of a wave packet $\psi = \int \psi(y)|y\rangle dy$. Since $\psi(y)|y\rangle$ is a vector in $\bar{H}_y$, the integrand has to be moved to a common point, x for the integral to make sense. This can be done by parallel transform maps which give

$$\psi_x = \int_x \psi(y)_x |y_x\rangle dy_x \tag{89}$$

or by correspondence maps which give

$$\psi_x = \int_x r_{y,x}\psi(y)_x |y_x\rangle dy_x. \tag{90}$$

Here the scaling factor, $r_{y,x}$, is the integral from x to y of the exponential of the gauge field, $\vec{A}(y)$, as in Eq. 57.

It was also seen that one can use both transform and correspondence maps to express the wave packet in a form that reflects exactly what one does in an experiment that measures either the spatial distribution or the position expectation value of a quantum particle. If the experiment setup consists of a collection of cube detectors of volume Δ^3 that fill 3 dimensional Euclidean space, the outcome of each of many repeated experiments is a triple of numbers, j,k,l that label the position, $j\Delta, k\Delta.l\Delta$ of the detector that fired.

As was seen in the discussion of mathematical and physical commerce, the outcomes of repeated experiments must be physically transported to a common point, x where they are interpreted as numbers in $\bar{R}_x$ for mathematical combination. It follows that the scaling factors are limited to integration over the volume of each detector. This results in the replacement of ψ_x by ψ_x' where, Eq. 70,

$$\psi_x' = \sum_{j,k,l} U_{x,z} \int_{V_{j,k,l}} r_{w,z}\psi(w)_z |w_z\rangle dw_z. \tag{91}$$

The sum is over all cubes. $z = z_{j,k,l}$ is a point on each cube surface. Each integral is over the volume, $V_{j,k,l}$, of each cube. The r factor, Eq. 72, scales the values of each integrand at point w to values at z. $U_{x,z}$ parallel transforms the integrals at different z to a common point x.

The state, ψ'_x, differs from ψ_x, or the usual quantum mechanical wave packet expression for ψ, in that it ties theory closer to experiment. It also reduces the effect of scaling to the sum of the effects for the volumes of each of the detectors. The effect is reduced because, for any point w in the sensitive volume of the experiment, the effect of $\vec{A}$ on the transform from w to x is limited to the part of the path in the detector volume containing w. As the detector volumes go to 0, so does the effect of $\vec{A}$. The increase in the number of detectors as the volume of each gets smaller does not remove this effect.

In this sense the usual quantum theory wave packet integral for ψ is a limit in that it is independent of experimental details. Unlike the case for the usual representation of ψ, use of ψ'_x to make predictions will give values that depend on experimental details. The fact that there is no indication, so far, of such dependence, at least to the accuracy of experiment, means that the effect of $\vec{A}(x)$ must be very small. Whether the small effect is due to small values of $\vec{A}$ itself or a small value of a coupling constant of $\vec{A}$ to states and matter fields, is not known at present.

The relation between ψ'_x and the usual integral for ψ is further clarified by the observation that $\psi_x \rightarrow \psi$ as the detector volume goes to 0. However, the Heisenberg uncertainty principle prevents experimental attainment of this limit.

It must be emphasized that the tying the wave packet integral to experiment details, as with ψ'_x has nothing to do with collapse of the wave packet during carrying out of the experiment. ψ'_x is just as coherent a state as is ψ.

The gauge field $\vec{A}$ also appears in the expression for the canonical momentum. The usual expression for momentum $\mathbf{p} = \sum_j i\hbar\partial_{j,x}$ is replaced by, Eq. 73

$$\mathbf{p}_{A,x} = i_x\hbar_x D_{j,x} \tag{92}$$

where

$$D_{j,x} = \partial'_{j,x} + \vec{A}. \tag{93}$$

$\partial'_{j,x}$, Eq. 47, accounts for the local availability of mathematics.

As a covariant derivative, $D_{\mu,x}$ appears in gauge theories with additional terms. It was seen that the limitation of Lagrangians to terms that are invariant under local gauge transformations, (Cheng & Li, 1984; Montvay & Münster, 1994), results in $\vec{A}$ appearing as a gauge boson for which mass is optional. This is the case for both Abelian and nonabelian gauge theories.

The physical nature of $\vec{A}$, if any, is unknown. What is known is that the great accuracy of QED requires that the coupling constant of $\vec{A}$ to matter fields must be very small.

It must be emphasized that this work is only a first step in combining "mathematics is local" with the freedom of choice of scaling factors for number structures. An example of work for the future is to determine the effect of number scaling factors on geometry. It is suspected that the scaling factors may induce conformal transformations into geometry. More work also needs to be done on the effects of number scaling on quantum mechanics. An interesting question here is whether scaling factors are needed at all in classical mechanics.

Finally, one may hope that this work provides a real entry into the description of a coherent theory of physics and mathematics together. Such a theory would be expected to describe

Effects on Quantum Physics of the Local Availability of
Mathematics and Space Time Dependent Scaling Factors for Number Systems

73

mathematics and physics together as part of a coherent whole instead of as two separate but closely related disciplines.

10. Acknowledgement

This work was supported by the U.S. Department of Energy, Office of Nuclear Physics, under Contract No. DE-AC02-06CH11357.

11. References

Albrecht, A. & Steinhardt, P. (1982). Cosmology for Grand Unified Theories with Radiatively Induced Symmetry Breaking, *Phys. Rev. Lett.* 48, 1220-1223.

Barwise, J. (1977). An Introduction to First Order Logic, in *Handbook of Mathematical Logic*, J. Barwise, Ed. North-Holland Publishing Co. New York, 5-46.

BenDaniel, D. (1999). Linking the foundations of physics and mathematics, arXiv:math-phy/9907004.

Benioff, P. (2011). New gauge fields from extension of parallel transport of vector spaces to underlying scalar fields, *Proceedings of SPIE, Quantum Information and Computation IX* 8057,80570X, Orlando, Florida, USA, April, 2011, SPIE, Bellingham, Washington.

Benioff, P. (2011). Representations of each number type that differ by scale factors, arXiv:1102.3658

Benioff, P. (2011). New gauge field from extension of space time parallel transport of vector spaces to the underlying number systems, *International Journal of theoretical physics* 50, 1887-1907 2011; arXiv:1008.3134

Benioff, P. (2005). Towards a coherent theory of physics and mathematics: the theory experiment connection, *Foundations of Physics*,35, 1825-1856, 2005; arXiv:quant-ph/0403209.

Benioff, P. (2002). Towards a coherent theory of physics and mathematics, *Foundations of Physics*,32, 989-1029, 2002; arXiv:quant-ph/0201093.

Bernal, A.; Sanchez, M.; Gil, F. (2008). Physics ftrom scratch. Letter on M. Tegmark's, 'The mathematical universe", arXiv:0803.0944.

Cheng, T. & Li, L. (1984). *Gauge Theory of Elementary Particle Physics*, Oxford University Press, New York, NY, Chapter 8.

Davies, P. (1990). Why is the Physical World so Comprehensible? in *Complexity, Entropy, and Physical Information*, Proceedings of the 1988 workshop on complexity, entropy, and the physics of information, may-june 1989, Santa Fe New Mexico, W. H. Zurek, Editor, Addison-Weseley Publishing Co. Redwood City CA, 61-70.

Hut,P.; Alford, M.; Tegmark, M. (2006). On math, matter, and mind, *Foundations of Physics*, 36, 765-794, 2006; arXiv: physics/ 0510188.

Jannes, G. (2009) Some comments on "The mathematical universe", *Foundations of Physics*, 39, 397-406, 2009.

Kadison, R. & Ringrose, J. (1983). *Fundamentals of the Theory of Operator Algebras: Elementary theory*, Academic Press, New York, Chap 2.

Kaye, R. (1991). *Models of Peano Arithmetic* Clarendon Press, Oxford, 1991, 16-21.

Keisler, H. (1977). Fundamentals of Model Theory, in *Handbook of Mathematical Logic*, J. Barwise, Ed. North-Holland Publishing Co. New York, 1977, 47-104.

Landauer, R. (1991). Information is Physical, *Physics Today*, 44, 23-29, 1991.

Linde, A. (1982). A new inflationary universe scenario: A possible solution of the horizon, flatness, homogeneity, isotropy and primordial monopole problems, *Phys. Letters B*, 108, 389-393, 1982.

Mack, G. (1981). Physical principles , geometrical aspects, and locality properties of gauge field theories, *Fortschritte der Physik*, 29, 135-185, 1981.

Montvay, I. & Münster, G. (1994). *Quantum Fields on a Lattice*, Cambrodge University Press, Cambridge, UK, Chapter 3.

Novaes, S. (2000). Standard model: an Introduction, in *Particles and Fields*, Proceedings, X Jorge Andre Swieca Summer School, Sao Paulo, February 1999, Editors, J. Barata, M. Begalli, R. Rosenfeld, World Scientific Publishing Co. Singapore; arXiv:hep-th/0001283.

Omnes, R. (2011). Wigner's "Unreasonable Effectiveness of Mathematics", Revisited, *Foundations of Physics*, 41, 1729-1739, 2011.

O'Raifeartaigh, L. (1997). *The Dawning of Gauge Theory*, Princeton Series in Physics, Princeton University Press, Princeton, N. J.

Randolph, J. (1968). *Basic Real and Abstract Analysis*, Academic Press, Inc. New York, NY, p 26.

Shoenfield, J. (1967). *Mathematical Logic*, Addison Weseley Publishing Co. Inc. Reading Ma, p. 86.

Tegmark, M. (2008). The mathematical universe, *Foundations of Physics*, 38, 101-150, 2008; arXiv:0704.0646.

Utiyama, R. (1956). Invariant Theoretical Interpretation of Interaction, *Phys. Rev.* 101, 1597-1607, 1956.

Welch, L. (2009). A possible mathematical structure for physics, arXiv:0908.2063.

Wigner, E. (1960). The unreasonable effectiveness of mathematics in the natural sciences, *Commum. Pure and Applied Math.* **13**, 001 (1960), Reprinted in E. Wigner, *Symmetries and Reflections*, (Indiana Univ. Press, Bloomington IN, pp. 222-237.

Yang , C. & Mills, R. (1954). Conservation of isotopic spin and isotopic gauge invariance, *Physical Review* 96, 191-195, 1954.

Part 2

Quantum Matter

4

Quantum Theory of Coherence and Polarization of Light

Mayukh Lahiri

Department of Physics and Astronomy, University of Rochester, Rochester, NY
USA

1. Introduction

When we open our eyes on earth for the first time, light generates the sensation of vision in our mind. If there was no light, our way of thinking would undoubtedly be quite different. Therefore, it is natural to ask the question, what is light? Our early inquisitiveness about light is documented in many stories and characters from almost all religions and cultures. Later, with the development of science, we tried to use more rigorous techniques to understand the nature of light. However, regardless of intense investigations for centuries on this subject, we are far from obtaining a convincing answer to the question "what is light?" Einstein seems to have put it best[1]:

"All the fifty years of conscious brooding have brought me no closer to answer the question, 'What are light quanta?' Of course today every rascal thinks he knows the answer, but he is deluding himself."

It is clear from the preceding remarks that this question represents one of the ever-lasting ones in physics. Although, at this stage one does not fully understand the nature of light, one can nevertheless study its properties. In this context, one may recall Feynman's illustration of the principle of scientific research (Feynman et al., 1985):

"...in trying to get some idea of what we're doing in trying to understand nature, is to imagine that the gods are playing some great game like chess, let's say, and you don't know the rules of the game, but you're allowed to look at the board, at least from time to time ..., and from those observations you try to figure out what the rules of the game are."

To understand the nature of light thus scientists have lead to investigate different properties of it. Coherence and polarization may be identified as the two important ones. In this chapter, we will give brief descriptions of them, and will show how the two apparently different phenomena can be described by analogous theoretical formulations, which involve incorporating the statistical fluctuations present in light.

Detailed descriptions of the history of the theories of coherence and polarization may be found in many scholarly articles [see, for example, (Born & Wolf, 1999; Brosseau, 2010; Mandel & Wolf, 1995). It may be said that the topic originated in Hooke's conjecture about the wave nature of light (Hooke, 1665), which was put in a sounder basis by Huygens (Huygens, 1690).

---

[1] From Einstein's letter to Michael Besso, written in 1954.

Later Young discovered that light waves may produce interference fringes (Young, 1802). Young's discovery lead to many investigations concerning interference properties of light[2] and it turned out that light from different sources may differ in their abilities to interfere. Traditionally "coherence" of an optical field is understood as the ability of light to interfere. Early relevant investigations on interfering properties of light may be found in the works of Fresnel and Arago (Arago & Fresnel, 1819)[3]. An important contribution was made by Verdet (Verdet, 1865) when he determined the size of the region on the earth-surface, where "sunlight vibrations are in unison". Michelson established relationships between visibility of interference fringes and intensity distribution on the surface of an extended primary source (Michelson, 1890; 1891a;b; 1902; 1927). He also elucidated the connection which exists between the visibility of interference fringes and energy distribution in spectral lines. However, he did not interpret his results in terms of field correlations.

It has become customary in traditional optics to represent an optical field by a deterministic function. Although a deterministic model provides simple solutions to some problems, it often suffers from lack of self-consistency and leaves out many questions which can only be answered by taking the random nature of the field into account. The necessity of developing a statistical theory of light arises from the fact that all optical fields, whether found in nature or generated in a laboratory, have some random fluctuations associated with them. Even though these fluctuations are too rapid to be observed directly, their existence can be experienced by various experiments which involve effects of correlations among the fluctuating fields at a point, or at several points in space. In the quantum mechanical description, in addition to that, one also needs to consider the presence of a detector, which is an atom or a collection of atoms.

The first quantitative measure of the correlations of light vibrations was introduced by Laue (Laue, 1906; 1907). Later Berek used this concept of correlation in his work on image formation in microscopes (Berek, 1926a;b;c;d). A new era in this subject began when van Cittert determined the joint probability distribution for the light disturbances at any two points on a plane illuminated by an extended primary source (van Cittert, 1934) and also determined the probability distribution for light disturbances at one point, at two different instances of time (van Cittert, 1939).

A simpler and more profound approach for addressing such problems was developed by Zernike (Zernike, 1938), who also introduced the concept of the "degree of coherence" in terms of visibility of interference fringes, which is a measurable quantity. Although Zernike's work brought new light to this subject, it had some limitations, because it did not take into consideration the time-difference that may exist between interfering beams. Wolf formulated a more general theory of optical coherence by introducing more generalized correlation functions into the analysis (Wolf, 1955). Analyzing the propagation of such correlation functions is often quite complicated due to time-retardation factors, and in most of the cases it does not lead to any useful solution. Also, for the same reason, it is rather difficult in this formulation to analyze many practical problems. Such problems are treated much

---

[2] For a detail discussion of the influence of Young's interference experiment on the development of coherence theory, see (Wolf, 2007a).

[3] Interpretation of the results obtained by Fresnel and Arago in terms of moderns coherence theory is given in (Mujat et al., 2004).

more conveniently by a different formulation, known as the space-frequency formulation of coherence theory (Wolf, 1981; 1982; 1986).

All the investigations mentioned above were carried out by the use of the scalar theory. To take into account another property of light, namely polarization, one needs to consider the vector nature of the fluctuating field. The credit of first notifying polarization properties of light is attributed to Erasmus Bartholinus, who studied the phenomenon of double refraction using calcite crystals. In 1672 Huygens (Huygens, 1690) provided an interpretation of the double refraction from the conception of spherical light waves. Use of double refraction became widely used in different fundamental and practical applications of optical sciences, which led to important investigations on this subject by several distinguished scientists. The foundation of the modern theory of polarization properties of light was laid down by Stokes (Stokes, 1852) as he formulated the theory of polarization in terms of certain parameters, now known as Stokes parameters [see also (Berry et al., 1977)]. Poincaré (Poincaré, 1892) provided a detailed mathematical treatment of the polarization properties of light, in which he introduced the concept of Poincaré sphere to specify any state of polarization of light. A matrix treatment of polarization was introduced by Wiener (Wiener, 1927; 1930; 1966), whose analysis related field correlations and polarization properties of light. Later, Wolf (Wolf, 1959) used a similar matrix formulation for systematic studies of polarization properties of statistically stationary light beams [see also, (Brosseau, 1995; Collett, 1993; Mandel & Wolf, 1995)]. A detailed description of the history of the theory of polarization can be found in a recently published article by Brosseau (Brosseau, 2010).

From the above discussion, it seems that coherence and polarization are two completely different phenomena with different histories and origins. However, according to classical theory, both of the can be interpreted as measures of correlations present between fluctuating electric field components. It is interesting to note that Verdet probably suspected, almost 140 years ago, that there is some analogy between the concepts of polarization and of coherence. The title of his paper (Verdet, 1865) states in loose translation: "Study of the nature of unpolarized and partially polarized light". Yet in spite of the stress on polarization, it is the very paper in which Verdet estimated the region of coherence of sunlight on the earth's surface[4]. This connection, became even more prominent in the works of Wiener and Wolf. However, for many years coherence and polarization have been considered as two independent branches of optics. The connection between them became evident with the introduction of a unified theory of coherence and polarization for stochastic electromagnetic beams (Wolf, 2003a;b).

All the investigations mentioned above are based on the classical theory of electromagnetic fields. In the early part of the last century, development of quantum mechanics began to provide deeper explanations of the intrinsic properties of light and its interaction with matter. Dirac's quantization of electromagnetic fields (Dirac, 1957) made it possible to analyze various properties of light by the use of quantum mechanical techniques. In 1963, Glauber introduced quantum mechanical formulation of coherence theory (Glauber, 1963), which has been followed by systematic investigations of the subject [see, for example, (Glauber, 2007; Mandel & Wolf, 1995)].

---

[4] It was pointed out by Wolf in a conference talk (Wolf, 2009b), which was presented by the author.

This chapter intends to provide a quantum mechanical description of coherence and polarization properties of light with emphasis on some recently obtained results. We will begin with a brief discussion on the classical theories of coherence and polarization of light. We will emphasize how the two apparently different properties of light may be described by analogous theoretical techniques. We will stress the importance of describing the subject in terms of observable quantities. We will then recall some basic results of quantum theory of optical coherence in the space-time domain, followed by detailed description of coherence properties of optical fields in the space-frequency domain. After that, we will discuss a formulation of optical coherence in the space-frequency domain. We will also provide a quantum mechanical description of polarization properties of optical beams. In the end, we will point out an interesting observation on the connection of Bohr's complementarity principle with partial coherence, and with partial polarization.

2. Classical theory of stochastic fields

We already mentioned in the Introduction that a deterministic model of optical fields leads to many discrepancies in the analysis of properties of light. A systematic development of the theory of coherence and polarization properties of light requires to take account the random fluctuations present in the field. We will show how this theory may be used to elucidate coherence and polarization properties of light.

From the classic theory of Maxwell, it is well known that an optical field can be represented by an electric field $\mathbf{E}(\mathbf{r};t)$ and a magnetic field $\mathbf{B}(\mathbf{r};t)$, which obey the four Maxwell's equations [see, for example, (Jackson, 2004)]. For a stochastic (randomly fluctuating) three-dimensional optical field, each of them will be represented by three random components $E_i(\mathbf{r};t)$ and $B_i(\mathbf{r};t)$, $i = x,y,z$, where x,y,z are three arbitrary mutually orthogonal directions in space. In our discussion we will neglect the effects due to magnetic fields.

2.1 Optical coherence in the space-time domain

The word "coherence" refers to the ability of light to interfere. Coherence properties of light may, therefore, be understood by analyzing the interference fringes produced in an Young's interference experiment (Fig. 1). Many years ago, Zernike (Zernike, 1938) defined the "degree of coherence" of a wave field by the maximum value of visibility in the interference pattern produced by it "under the best circumstances"[5]. However, as already mentioned in the Introduction that Zernike did not take into account the time-delay between the fields arriving from different pinholes. Consequently, his theory could not address some interesting aspects of coherence, which was later generalized by Wolf (Wolf, 1955). We will now briefly discuss the main results of Wolf's theory in the space-time domain. For simplicity we begin with scalar description of fields, or in other words we assume that all the electric field components behave in the same way.

A randomly fluctuating generally complex optical scalar field[6], at a point $P(\mathbf{r})$, at a time t, may be represented by a statistical ensemble $\{V(\mathbf{r};t)\}$ of realizations. The second-order

---

[5] By the term "best circumstances" Zernike meant that the intensities of the two interfering beams were equal and that only small path difference was introduced between them.
[6] The corresponding technical term is the "complex analytic signal" of a real electric field (Gabor, 1946).

cross-correlation function $\Gamma(\mathbf{r}_1, \mathbf{r}_2; t_1, t_2)$ of the random fields at two different space-time points $(\mathbf{r}_1; t_1)$ and $(\mathbf{r}_2; t_2)$ is defined by the expression

$$\Gamma(\mathbf{r}_1, \mathbf{r}_2; t_1, t_2) \equiv \langle V^*(\mathbf{r}_1; t_1) V(\mathbf{r}_2; t_2) \rangle, \tag{1}$$

where the asterisk denotes the complex conjugate and the angular brackets denote ensemble average. Now if the field is statistically stationary, at least in the wide sense (Mandel & Wolf, 1995; Wolf, 2007b), the expression on the right hand side of Eq. (1) then depends only on the time difference $\tau \equiv t_2 - t_1$. Therefore, for statistically stationary fields, the cross-correlation function takes the form

$$\Gamma(\mathbf{r}_1, \mathbf{r}_2; \tau) \equiv \langle V^*(\mathbf{r}_1; t) V(\mathbf{r}_2; t + \tau) \rangle. \tag{2}$$

The cross-correlation function $\Gamma(\mathbf{r}_1, \mathbf{r}_2; \tau)$ is known as the *mutual coherence function*. It characterizes the second-order correlation properties of such fields in the space-time domain.

The average intensity $I(\mathbf{r})$ of light at a point $P(\mathbf{r})$, apart from a constant factor depending on the choice of units, is given by $\langle |V(\mathbf{r}; t)|^2 \rangle$. From Eq. (2) it follows that

$$I(\mathbf{r}) \equiv \langle |V(\mathbf{r}; t)|^2 \rangle = \Gamma(\mathbf{r}, \mathbf{r}; 0). \tag{3}$$

Evidently for statistically stationary light the average intensity *does not depend on time*.

Let us now consider a Young's two pinhole interference experiment (Fig. 1). Suppose that a light beam is incident from the half-space $z < 0$ onto an opaque screen A, placed in the plane $z = 0$ containing two pinholes $Q_1(\mathbf{r}_1)$ and $Q_2(\mathbf{r}_2)$. For the sake of simplicity, we assume that

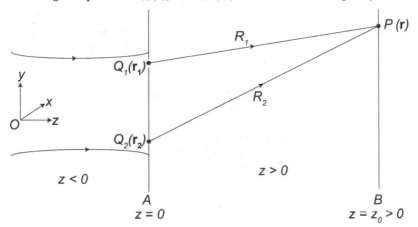

Fig. 1. Illustrating the notation relating to Young's interference experiment.

the beam is incident normally on the screen A. In general, interference fringes will be formed on a screen B, placed in a plane $z = z_0 > 0$, some distance behind the screen A (see Fig. 1). If we assume that the contributions to the total intensity from the two pinholes are equal to each other, i.e., that $I^{(1)}(\mathbf{r}) = I^{(2)}(\mathbf{r}) \equiv I^{(0)}(\mathbf{r})$ (say), then we find that

$$I(\mathbf{r}) = 2I^{(0)}(\mathbf{r}) \left\{ 1 + \left| \frac{\Gamma(\mathbf{r}_1, \mathbf{r}_2; \tau)}{\sqrt{I(\mathbf{r}_1)} \sqrt{I(\mathbf{r}_2)}} \right| \cos\left[\alpha(\mathbf{r}_1, \mathbf{r}_2; \tau) \right] \right\}, \tag{4}$$

where

$$\tau \equiv t_2 - t_1 = \frac{R_2 - R_1}{c} \tag{5}$$

is the time delay between the fields arriving at $P(\mathbf{r})$ from pinholes $Q(\mathbf{r}_1)$ and $Q(\mathbf{r}_2)$, c being the speed of light in free space, and $\alpha(\mathbf{r}_1, \mathbf{r}_2; \tau) = \arg\{\Gamma(\mathbf{r}_1, \mathbf{r}_2; \tau)\}$. The visibility $\mathcal{V}$ of the fringes is defined by the famous formula due to Michelson, viz.,

$$\mathcal{V} \equiv \frac{I_{\max} - I_{\min}}{I_{\max} + I_{\min}}, \qquad 0 \le \mathcal{V} \le 1. \tag{6}$$

One can readily show from Eq. (4) and (6) that the visibility of the fringes at the point $P(\mathbf{r})$ is given by ((Wolf, 2007b), Sec. 3.1, Eq. (19))

$$\mathcal{V} = \left| \frac{\Gamma(\mathbf{r}_1, \mathbf{r}_2; \tau)}{\sqrt{I(\mathbf{r}_1)}\sqrt{I(\mathbf{r}_2)}} \right|. \tag{7}$$

The normalized cross-correlation function in this expression is defined as the *degree of coherence* $\gamma(\mathbf{r}_1, \mathbf{r}_2; \tau)$, i.e.,

$$\gamma(\mathbf{r}_1, \mathbf{r}_2; \tau) \equiv \frac{\Gamma(\mathbf{r}_1, \mathbf{r}_2; \tau)}{\sqrt{I(\mathbf{r}_1)}\sqrt{I(\mathbf{r}_2)}}, \tag{8}$$

Since the visibility is always bounded by zero and by unity, so is the modulus of degree of coherence. It can also be proved explicitly by use of the Cauchy-Schwarz inequality that

$$0 \le |\gamma(\mathbf{r}_1, \mathbf{r}_2; \tau)| \le 1. \tag{9}$$

When $|\gamma(\mathbf{r}_1, \mathbf{r}_2; \tau)| = 1$, sharpest possible fringes are obtained and the field is said to be *completely coherent*, for the time delay τ, at the pair of points $Q(\mathbf{r}_1)$ and $Q(\mathbf{r}_2)$. In the other extreme case, when $|\gamma(\mathbf{r}_1, \mathbf{r}_2; \tau)| = 0$, no fringe is obtained and the field is said to be *incoherent*, for the time delay τ, at the two points. In the intermediate case $0 < |\gamma(\mathbf{r}_1, \mathbf{r}_2; \tau)| < 1$, the field is said to be *partially coherent*. It is to be noted that the degree of coherence is, in general, a complex quantity. Its phase is also a meaningful physical quantity and can be determined from measurements of positions of maximum and minimum in the fringe pattern [see, for example, (Mandel & Wolf, 1995), p-167]. It must be noted that the mutual coherence function $\Gamma(\mathbf{r}_1, \mathbf{r}_2; \tau)$ obeys certain propagation laws which make it possible to determine changes in correlation properties of light on propagation. These propagation laws are often called Wolf's equations (see, for example, (Wolf, 2007b), Sec. 3.5).

The theory can be immediately generalized to vector fields. If we restrict ourselves to a stationary stochastic light beam propagating along positive z direction, then the coherence properties can be described by a 2×2 matrix $\overleftrightarrow{\Gamma}(\mathbf{r}_1, \mathbf{r}_2; \tau)$, which is defined by the formula

$$\overleftrightarrow{\Gamma}(\mathbf{r}_1, \mathbf{r}_2, \tau) \equiv \left[\Gamma_{ij}(\mathbf{r}_1, \mathbf{r}_2; \tau) \right] \equiv \left[\langle E_i^*(\mathbf{r}_1; t) E_j(\mathbf{r}_2; t + \tau) \rangle \right], \quad (i = x, y; \; j = x, y), \tag{10}$$

where E_i is a component of the electric field vector. In this case, the average intensity at a point $P(\mathbf{r})$ is given by

$$I(\mathbf{r}) \equiv \mathrm{Tr}\, \overleftrightarrow{\Gamma}(\mathbf{r}, \mathbf{r}; 0), \tag{11}$$

where Tr represents trace of a matrix. The degree of coherence can be shown to be given by the formula (Karczewski, 1963)

$$\gamma(\mathbf{r}_1, \mathbf{r}_2, \tau) \equiv \frac{\text{Tr}\ \overleftrightarrow{\Gamma}(\mathbf{r}_1, \mathbf{r}_2, \tau)}{\sqrt{I(\mathbf{r}_1)I(\mathbf{r}_2)}}. \tag{12}$$

2.2 Optical coherence in the space-frequency domain

The space-time formulation of the coherence theory, discussed in the previous section, is quite natural, intuitive and often useful. However, as mentioned in the Introduction, there are some problems in statistical optics which turn out to be almost impossible to solve by the use of this formulation. For example, attempts to solve problems involving change in coherence properties of light on propagation through various media, or the problems involving scattering of partially coherent light, presents considerable difficulties in this formulation. These types of problems can be much more conveniently addressed by the use of a somewhat different formulation, known as the *space-frequency formulation* of coherence theory (Wolf, 1981; 1982; 1986). This space-frequency formulation has led to discoveries and understanding of some new physical phenomena, such as correlation-induced spectral changes (Wolf, 1987) and changes in polarization properties of light on propagation (James, 1994). Recent studies have also revealed a great usefulness of this theory in connection with determining the structure of objects by inverse scattering technique [see, for example, Refs. (Lahiri et al., 2009; Wolf, 2009a; 2010a; 2011)]. In this section we will briefly present some basic results in the theory of optical coherence in the space-frequency domain for scalar fields which are statistically stationary, at least in the wide sense.

A stationary random function $V(\mathbf{r}; t)$ is not square integrable and, consequently, its Fourier transform does not exist. However, for most statistically stationary optical fields, it is reasonable to assume that the mutual coherence function $\Gamma(\mathbf{r}_1, \mathbf{r}_2; \tau)$ exists and is a square integrable function of τ. One can then define a function $W(\mathbf{r}_1, \mathbf{r}_2; \omega)$ which together with $\Gamma(\mathbf{r}_1, \mathbf{r}_2; \tau)$ form a Fourier transform pair, i.e.,

$$W(\mathbf{r}_1, \mathbf{r}_2; \omega) = \frac{1}{2\pi} \int_{-\infty}^{\infty} \Gamma(\mathbf{r}_1, \mathbf{r}_2; \tau) e^{i\omega\tau}\, d\tau, \tag{13a}$$

$$\Gamma(\mathbf{r}_1, \mathbf{r}_2; \tau) = \int_{0}^{\infty} W(\mathbf{r}_1, \mathbf{r}_2; \omega) e^{-i\omega\tau}\, d\omega, \tag{13b}$$

where ω denotes the frequency. The quantity $W(\mathbf{r}_1, \mathbf{r}_2; \omega)$ is called the cross-spectral density function (to be abbreviated by CSDF) of the field. It can be shown that CSDF is also a correlation function, i.e., that it can be represented in the form

$$W(\mathbf{r}_1, \mathbf{r}_2; \omega) = \langle U^*(\mathbf{r}_1; \omega) U(\mathbf{r}_2; \omega) \rangle_\omega, \tag{14}$$

where $U(\mathbf{r}; \omega)$ is a typical member of a suitably constructed ensemble of monochromatic realizations[7], all of frequency ω ((Wolf, 2007b), Sec. 4.1). In the special case, when the two points $\mathbf{r}_1$ and $\mathbf{r}_2$ coincide, it follows from generalized Wiener-Khinchin theorem (see, for example, (Mandel & Wolf, 1995), sec. 2.4.4) that the CSDF represents the spectral density

---

[7] It is important to note that $U(\mathbf{r}; \omega)$ is not the Fourier transform of the field $V(\mathbf{r}; t)$.

$S(\mathbf{r}, \omega)$ of the field, i.e., that

$$S(\mathbf{r}, \omega) = W(\mathbf{r}, \mathbf{r}; \omega). \tag{15}$$

The spectral density $S(\mathbf{r}, \omega)$ is a physically meaningful quantity which represents the average intensity at a particular frequency.

The CSDF is the key quantity in the second-order coherence theory in space-frequency domain. As Eq. (15) shows, one can obtain the spectral density directly from it by taking the two spatial arguments to be equal. The spectral coherence properties of scalar fields are also completely described by the CSDF. To see that let us consider, once again, an Young's interference experiment (Fig. 1), but now we consider the fringe pattern produced by each frequency component present in the spectrum of the light. This situation can be realized by imagining that the incident light is filtered around the frequency ω before reaching the pinholes. The distribution of the spectral density $S(\mathbf{r}; \omega)$ on the screen B is given by the expression [(Wolf, 2007b), Sec. 4.2]

$$S(\mathbf{r}; \omega) = S^{(1)}(\mathbf{r}; \omega) \left\{ 1 + |\mu(\mathbf{r}_1, \mathbf{r}_2; \omega)| \cos\left[\beta(\mathbf{r}_1, \mathbf{r}_2; \omega) - \delta\right] \right\}. \tag{16}$$

Here $S^{(1)}(\mathbf{r}; \omega)$ is the contribution of light reaching at $P(\mathbf{r})$ from either of the two pinholes, $\delta = \omega(R_2 - R_1)/c$, and $\beta(\mathbf{r}_1, \mathbf{r}_2; \omega)$ is the phase of the so-called *spectral degree of coherence* $\mu(\mathbf{r}_1, \mathbf{r}_2; \omega)$ which is given by the expression ((Wolf, 2007b), Sec. 4.2)

$$\mu(\mathbf{r}_1, \mathbf{r}_2; \omega) \equiv \frac{W(\mathbf{r}_1, \mathbf{r}_2; \omega)}{\sqrt{S(\mathbf{r}_1, \omega)}\sqrt{S(\mathbf{r}_2, \omega)}}. \tag{17}$$

The formula (16) is known as the *spectral intensity law*. By analogy with the space-time formulation, one can readily show that $|\mu(\mathbf{r}_1, \mathbf{r}_2; \omega)|$ is equal to the fringe visibility associated with the frequency component ω, in the experiment sketched out in Fig. 1. It should be noted that in this case, unlike in the case of the space-time formulation, the fringe visibility is constant over the screen B. It can be shown that [(Mandel & Wolf, 1995), Sec. 4.3.2]

$$0 \le |\mu(\mathbf{r}_1, \mathbf{r}_2; \omega)| \le 1. \tag{18}$$

When $|\mu(\mathbf{r}_1, \mathbf{r}_2; \omega)| = 1$, the field at the two points $Q_1(\mathbf{r}_1)$ and $Q_2(\mathbf{r}_2)$ is said to be *spectrally completely coherent* at the frequency ω. If $\mu(\mathbf{r}_1, \mathbf{r}_2; \omega) = 0$, the field is said to be *spectrally completely incoherent* at the two points, at that frequency. In the intermediate case, it is said to be spectrally partially coherent at frequency ω. Like the mutual coherence function, the cross-spectral density function also obey certain propagation laws [see, for example, (Mandel & Wolf, 1995), Sec. 4.4.1].

The theory can also be generalized to the vector fields. For an optical beam propagating along positive direction of z axis, one can define a 2×2 matrix, known as the cross-spectral density matrix (CSDM), which is the Fourier transform of mutual coherence matrix [(Wolf, 2007b), Chapter 9, Eqs. (1) and (2)]:

$$\overleftrightarrow{W}(\mathbf{r}_1, \mathbf{r}_2; \omega) \equiv \frac{1}{2\pi} \int_{-\infty}^{\infty} \overleftrightarrow{\Gamma}(\mathbf{r}_1, \mathbf{r}_2; \tau) \exp[i\omega\tau] \, d\tau. \tag{19}$$

As was in the scalar case, it can be shown that each element of the CSDM is a correlation function [see, (Wolf, 2007b), Chapter 9]. In this case, the spectral density $S(\mathbf{r}, \omega)$ is given by the expression [(Wolf, 2007b), Sec. 9.2, Eq. (2)]

$$S(\mathbf{r}, \omega) = \mathrm{Tr}\ \overleftrightarrow{W}(\mathbf{r}, \mathbf{r}; \omega),\tag{20}$$

and the spectral degree of coherence is given by the expression [(Wolf, 2007b), Sec. 9.2, Eq. (8)]

$$\mu(\mathbf{r}_1, \mathbf{r}_2; \omega) = \frac{\mathrm{Tr}\ \overleftrightarrow{W}(\mathbf{r}_1, \mathbf{r}_2; \omega)}{\sqrt{S(\mathbf{r}_1; \omega)}\sqrt{S(\mathbf{r}_2; \omega)}}.\tag{21}$$

The relationship between coherence properties of light in the space-time and in the space-frequency domains have been subject of interest [for details see, for example, (Friberg & Wolf, 1995; Lahiri & Wolf, 2010a;d; Wolf, 1983)].

2.3 Polarization properties of stochastic beams in the space-time domain

In this section, we will briefly discuss some basic results in the matrix theory of polarization of electromagnetic beams, following the work of Wolf. For detailed discussions on this topic see any standard textbook, for example, (Born & Wolf, 1999; Brosseau, 1995; Collett, 1993; Mandel & Wolf, 1995). Let us consider a statistically stationary light beam characterized by a randomly fluctuating electric field vector $\mathbf{E}(\mathbf{r}, t)$. Without any loss of generality, we assume that the beam propagation direction is along positive z axis. Therefore, $\mathbf{E}(\mathbf{r}, t)$ may be represented by the two mutually orthogonal random components $E_x(\mathbf{r}, t)$ and $E_y(\mathbf{r}, t)$. Suppose that these components are represented by the ensembles of realizations $\{E_x(\mathbf{r}, t)\}$ and $\{E_y(\mathbf{r}, t)\}$, respectively. One can construct a 2×2 correlation matrix, known as coherency matrix (Wiener, 1927), which is given by [see, for example, (Mandel & Wolf, 1995), Sec. 6.2, Eq. 6.2-6]

$$\overleftrightarrow{J}(\mathbf{r}) \equiv \overleftrightarrow{\Gamma}(\mathbf{r}, \mathbf{r}; 0) \equiv \left[\left\langle E_i^*(\mathbf{r}; t) E_j(\mathbf{r}, t) \right\rangle \right], \qquad i = x, y, \quad j = x, y.\tag{22}$$

The elements of this matrix are *equal-time* correlation functions; consequently for statistically stationary fields they are time independent. This matrix contains all information about polarization properties of a stationary stochastic light beam at a point. Each element of this matrix can be determined from a canonical experiment, which involves passing the beam thought a compensator plate and polarizer, and then measuring the intensities for different values of the polarizer angle and of different values time delays introduced by the compensator plate among the components of electric field (Born & Wolf, 1999). A similar experiment will be discussed in detail in section 7. The Stokes parameters can be expressed in terms of the elements of a coherency matrix [see, for example, (Born & Wolf, 1999), section 10.9.3].

2.3.1 Unpolarized light beam

It can be shown that if a light beam is unpolarized at a point $P(\mathbf{r})$, then at that point the coherency matrix is proportional to a unit matrix, i.e., it has the form ((Mandel & Wolf, 1995), Sec. 6.3.1)

$$\overleftrightarrow{J}^{(u)}(\mathbf{r}) = \mathscr{A}(\mathbf{r}) \begin{pmatrix} 1 & 0 \\ 0 & 1 \end{pmatrix}.\tag{23}$$

It implies that in this case x and y components of the electric field are completely uncorrelated $[J_{xy}(\mathbf{r}) = J_{yx}(\mathbf{r}) = 0]$, and intensities of x and y components of the field are equal $[J_{xx}(\mathbf{r}) = J_{yy}(\mathbf{r}) \equiv \mathscr{A}(\mathbf{r})]$. It immediately follows that in this case the normalized correlation function

$$j_{xy}(\mathbf{r}) \equiv J_{xy}(\mathbf{r})/\sqrt{J_{xx}(\mathbf{r})J_{yy}(\mathbf{r})} = 0 \tag{24}$$

2.3.2 Polarized light beam

In the other extreme case when $|j_{xy}(\mathbf{r})| = 1$, one can readily show that the elements of the coherency matrix factorize in the form

$$J_{ij}^{(\mathrm{p})}(\mathbf{r}) = \mathcal{E}_i^*(\mathbf{r})\mathcal{E}_j(\mathbf{r}), \quad (i = x, y; \ j = x, y), \tag{25}$$

where $\mathcal{E}_i(\mathbf{r})$ is a time-independent deterministic function of position. This coherency matrix is identical with that of a monochromatic field which is given by the expression

$$\mathbf{E}(\mathbf{r}; t) = \mathcal{E}_i(\mathbf{r})e^{-i\omega t}. \tag{26}$$

In analogy with monochromatic beams, completely polarized beams are traditionally defined by the the coherency matrices which can be expressed in the form (25). However, a polarized light beam must not be confused with a monochromatic light beam. Recently, the distinction between the two has been clearly pointed out (Lahiri & Wolf, 2009; 2010a).

2.3.3 Partially polarized light beam

Any optical beam, which is neither unpolarized, nor polarized, is said to be partially polarized. Evidently, any coherency matrix which cannot be expressed in the form (23), or in the form (25) represents a partially polarized beam. However, it is remarkable that any such coherency matrix, can always be uniquely decomposed into the sum of two matrices representing a polarized beam and an unpolarized beam (Wolf, 1959) (see also, (Mandel & Wolf, 1995), Sec. 6.3.3), i.e., that

$$\overleftrightarrow{J}(\mathbf{r}) = \overleftrightarrow{J}^{(\mathrm{u})}(\mathbf{r}) + \overleftrightarrow{J}^{(\mathrm{p})}(\mathbf{r}). \tag{27}$$

consequently, the average intensity $[I(\mathbf{r}) \equiv \mathrm{Tr}\,\overleftrightarrow{J}(\mathbf{r})]$ of any light beam, at a point, has contributions from a completely polarized and a completely unpolarized beam. The degree of polarization at a point $P(\mathbf{r})$ is defined as the ratio of the average intensity of the polarized part to the total average intensity at that point (Wolf, 1959). One can show that it is given by the expression (Wolf, 1959) (see also, Ref. (Born & Wolf, 1999), Sec. 10.9.2, Eq. (52))

$$P(\mathbf{r}) \equiv \frac{I^{(\mathrm{p})}(\mathbf{r})}{I(\mathbf{r})} = \sqrt{1 - \frac{4\,\mathrm{Det}\,\overleftrightarrow{J}(\mathbf{r})}{[\mathrm{Tr}\,\overleftrightarrow{J}(\mathbf{r})]^2}}, \tag{28}$$

where Det denotes the determinant and Tr denotes the trace. One can show that the degree of polarization is always bounded between zero and unity:

$$0 \leq P(\mathbf{r}) \leq 1. \tag{29}$$

If the degree of polarization is unity at a point $\mathbf{r}$, then the beam is said to be *completely polarized* at that point and if it has zero value, the light is said to be *completely unpolarized* at that point.

2.4 Polarization properties of stochastic beams in the space-frequency domain

As was in the case of coherence, polarization properties of light can also be analyzed in the space-frequency domain. Such a theory has certain advantages, because it makes it easier to study the change in polarization properties of light on propagation and scattering. The theory involves analyzing polarization properties of light at each frequency present in its spectrum. It is similar to the theory in the space-time domain, except that the coherency matrix $\overleftrightarrow{J}(\mathbf{r})$ is replaced by equal-point CSDM $\overleftrightarrow{W}(\mathbf{r},\mathbf{r};\omega)$. In this case the spectral degree of polarization is given by the expression ((Wolf, 2007b), Sec. 9.2, Eq. (14))

$$\mathcal{P}(\mathbf{r};\omega) \equiv \sqrt{1 - \frac{4 \operatorname{Det} \overleftrightarrow{W}(\mathbf{r},\mathbf{r};\omega)}{[\operatorname{Tr} \overleftrightarrow{W}(\mathbf{r},\mathbf{r};\omega)]^2}}. \tag{30}$$

The relationship between space-time and space-frequency description of polarization properties of light are being investigated recently [for details see, for example, (Lahiri, 2009; Lahiri & Wolf, 2010a; Setälä et al., 2009)].

2.5 Unified theory of coherence and polarization

For many years coherence and polarization properties of light have been considered as independent subjects. However, the matrix formulation, especially in the space-frequency domain, shows that they are intimately related. This fact was firmly established by the introduction of unified theory of coherence and polarization (Wolf, 2003b). According to the unified theory, both the coherence and polarization properties of a stochastic electromagnetic beam, can be described in terms of the 2×2 cross-spectral density matrix (CSDM) $\overleftrightarrow{W}(\mathbf{r}_1,\mathbf{r}_2;\omega)$. The spectral density, spectral degree of coherence, and the spectral degree of polarization are described by Eqs. (20), (21), and (30) respectively.

Elements of the CSDM obey definite propagation laws [see, for example, (Wolf, 2007b), Sec. 9.4.1, Eq. (3)]. If the CSDM is specified at all pairs of points at any cross-sectional plane of a beam, then it is possible to determine the CSDM at all pairs of points on any other cross-section of that beam, both in free space and in a medium. Therefore, it is possible to study the changes in spectral, coherence and polarization properties of a beam on propagation.

3. Optics in terms of observable quantities

At this point, we would like to emphasize that every optical phenomenon, which will be addressed in this chapter, will be illustrated and interpreted in terms of *observable quantities*. Formulating optical physics in terms of observable quantities is due to valuable work of Emil Wolf (Wolf, 1954). His effort was highly appreciated by Danis Gabor in a lecture on "Light and Information" (Gabor, 1955), as he mentioned "perhaps the most satisfactory feature of the theory is that it operates entirely with quantities which are in principle observable, in line with the valuable efforts of E. Wolf to rid optics of its metaphysical residues."

The analysis of coherence and polarization properties of light are based on the theory of electromagnetic fields. However, one must note that even today, one is not able to "directly" detect such a field at an optical frequency, or at a higher frequency. Any optical phenomenon that one observes in a laboratory, or in nature, is the result of generation of electric currents in detectors. Such currents originate from the so-called *destruction of photons* by light-matter interactions. An example of commonly available sophisticated optical detectors is a human eye. Another example of an optical detector is an EM-CCD camera, which is often used in today's laboratories. From the measurement of current in a detector one may predict the so-called photon detection rate, or intensity of light.

In section 2, we interpreted spatial coherence properties of light in terms of correlation between classical electric field at a pair of points. However, since electric field in an optical frequency is not observable, one must not be too much carried away with such an interpretation. It must be kept in mind that coherence is the ability of light to interfere and a physical measure of coherence is the visibility of fringes in an interference experiment, not a correlation function. On the other hand, polarization properties of light beams were interpreted as correlation between electric field components at a particular point. However, the physical phenomenon which leads to such a mathematical formulation is the modulation of intensity of the beam, as it is passed through polarization controlling devices, such as polarizers, compensator plates etc.

4. Quantum theory of optical fields

We begin by recalling some basic properties of quantized electromagnetic fields (Dirac, 1957). A quantized electric field may be represented by a Hermitian operator (Mandel & Wolf, 1995)

$$\widehat{\mathbf{E}}(\mathbf{r}, t) = i \sum_{\mathbf{k}} \sum_{s} \left(\frac{1}{2} \hbar \omega \right)^{\frac{1}{2}} \left[\widehat{a}_{\mathbf{k},s} C_{\mathbf{k},s} \boldsymbol{\epsilon}_{\mathbf{k},s} e^{i(\mathbf{k} \cdot \mathbf{r} - \omega t)} - \widehat{a}_{\mathbf{k},s}^{\dagger} C_{\mathbf{k},s}^{*} \boldsymbol{\epsilon}_{\mathbf{k},s}^{*} e^{-i(\mathbf{k} \cdot \mathbf{r} - \omega t)} \right], \tag{31}$$

where the wave vectors $\mathbf{k}$ labels plane wave modes, $|\mathbf{k}| = k = \omega/c$, c is the speed of light in free space, $C_{\mathbf{k},s}$ is a constant, and $\boldsymbol{\epsilon}_{\mathbf{k},s}$ ($s = 1, 2$), are mutually orthonormal base vectors, which obey the conditions[8]

$$\mathbf{k} \cdot \boldsymbol{\epsilon}_{\mathbf{k},s} = 0, \qquad \boldsymbol{\epsilon}_{\mathbf{k},s}^{*} \cdot \boldsymbol{\epsilon}_{\mathbf{k},s'} = \delta_{ss'}, \qquad \boldsymbol{\epsilon}_{\mathbf{k},1} \times \boldsymbol{\epsilon}_{\mathbf{k},2} = \mathbf{k}/k. \tag{32}$$

In the expansion (31), $\widehat{a}_{\mathbf{k},s}$ and $\widehat{a}_{\mathbf{k},s}^{\dagger}$ are the photon annihilation and the photon creation operators respectively, for the mode labeled by $(\mathbf{k}, s)$. These operators obey the well known commutation relations [see, for example, (Mandel & Wolf, 1995), Sec. 10.3]

$$\left[\widehat{a}_{\mathbf{k},s}, \widehat{a}_{\mathbf{k}',s'}^{\dagger} \right] = \delta_{\mathbf{k}\mathbf{k}'}^{3} \delta_{ss'},$$
$$\left[\widehat{a}_{\mathbf{k},s}, \widehat{a}_{\mathbf{k}',s'} \right] = 0, \tag{33}$$
$$\left[\widehat{a}_{\mathbf{k},s}^{\dagger}, \widehat{a}_{\mathbf{k}',s'}^{\dagger} \right] = 0,$$

---

[8] The unit base vectors $\boldsymbol{\epsilon}_{k1}$, $\boldsymbol{\epsilon}_{k2}$ may be chosen to be complex for general expansion of the field into two orthogonal polarization components, for example, in connection with elliptic polarization.

where δ_{ij} is the Kronecker symbol. As evident from Eq. (31), the electric field operators consists of a positive frequency part[9]

$$\widehat{\mathbf{E}}^{(+)}(\mathbf{r},t) = i \sum_{\mathbf{k}} \sum_s \left(\frac{1}{2}\hbar\omega\right)^{\frac{1}{2}} \widehat{a}_{\mathbf{k},s} C_{\mathbf{k},s} \boldsymbol{\epsilon}_{\mathbf{k},s} e^{i(\mathbf{k}\cdot\mathbf{r}-\omega t)}, \tag{34}$$

and a negative frequency part $\widehat{\mathbf{E}}^{(-)}(\mathbf{r},t)$, which is the Hermitian adjoint of the positive frequency part. The expansion (31) represents the field in discrete modes. Such a representation is appropriate, for example, when treating an electric field in a cavity. In more general situations, a continuous mode representation may be more appropriate [for discussion on such representation see (Mandel & Wolf, 1995), Sec. 10.10].

The state of the field is described by a state vector $|i\rangle$ in the Fock space, or, more generally, by a density operator $\widehat{\rho} = \{\langle i|\,|i\rangle\}_{\text{average}}$, where the average is taken over an appropriate ensemble. The expectation value $\langle\widehat{O}\rangle$ of any operator $\widehat{O}$ is given by the well known expression

$$\langle\widehat{O}\rangle = \text{Tr}\{\widehat{\rho}\,\widehat{O}\}, \tag{35}$$

where Tr denotes the trace. An informative description of how the measurable quantities may be interpreted in terms of quantized field and density operators, was given by Glauber (Glauber, 1963). In the following section, we will briefly go over some basic concepts from the Glauber's interpretation of quantum theory of optical coherence.

5. Summary of some basic results of the quantum theory of optical coherence in the space-time domain

In the quantum-mechanical interpretation, a photon can be detected only by destroying it. The photo-detector is assumed to be ideal in the sense that it is of negligible size and has a frequency-independent photo absorbtion probability. Let us now consider absorption (detection) of a photon by an ideal detector at a space-time point $(\mathbf{r};t)$. Suppose that due to this absorption the field goes from the initial state $|i\rangle$ to a final state $|f\rangle$. The probability of the detector for absorbing a photon in final state $|f\rangle$ is $\left|\langle f|\,\widehat{\mathbf{E}}^{(+)}(\mathbf{r};t)\,|i\rangle\right|^2$. The counting rate in the detector is obtained by summing over all the final states which can be reached from $|i\rangle$, by absorption of a photon. One can extend the summation over a complete set of final states, since the states which cannot be reached in this process will not contribute to the result [for details see (Glauber, 2007)]. The counting rate of the detector then becomes proportional to

$$\sum_f \left|\langle f|\,\widehat{\mathbf{E}}^{(+)}(\mathbf{r};t)\,|i\rangle\right|^2 = \langle i|\,\widehat{\mathbf{E}}^{(-)}(\mathbf{r};t)\cdot\widehat{\mathbf{E}}^{(+)}(\mathbf{r};t)\,|i\rangle. \tag{36}$$

---

[9] The classical analogue of the positive frequency part of electric field operator is the so-called complex analytic signal of a real electric field, introduced by Gabor (Gabor, 1946). For a discussion of the physical interpretation of the positive and negative frequency parts of the quantized electric field operator see (Glauber, 1963).

If one considers the random fluctuations associated with light, one leads to a more general expression involving the density operator. The average counting rate of a photo-detector placed at a position $\mathbf{r}$ at time t then becomes proportional to [c.f (Glauber, 1963), Eq. (3.3)]

$$\mathscr{R}(\mathbf{r}, t) \equiv \mathrm{Tr}\left\{\hat{\rho}\,\hat{\mathbf{E}}^{(-)}(\mathbf{r}, t) \cdot \hat{\mathbf{E}}^{(+)}(\mathbf{r}, t)\right\}. \tag{37}$$

Here the dot $(\cdot)$ denotes scalar product.

Using similar arguments, one can obtain quantum-mechanical analogues of all correlation functions (such as mutual coherence function etc.) used in the classical theory. The first-order correlation properties (corresponds to second-order properties in the classical theory) of the field may be specified by a 3×3 correlation matrix (Glauber, 1963)

$$\begin{aligned}
\overleftrightarrow{G}^{(1)}(\mathbf{r}, t; \mathbf{r}', t') &\equiv \left[G_{\mu\nu}^{(1)}(\mathbf{r}, t; \mathbf{r}', t')\right] \\
&\equiv \left[\mathrm{Tr}\left\{\hat{\rho}\,\hat{E}_{\mu}^{(-)}(\mathbf{r}, t)\hat{E}_{\nu}^{(+)}(\mathbf{r}', t')\right\}\right],
\end{aligned} \tag{38}$$

where μ, ν label, mutually orthogonal components of the electric field operator. For the sake of simplicity, let us neglect the polarization properties of the light, restricting our analysis to scalar fields. Hence, with a suitable choice of axes, *only one* element $G_{\mu\mu}^{(1)}(\mathbf{r}, t; \mathbf{r}', t')$ [no summation over repeated indices] of $\overleftrightarrow{G}^{(1)}(\mathbf{r}, t; \mathbf{r}', t')$ will completely characterize all first-order correlation properties of the field in the space-time domain. We omit the suffix μ and write

$$G^{(1)}(\mathbf{r}, t; \mathbf{r}', t') = \mathrm{Tr}\left\{\hat{\rho}\,\hat{E}^{(-)}(\mathbf{r}; t)\hat{E}^{(+)}(\mathbf{r}'; t')\right\}. \tag{39}$$

The simplest coherence properties of light, in the space-time domain, are characterized by the first-order correlation function $\overleftrightarrow{G}^{(1)}(\mathbf{r}, t; \mathbf{r}', t')$. In terms of it one can define the first-order degree of coherence by the formula [(Glauber, 1963), Eq. (4)]

$$g^{(1)}(\mathbf{r}, t; \mathbf{r}', t') \equiv \frac{G^{(1)}(\mathbf{r}, t; \mathbf{r}', t')}{\sqrt{G^{(1)}(\mathbf{r}; t; \mathbf{r}; t)}\sqrt{G^{(1)}(\mathbf{r}'; t'; \mathbf{r}'; t')}}. \tag{40}$$

The modulus of $g^{(1)}(\mathbf{r}, t; \mathbf{r}', t')$ may be shown to be bounded by zero and unity [(Glauber, 1963), Eq. (4.2)], i.e.,

$$0 \leq |g^{(1)}(\mathbf{r}, t; \mathbf{r}', t')| \leq 1. \tag{41}$$

It can be shown that this quantity is related to fringe visibility in interference experiments [see, for example, (Glauber, 2007), Sec. 2.7.2]. Complete first-order coherence (corresponding to second-order coherence in classical theory) is characterized by the condition $|g^{(1)}(\mathbf{r}, t; \mathbf{r}', t')| = 1$, and complete first-order incoherence by the other extreme, $g^{(1)}(\mathbf{r}, t; \mathbf{r}', t') = 0$. Equations (8), and (40) may look similar, but one must appreciate the fact that the quantum mechanical interpretation is much more effective as one goes to a low-intensity domain, where absorption or emission of one, or, few numbers of photons may affect the experimental observations.

Detailed descriptions on this topic have been discussed in many scholarly articles [see, for example, (Glauber, 1963; 2007; Mandel & Wolf, 1995)].

6. Quantum theory of optical coherence in space-frequency domain

In this section we present a quantum-mechanical theory of first-order optical coherence for statistically non-stationary light in the space-frequency domain. We discuss some relevant correlation functions, associated with the quantized field and the density operator, which can be introduced in the space-frequency representation. We also show, by use of the technique of linear filtering, that these new correlation functions may be related to the photo-counting rate and to well known coherence-functions of space-time domain. We consider non-stationary light[10], because the assumption of statistical stationarity is not always appropriate, especially in many situations encountered in quantum optics where light emission may take place over a finite time-interval. A detailed description of the main material presented in this section has been recently published (Lahiri & Wolf, 2010c).

6.1 Correlation functions in the space-frequency domain

Let us first note some properties of the operator $\widehat{e}(r, \omega)$, which is the Fourier transform of $\widehat{E}(r; t)$, i.e., which is given by

$$\widehat{e}(r, \omega) = \frac{1}{2\pi} \int_{-\infty}^{\infty} \widehat{E}(r; t) e^{i\omega t} \, dt. \tag{42}$$

Using the fact that $\widehat{E}(r; t) = \widehat{E}^{(+)}(r; t) + \widehat{E}^{(-)}(r; t)$, one may express $\widehat{e}(r, \omega)$ in the form

$$\widehat{e}(r, \omega) = \widehat{e}^{(+)}(r, \omega) + \widehat{e}^{(-)}(r, \omega), \tag{43}$$

where

$$\widehat{e}^{(+)}(r, \omega) = \frac{1}{2\pi} \int_{-\infty}^{\infty} \widehat{E}^{(+)}(r; t) e^{i\omega t} \, dt, \tag{44a}$$

$$\widehat{e}^{(-)}(r, \omega) = \frac{1}{2\pi} \int_{-\infty}^{\infty} \widehat{E}^{(-)}(r; t) e^{i\omega t} \, dt. \tag{44b}$$

Using the property $\left\{ \widehat{E}^{(+)}(r; t) \right\}^{\dagger} = \widehat{E}^{(-)}(r; t)$, it follows from Eqs. (44) that $\left\{ \widehat{e}^{(+)}(r, \omega) \right\}^{\dagger} = \widehat{e}^{(-)}(r, -\omega)$.

Let us now consider the following 3×3 correlation matrix

$$\overleftrightarrow{\mathscr{W}}^{(1)}(r, \omega; r', \omega') \equiv \left[\mathscr{W}_{\mu\nu}^{(1)}(r, \omega; r', \omega') \right] \equiv \mathrm{Tr} \left\{ \widehat{\rho} \, \widehat{e}_{\mu}^{(-)}(r, -\omega) \widehat{e}_{\nu}^{(+)}(r', \omega') \right\}. \tag{45}$$

On using Eqs. (44) and (45), one can readily show that the elements of the correlation matrices $\overleftrightarrow{G}^{(1)}(r, t; r', t')$ and $\overleftrightarrow{\mathscr{W}}^{(1)}(r, \omega; r', \omega')$ are Fourier transforms of each other, i.e., that

$$\mathscr{W}_{\mu\nu}^{(1)}(r, \omega; r', \omega') = \left(\frac{1}{2\pi} \right)^2 \iint_{-\infty}^{\infty} G_{\mu\nu}^{(1)}(r, t; r', t') e^{i(-\omega t + \omega' t')} \, dt \, dt', \tag{46a}$$

$$G_{\mu\nu}^{(1)}(r, t; r', t') = \iint_{0}^{\infty} \mathscr{W}_{\mu\nu}^{(1)}(r, \omega; r', \omega') e^{i(\omega t - \omega' t')} \, d\omega \, d\omega'. \tag{46b}$$

---

[10] Attempts to formulate coherence theory for classical non-stationary fields have been made (Bertolotti et al., 1995; Sereda et al., 1998).

We will restrict our analysis to scalar fields. Therefore, the 3×3 correlation matrix $\overleftrightarrow{\mathscr{W}}^{(1)}(\mathbf{r}, \omega; \mathbf{r}', \omega')$ may now be replaced by the correlation function

$$\mathscr{W}^{(1)}(\mathbf{r}, \omega; \mathbf{r}', \omega') = \mathrm{Tr}\left\{ \hat{\rho}\, \hat{e}^{(-)}(\mathbf{r}, -\omega)\hat{e}^{(+)}(\mathbf{r}', \omega') \right\}, \tag{47}$$

which, in analogy with the relations (46), is the Fourier transform of $G^{(1)}(\mathbf{r}, t; \mathbf{r}', t')$; viz,

$$\mathscr{W}^{(1)}(\mathbf{r}, \omega; \mathbf{r}', \omega') = \left(\frac{1}{2\pi}\right)^2 \iint\limits_{-\infty}^{\infty} G^{(1)}(\mathbf{r}, t; \mathbf{r}', t')e^{i(-\omega t + \omega' t')}\, dt\, dt', \tag{48a}$$

$$G^{(1)}(\mathbf{r}, t; \mathbf{r}', t') = \iint\limits_{0}^{\infty} \mathscr{W}^{(1)}(\mathbf{r}, \omega; \mathbf{r}', \omega')e^{i(\omega t - \omega' t')}\, d\omega\, d\omega'. \tag{48b}$$

We will refer to $\mathscr{W}^{(1)}(\mathbf{r}, \omega; \mathbf{r}', \omega')$ as the *two-frequency cross-spectral density function* (to be abbreviated by two-frequency CSDF) and the single-frequency correlation function $\mathscr{W}^{(1)}(\mathbf{r}, \omega; \mathbf{r}', \omega)$ as the *cross-spectral density function* (CSDF) of the field, in analogy with terminology used in the classical theory.

6.2 Physical interpretation of correlation functions in the space-frequency domain

Let us now assume that a light beam is transmitted by a linear filter which allows only a narrow frequency band to pass through it. Suppose that the light, emerging from the filter, has mean frequency $\bar{\omega}$ and effective bandwidth $\Delta\omega \ll \bar{\omega}$. The field operators representing this filtered narrow-band light in the space-frequency domain, may then be represented by the formulas

$$\hat{e}^{(+)}_{(\bar{\omega})}(\mathbf{r}, \omega) = T_{(\bar{\omega})}(\omega)\hat{e}^{(+)}(\mathbf{r}, \omega), \tag{49a}$$

$$\hat{e}^{(-)}_{(\bar{\omega})}(\mathbf{r}, -\omega) = T^*_{(\bar{\omega})}(\omega)\hat{e}^{(-)}(\mathbf{r}, -\omega), \tag{49b}$$

where $T_{(\bar{\omega})}(\omega)$ is the transmission function of the filter, whose modulus is negligible outside the pass-band $\bar{\omega} - \Delta\omega/2 \leq \omega \leq \bar{\omega} + \Delta\omega/2$ of the filter. Using Eqs. (47) and (49), it follows that for the filtered light,

$$\mathscr{W}^{(1)}_{(\bar{\omega})}(\mathbf{r}, \omega; \mathbf{r}', \omega') = T^*_{(\bar{\omega})}(\omega)T_{(\bar{\omega})}(\omega')\mathscr{W}^{(1)}(\mathbf{r}, \omega; \mathbf{r}', \omega'). \tag{50}$$

Here, the function $\mathscr{W}^{(1)}_{(\bar{\omega})}(\mathbf{r}, \omega; \mathbf{r}', \omega')$ is the two-frequency CSDF of the filtered light, of mean frequency $\bar{\omega}$, and $\mathscr{W}^{(1)}(\mathbf{r}, \omega; \mathbf{r}', \omega')$ on the right hand side is the two-frequency CSDF of the unfiltered light incident on the filter.

From Eqs. (48b) and (50), one readily finds that the space-time correlation function $G^{(1)}_{(\bar{\omega})}(\mathbf{r}; t; \mathbf{r}'; t')$ of the filtered narrow-band light is given by the expression

$$G^{(1)}_{(\bar{\omega})}(\mathbf{r}; t; \mathbf{r}'; t') = \iint\limits_{\bar{\omega}-\Delta\omega/2}^{\bar{\omega}+\Delta\omega/2} T^*_{(\bar{\omega})}(\omega)T_{(\bar{\omega})}(\omega')\mathscr{W}^{(1)}(\mathbf{r}, \omega; \mathbf{r}', \omega')e^{i(\omega t - \omega' t')}\, d\omega\, d\omega'. \tag{51}$$

Because of the assumption that $\Delta\omega/\bar{\omega} \ll 1$, the function $\mathscr{W}^{(1)}(\mathbf{r}, \omega; \mathbf{r}', \omega')$ in Eq. (51) does not change appreciably as function of ω and ω' over the ranges $\bar{\omega} - \Delta\omega/2 \leq \omega \leq \bar{\omega} + \Delta\omega/2$, $\bar{\omega} - \Delta\omega/2 \leq \omega' \leq \bar{\omega} + \Delta\omega/2$ and is approximately equal to $\mathscr{W}^{(1)}(\mathbf{r}, \bar{\omega}; \mathbf{r}', \bar{\omega})$. From Eq. (51) it then follows that

$$G_{(\bar{\omega})}^{(1)}(\mathbf{r}; t; \mathbf{r}'; t') \approx \mathscr{W}^{(1)}(\mathbf{r}, \bar{\omega}; \mathbf{r}', \bar{\omega}) \iint\limits_{\bar{\omega}-\Delta\omega/2}^{\bar{\omega}+\Delta\omega/2} T_{(\bar{\omega})}^*(\omega) T_{(\bar{\omega})}(\omega') e^{i(\omega t - \omega' t')} \, d\omega \, d\omega'. \tag{52}$$

6.3 Physical significance of $\mathscr{W}^{(1)}(\mathbf{r}, \omega; \mathbf{r}, \omega)$

Let us consider light generated by some physical process which begins at time $t = 0$ and ceases at time $t = \mathcal{T}$, say; for example light emitted by a collection of exited atoms. Clearly, such light is not statistically stationary. Suppose now that this light is filtered and is then incident on a photo-detector. It follows from Eq. (52) that the average counting rate $\mathscr{R}_{(\bar{\omega})}(\mathbf{r}; t) \equiv G_{(\bar{\omega})}^{(1)}(\mathbf{r}; t; \mathbf{r}; t)$ of the detector placed at a point $\mathbf{r}$, at time t, is given by the expression

$$\mathscr{R}_{(\bar{\omega})}(\mathbf{r}; t) \approx \mathscr{W}^{(1)}(\mathbf{r}, \bar{\omega}; \mathbf{r}, \bar{\omega}) \iint\limits_{\bar{\omega}-\Delta\omega/2}^{\bar{\omega}+\Delta\omega/2} T_{(\bar{\omega})}^*(\omega) T_{(\bar{\omega})}(\omega') e^{i(\omega - \omega')t} \, d\omega \, d\omega'. \tag{53}$$

Since, $\mathscr{R}_{(\bar{\omega})}(\mathbf{r}; t) = 0$, unless $0 < t < \mathcal{T}$, the total energy $\mathscr{E}(\mathbf{r}; \bar{\omega})$ detected (total counts) by the photo-detector will be proportional to

$$\mathscr{E}(\mathbf{r}; \bar{\omega}) \equiv \int_0^{\mathcal{T}} \mathscr{R}_{(\bar{\omega})}(\mathbf{r}; t) \, dt = \int_{-\infty}^{\infty} \mathscr{R}_{(\bar{\omega})}(\mathbf{r}; t) \, dt$$

$$\approx \mathscr{W}^{(1)}(\mathbf{r}, \bar{\omega}; \mathbf{r}, \bar{\omega}) \left\{ 2\pi \int_{\bar{\omega}-\Delta\omega/2}^{\bar{\omega}+\Delta\omega/2} \left| T_{(\bar{\omega})}(\omega) \right|^2 \, d\omega \right\} \tag{54}$$

Thus, we may conclude that the total energy $\mathscr{E}(\mathbf{r}; \omega)$ detected by the photo-detector, placed at a point $\mathbf{r}$ in the path of a narrow-band light beam of mean frequency ω, is proportional to $\mathscr{W}^{(1)}(\mathbf{r}, \omega; \mathbf{r}, \omega)$, i.e., that

$$\mathscr{E}(\mathbf{r}; \omega) \equiv \int_0^{\mathcal{T}} \mathscr{R}_{(\bar{\omega})}(\mathbf{r}; t) \, dt \propto \mathscr{W}^{(1)}(\mathbf{r}, \omega; \mathbf{r}, \omega), \tag{55}$$

the proportionality constant being dependent on the filter and on the choice of units. In other words, the correlation function $\mathscr{W}^{(1)}(\mathbf{r}, \omega; \mathbf{r}, \omega)$ provides a measure of the energy density, at a point $\mathbf{r}$, associated with the frequency ω of light. It is evident that $\mathscr{W}^{(1)}(\mathbf{r}, \omega; \mathbf{r}, \omega) \geq 0$.

$\mathscr{W}^{(1)}(\mathbf{r}, \omega; \mathbf{r}, \omega)$ must not be confused with the well-known Wiener's spectral density (Wiener, 1930) of statistically stationary light. In the quantum mechanical interpretation, Wiener's spectral density is equivalent to the *counting rate* of the photo-detector, associated with a frequency. On the other hand, the quantity $\mathscr{W}^{(1)}(\mathbf{r}, \omega; \mathbf{r}, \omega)$ represents the *total counts* in the photo-detector associated with the frequency ω. Therefore, $\mathscr{W}^{(1)}(\mathbf{r}, \omega; \mathbf{r}, \omega)$ has a different dimension than Wiener's spectral density. Even in the stationary limit, $\mathscr{W}^{(1)}(\mathbf{r}, \omega; \mathbf{r}, \omega)$ does not reduce to the Wiener's spectral density. In fact, in such a case, the energy density $\mathscr{E}(\mathbf{r}; \omega)$ will be infinitely large and hence, in that limit, the quantity $\mathscr{W}^{(1)}(\mathbf{r}, \omega; \mathbf{r}, \omega)$ will not be useful.

For a light that is not statistically stationary, defining spectral density is a nontrivial problem which is still a subject of an active discussion. Many publications have been dedicated to it and several different definitions have been proposed [see, for example, (Eberly & Wódkievicz, 1977; Lampard, 1954; Mark, 1970; Page, 1952; Ponomarenko et al., 2004; Silverman, 1957)]. Nevertheless, for statistically non-stationary processes, $\mathscr{W}^{(1)}(\mathbf{r}, \omega; \mathbf{r}, \omega)$ is measurable and a physically meaningful quantity. In the following sections we show that it has an important role in defining the spectral-degree of coherence of non-stationary light.

6.4 First-order coherence

We will now consider the first-order coherence properties of non-stationary light, again restricting ourselves to scalar fields. As in the previous Section, we will consider a filtered narrow-band light. On using Eqs. (40) and (52), one then has

$$g_{(\bar{\omega})}^{(1)}(\mathbf{r}, t; \mathbf{r}', t') \approx \frac{\mathscr{W}^{(1)}(\mathbf{r}, \bar{\omega}; \mathbf{r}', \bar{\omega})}{\sqrt{\mathscr{W}^{(1)}(\mathbf{r}, \bar{\omega}; \mathbf{r}, \bar{\omega})}\sqrt{\mathscr{W}^{(1)}(\mathbf{r}', \bar{\omega}; \mathbf{r}', \bar{\omega})}} \Theta(t, t'), \tag{56}$$

where

$$\Theta(t, t') = \frac{\iint_{\bar{\omega}-\Delta\omega/2}^{\bar{\omega}+\Delta\omega/2} T_{(\bar{\omega})}^*(\omega) T_{(\bar{\omega})}(\omega') e^{i(\omega t - \omega' t')}\, d\omega\, d\omega'}{\left| \int_{\bar{\omega}-\Delta\omega/2}^{\bar{\omega}+\Delta\omega/2} T_{(\bar{\omega})}(\omega) e^{-i\omega t}\, d\omega \right| \left| \int_{\bar{\omega}-\Delta\omega/2}^{\bar{\omega}+\Delta\omega/2} T_{(\bar{\omega})}(\omega) e^{-i\omega t'}\, d\omega \right|}. \tag{57}$$

Since, the numerator on the right hand side of Eq. (57) factorizes into a product of the two integrals which appear in the denominator, it is evident that

$$|\Theta(t, t')| = 1, \tag{58}$$

for all values of t and t'. From Eqs. (56) and (58), one readily finds that the modulus of the first-order "space-time" degree of coherence of the filtered narrow-band light of mean frequency $\bar{\omega}$, is given by the formula

$$\left| g_{(\bar{\omega})}^{(1)}(\mathbf{r}, t; \mathbf{r}', t') \right| \approx \left| \frac{\mathscr{W}^{(1)}(\mathbf{r}, \bar{\omega}; \mathbf{r}', \bar{\omega})}{\sqrt{\mathscr{W}^{(1)}(\mathbf{r}, \bar{\omega}; \mathbf{r}, \bar{\omega})}\sqrt{\mathscr{W}^{(1)}(\mathbf{r}', \bar{\omega}; \mathbf{r}', \bar{\omega})}} \right|. \tag{59}$$

The expression within the modulus signs on the right-hand side is the normalized CSDF at the frequency $\bar{\omega}$.

It may be concluded from Eq. (59) that this normalized CSDF provides a measure of first-order coherence of the filtered light of mean frequency $\bar{\omega}$. Consequently, one may define the *first-order spectral degree of coherence* at frequency ω by the formula

$$\eta^{(1)}(\mathbf{r}, \omega; \mathbf{r}', \omega) \equiv \frac{\mathscr{W}^{(1)}(\mathbf{r}, \omega; \mathbf{r}', \omega)}{\sqrt{\mathscr{W}^{(1)}(\mathbf{r}, \omega; \mathbf{r}, \omega)}\sqrt{\mathscr{W}^{(1)}(\mathbf{r}', \omega; \mathbf{r}', \omega)}}. \tag{60}$$

It can be immediately shown that (Lahiri & Wolf, 2010c)

$$0 \leq |\eta^{(1)}(\mathbf{r}, \omega; \mathbf{r}', \omega)| \leq 1. \tag{61}$$

When $|\eta^{(1)}(\mathbf{r},\omega;\mathbf{r}',\omega)| = 1$, the light may be said to be completely coherent at the frequency ω, and when $\eta^{(1)}(\mathbf{r},\omega;\mathbf{r}',\omega) = 0$, it may be said to be completely incoherent at that frequency. We stress, once again, that $\mathcal{W}^{(1)}(\mathbf{r},\omega;\mathbf{r}',\omega)$ and $\eta^{(1)}(\mathbf{r},\omega;\mathbf{r}',\omega)$ being *single-frequency* quantities, provide measures of the ability of a particular frequency-component of the light to interfere.

We will now return to the two-frequency correlation function $\mathcal{W}^{(1)}(\mathbf{r},\omega;\mathbf{r}',\omega')$ given by formula (47). It characterizes correlations between different frequency components of light, i.e., the ability of two different frequency components to interfere. Interference of fields of different frequencies has generally not been considered in the literature, probably because it is not a commonly observed phenomenon. According to the classical theory, different frequency components of *statistically stationary* light do not interfere (see, for example, (Wolf, 2007b), Sec. 2.5). For the sake of completeness, we will now show that this fact is also true in the non-classical domain. The proof is similar to that for classical fields. If one considers light whose fluctuations are statistically stationary in the wide sense, i.e., if $G^{(1)}(\mathbf{r},t;\mathbf{r}',t') = G^{(1)}(\mathbf{r},\mathbf{r}';t'-t)$ then, ignoring some mathematical subtleties, one finds from Eq. (48a) that the two-frequency CSDF $\mathcal{W}^{(1)}(\mathbf{r},\omega;\mathbf{r}',\omega')$ has the form

$$\mathcal{W}^{(1)}(\mathbf{r},\omega;\mathbf{r}',\omega') = f[\mathbf{r},\mathbf{r}';(\omega+\omega')/2]\delta(\omega-\omega'), \tag{62}$$

where f is, in general, a complex function of its arguments and δ denotes the Dirac delta function. This formula shows that when $\omega \neq \omega'$, the two-frequency CSDF $\mathcal{W}^{(1)}(\mathbf{r},\omega;\mathbf{r}',\omega') = 0$, implying that different frequency components of statistically stationary light do not interfere. However, for non-stationary random processes, this correlation function may have a non-zero value and, consequently, interference among different frequency components may take place. One may define a normalized two-frequency correlation function

$$\eta^{(1)}(\mathbf{r},\omega;\mathbf{r}',\omega') = \frac{\mathcal{W}^{(1)}(\mathbf{r},\omega;\mathbf{r}',\omega')}{\sqrt{\mathcal{W}^{(1)}(\mathbf{r},\omega;\mathbf{r},\omega)\mathcal{W}^{(1)}(\mathbf{r}',\omega';\mathbf{r}',\omega')}} \tag{63}$$

as a generalized first-order spectral degree of coherence for a pair of frequencies ω and ω'. It can also be shown that (Lahiri & Wolf, 2010c)

$$0 \leq |\eta^{(1)}(\mathbf{r},\omega;\mathbf{r}',\omega')| \leq 1. \tag{64}$$

The extreme value $|\eta^{(1)}(\mathbf{r},\omega;\mathbf{r}',\omega')| = 1$, which corresponds to maximum possible fringe visibility observed in interference experiments, represents complete coherence in the space-frequency domain. The other extreme value, $\eta^{(1)}(\mathbf{r},\omega;\mathbf{r}',\omega') = 0$, implies that no interference fringes will be present, i.e that there is complete incoherence between different frequency components.

7. Polarization properties of optical beams

In this section we will discuss the polarization properties of light. As is clear from previous discussions that a scalar treatment will no more be sufficient for his purpose, and we have to consider the vector nature of the field. However, since, we will restrict our analysis to beam-like optical fields, we will encounter 2×2 correlation matrices.

We consider the canonical experiment depicted in Fig. 2. Suppose that a light beam, propagating along the positive z axis, passes through a compensator, followed by a polarizer (see Fig. 2). Light emerging from the polarizer is linearly polarized along some direction,

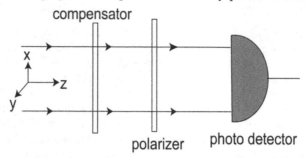

Fig. 2. Illustrating notation relating to the canonical polarization experiment.

which makes an angle θ, say, with the positive direction of a chosen x axis. We call θ the polarizer angle. Effects of the compensator may be described by introducing phase delays α_x and α_y in the x and the y components of the field operator $\widehat{\mathbf{E}}^{(+)}$ respectively. Suppose now that a photodetector is placed behind the polarizer (see Fig. 2), which detects photons that emerge from the polarizer. From Eq. (37), it follows that the counting rate $\mathscr{R}_{\theta,\alpha}(\mathbf{r};t)$ in the detector will be given by the formula

$$\mathscr{R}_{\theta,\alpha}(\mathbf{r};t) = \mathrm{Tr}\left\{\widehat{\rho}\,\widehat{E}_{\theta,\alpha}^{(-)}(\mathbf{r};t)\widehat{E}_{\theta,\alpha}^{(+)}(\mathbf{r};t)\right\}, \tag{65}$$

where

$$\widehat{E}_{\theta,\alpha}^{(\pm)} = \widehat{E}_x^{(\pm)}e^{i\alpha_x}\cos\theta + \widehat{E}_y^{(\pm)}e^{i\alpha_y}\sin\theta. \tag{66}$$

Using Eqs. (65) and (66), one readily finds that the average counting rate of the photo-detector is given by the expression

$$\mathscr{R}_{\theta,\alpha}(\mathbf{r};t) = G_{xx}^{(1)}(\mathbf{r},t;\mathbf{r},t)\cos^2\theta + G_{yy}^{(1)}(\mathbf{r},t;\mathbf{r},t)\sin^2\theta$$
$$+ 2\sqrt{G_{xx}^{(1)}(\mathbf{r},t;\mathbf{r},t)}\sqrt{G_{yy}^{(1)}(\mathbf{r},t;\mathbf{r},t)}\sin\theta\cos\theta|g_{xy}^{(1)}(\mathbf{r};t)|\cos\left[\beta_{xy}(\mathbf{r};t) - \alpha\right], \tag{67}$$

where $\alpha = \alpha_y - \alpha_x$ and

$$g_{xy}^{(1)}(\mathbf{r};t) \equiv \frac{G_{xy}^{(1)}(\mathbf{r},t;\mathbf{r},t)}{\sqrt{G_{xx}^{(1)}(\mathbf{r},t;\mathbf{r},t)}\sqrt{G_{yy}^{(1)}(\mathbf{r},t;\mathbf{r},t)}} \equiv |g_{xy}^{(1)}(\mathbf{r};t)|e^{i\beta_{xy}(\mathbf{r};t)}. \tag{68}$$

It is important to note that only the "equal-point" ($\mathbf{r}_1 = \mathbf{r}_2 \equiv \mathbf{r}$) and "equal-time" ($t_1 = t_2 \equiv t$) correlation matrix $\overleftrightarrow{G}^{(1)}(\mathbf{r},t;\mathbf{r},t)$ contributes to the photon counting rate. We will refer to this matrix as the *quantum polarization matrix*. Equation (67) makes it possible to determine the elements of the quantum polarization matrix $\overleftrightarrow{G}^{(1)}(\mathbf{r},t;\mathbf{r},t)$ in a similar way as is done for the elements of the analogous correlation matrix for a classical field, introduced in Ref. (Wolf,

1959). One can readily show that

$$G_{xx}^{(1)}(\mathbf{r},t;\mathbf{r},t) = \mathcal{R}_{0,0}(\mathbf{r};t), \tag{69a}$$

$$G_{yy}^{(1)}(\mathbf{r},t;\mathbf{r},t) = \mathcal{R}_{\pi/2,0}(\mathbf{r};t), \tag{69b}$$

$$G_{xy}^{(1)}(\mathbf{r},t;\mathbf{r},t) = \frac{1}{2}\left[\mathcal{R}_{\pi/4,0}(\mathbf{r};t) - \mathcal{R}_{3\pi/4,0}(\mathbf{r};t)\right] + \frac{i}{2}\left[\mathcal{R}_{\pi/4,\pi/2}(\mathbf{r};t) - \mathcal{R}_{3\pi/4,\pi/2}(\mathbf{r};t)\right], \tag{69c}$$

$$G_{yx}^{(1)}(\mathbf{r},t;\mathbf{r},t) = \frac{1}{2}\left[\mathcal{R}_{\pi/4,0}(\mathbf{r};t) - \mathcal{R}_{3\pi/4,0}(\mathbf{r};t)\right] - \frac{i}{2}\left[\mathcal{R}_{\pi/4,\pi/2}(\mathbf{r};t) - \mathcal{R}_{3\pi/4,\pi/2}(\mathbf{r};t)\right]. \tag{69d}$$

Let us now briefly examine some properties of the quantum polarization matrix $\overleftrightarrow{G}^{(1)}(\mathbf{r},t;\mathbf{r},t)$. Each element of this matrix, being a correlation function, has the properties of a scalar product[11]. In particular, the elements of any such matrix satisfy the constraints

$$G_{\mu\mu}^{(1)}(\mathbf{r},t;\mathbf{r},t) \geq 0, \tag{70a}$$

$$G_{\mu\nu}^{(1)}(\mathbf{r},t;\mathbf{r},t) = \left\{G_{\nu\mu}^{(1)}(\mathbf{r},t;\mathbf{r},t)\right\}^*, \tag{70b}$$

$$\mathrm{Det}\,\overleftrightarrow{G}^{(1)}(\mathbf{r},t;\mathbf{r},t) \geq 0, \tag{70c}$$

where $\mu = x,y$; $\nu = x,y$ and Det denotes the determinant. Conditions (70a) and (70c) imply that a quantum polarization matrix is always *non-negative definite*. Formula (70c) follows from the Cauchy-Schwarz inequality. From Eqs. (68) and (70c), it follows that

$$0 \leq |g_{xy}^{(1)}(\mathbf{r};t)| \leq 1. \tag{71}$$

7.1 Unpolarized light beam

We will now assume that an *unpolarized* photon-beam[12] is used in the experiment depicted in Fig. 2. For such a beam, the photon detection rate $\mathcal{R}_{\theta,\alpha}(\mathbf{r};t)$, has to be independent of θ and α and consequently, one has from Eq. (67) that

$$g_{xy}^{(1)}(\mathbf{r};t) = 0, \tag{72a}$$

$$\text{and} \quad G_{xx}^{(1)}(\mathbf{r},t;\mathbf{r},t) = G_{yy}^{(1)}(\mathbf{r},t;\mathbf{r},t) \equiv A(\mathbf{r};t), \quad \text{say}, \tag{72b}$$

where $A(\mathbf{r};t)$ is a real function of space and time and $A(\mathbf{r};t) \geq 0$. From Eqs. (72) it is evident that for an unpolarized beam, the quantum polarization matrix $\overleftrightarrow{G}^{(1)}_{(u)}(\mathbf{r};t;\mathbf{r};t)$ is proportional to a unit matrix, i.e., that

$$\overleftrightarrow{G}^{(1)}_{(u)}(\mathbf{r};t;\mathbf{r};t) = A(\mathbf{r};t)\begin{pmatrix}1 & 0 \\ 0 & 1\end{pmatrix}. \tag{73}$$

---

[11] The proof is similar to that given for scalar field operators (Titulaer & Glauber, 1965).
[12] For discussion on unpolarized radiation see, for example, Refs. (Agarwal, 1971; Prakash et al., 1971)

7.2 Polarized light beam

Let us now consider the other extreme case, namely when the beam is fully polarized[13], i.e., when $|g_{xy}^{(1)}(\mathbf{r};t)| = 1$. In this case, it follows from Eq. (68) that the elements of $\overleftrightarrow{G}_{(p)}^{(1)}(\mathbf{r};t;\mathbf{r};t)$ can be expressed in the factorized form

$$\left\{ \overleftrightarrow{G}_{(p)}^{(1)}(\mathbf{r};t;\mathbf{r};t) \right\}_{\mu\nu} = \mathcal{E}_{\mu}^{*}(\mathbf{r};t)\mathcal{E}_{\nu}(\mathbf{r};t), \quad (\mu = x,y; \ \nu = x,y), \tag{74}$$

where

$$\mathcal{E}_x(\mathbf{r};t) = \sqrt{G_{xx}^{(1)}(\mathbf{r},t;\mathbf{r},t)} \exp\left[i\phi_{xx}(\mathbf{r};t)\right], \tag{75a}$$

$$\mathcal{E}_y(\mathbf{r};t) = \sqrt{G_{yy}^{(1)}(\mathbf{r},t;\mathbf{r},t)} \exp\left[i\phi_{yy}(\mathbf{r};t)\right]. \tag{75b}$$

Here $\phi_{yy}(\mathbf{r};t) - \phi_{xx}(\mathbf{r};t) = \beta_{xy}(\mathbf{r};t)$, where $\beta_{xy}(\mathbf{r};t)$ is the phase of $g_{xy}^{(1)}(\mathbf{r};t)$, defined in Eq. (68). The fact that the elements of $\overleftrightarrow{G}_{(p)}^{(1)}(\mathbf{r};t;\mathbf{r};t)$ can be factorized, was also noted in Ref. (Glauber, 1963). The condition (74) may readily shown to imply that for a completely polarized light beam

$$\text{Det}\, \overleftrightarrow{G}_{(p)}^{(1)}(\mathbf{r};t;\mathbf{r};t) = 0. \tag{76}$$

The converse of this statement is also true, i.e., if condition (76) holds at a point $\mathbf{r}$, at time t, one can readily show by use of properties (70a) and (70b) that $|g_{xy}^{(1)}(\mathbf{r};t)| = 1$; hence the light is then completely polarized at that point at time t.

We will next establish a necessary and sufficient condition of complete polarization[14]. We express it in the form of the following theorem:

Theorem 7.1. *In order that a beam is completely polarized, it is necessary and sufficient that the quantized field components and the density operator satisfy the condition*

$$\hat{E}_x^{(+)}(\mathbf{r};t)\hat{\rho} = A(\mathbf{r};t)\hat{E}_y^{(+)}(\mathbf{r};t)\hat{\rho}, \tag{77a}$$

$$where \quad A(\mathbf{r};t) = \frac{G_{yx}^{(1)}(\mathbf{r},t;\mathbf{r},t)}{G_{yy}^{(1)}(\mathbf{r},t;\mathbf{r},t)}, \tag{77b}$$

Proof. To prove that Eq. (77) is a necessary condition, let us introduce the operator

$$\widehat{M}(\mathbf{r};t) = \hat{E}_x^{(+)}(\mathbf{r};t) - A(\mathbf{r};t)\hat{E}_y^{(+)}(\mathbf{r};t), \tag{78}$$

where $A(\mathbf{r};t)$ is given by Eq. (77b). On using Eqs. (77b) and (78), one readily finds that

$$\text{Tr}\left\{ \hat{\rho}\,\widehat{M}^{\dagger}\widehat{M} \right\} = \frac{\text{Det}\, \overleftrightarrow{G}^{(1)}(\mathbf{r},t;\mathbf{r},t)}{G_{yy}^{(1)}(\mathbf{r},t;\mathbf{r},t)}. \tag{79}$$

---

[13] For discussion on polarized radiation see Ref. (Mehta et al., 1974)

[14] An analogous condition hold for complete first-order coherence in the space-time domain (Titulaer & Glauber, 1965). In the classical limit, these conditions resembles the recently introduced concept of statistical similarity (Roychowdhury & Wolf, 2005; Wolf, 2010b) for statistically stationary beams.

If the field is completely polarized, Eqs. (76) and (79) imply that

$$\text{Tr}\left\{\hat{\rho}\,\hat{M}^\dagger \hat{M}\right\} = 0, \tag{80}$$

and, consequently,

$$\hat{M}\hat{\rho} = \hat{\rho}\hat{M}^\dagger = 0. \tag{81}$$

Substituting for $\hat{M}$ from Eq. (78) into Eq. (81), one readily find that condition (77) is satisfied. Thus we have proved that Eq. (77) *is a necessary condition for complete polarization* of a light beam at a point $\mathbf{r}$, at time t.

On the other hand, if one imposes condition (77) on the components of the field operator and evaluates $\text{Det}\,\overleftrightarrow{G}_{(p)}^{(1)}(\mathbf{r};t;\mathbf{r};t)$, one readily obtains Eq. (76), i.e., one finds that if condition (77) holds, the beam is completely polarized. Hence this condition is also a sufficiency condition.
□

7.3 Partially polarized light beam

We have discussed the properties of completely polarized and completely unpolarized beams. Next we consider beams which are partially polarized, i.e., beams for which $0 < |g_{xy}^{(1)}(\mathbf{r};t)| < 1$.

We will first establish the following result: *If a beam of partially polarized photons is incident on a photo-detector, the average counting rate of the detector, at any time t, can always be decomposed into two parts, one which represents the counting rate for a polarized beam and the other the counting rate for an unpolarized beam.* It can be proved that any quantum polarization matrix $\overleftrightarrow{G}^{(1)}(\mathbf{r},t;\mathbf{r},t)$ can be uniquely decomposed in the form (Lahiri & Wolf, 2010b)

$$\overleftrightarrow{G}^{(1)}(\mathbf{r},t;\mathbf{r},t) = \overleftrightarrow{G}_{(p)}^{(1)}(\mathbf{r};t;\mathbf{r};t) + \overleftrightarrow{G}_{(u)}^{(1)}(\mathbf{r};t;\mathbf{r};t), \tag{82}$$

where the elements of $\overleftrightarrow{G}_{(p)}^{(1)}(\mathbf{r};t;\mathbf{r};t)$ and $\overleftrightarrow{G}_{(u)}^{(1)}(\mathbf{r};t;\mathbf{r};t)$ can be uniquely expressed in terms of the elements of $\overleftrightarrow{G}^{(1)}(\mathbf{r},t;\mathbf{r},t)$. From Eqs. (37) and (82), one can at once deduce that the counting rate of the photo-detector has contribution from an unpolarized part and from a polarized part; i.e., that

$$\mathscr{R}(\mathbf{r},t) = \text{Tr}\,\overleftrightarrow{G}^{(1)}(\mathbf{r},t;\mathbf{r},t) = \mathscr{R}_{(p)}(\mathbf{r},t) + \mathscr{R}_{(u)}(\mathbf{r},t). \tag{83}$$

By analogy with the classical theory of stochastic electromagnetic beams, we define the degree of polarization $\mathscr{P}$, at a point $P(\mathbf{r})$, at time t, as the ratio of the photon counting rate for the polarized part to the total counting rate:

$$\mathscr{P}(\mathbf{r};t) \equiv \frac{\mathscr{R}_{(p)}(\mathbf{r},t)}{\mathscr{R}(\mathbf{r},t)} = \frac{\text{Tr}\,\overleftrightarrow{G}_{(p)}^{(1)}(\mathbf{r};t;\mathbf{r};t)}{\text{Tr}\,\overleftrightarrow{G}^{(1)}(\mathbf{r},t;\mathbf{r},t)}. \tag{84}$$

On expressing the elements of $\overset{\leftrightarrow}{G}{}^{(1)}_{(p)}(\mathbf{r}; t; \mathbf{r}; t)$ in terms of the elements of $\overset{\leftrightarrow}{G}{}^{(1)}(\mathbf{r}, t; \mathbf{r}, t)$, one can show that (Lahiri & Wolf, 2010b)

$$\mathscr{P}(\mathbf{r}; t) = \sqrt{1 - \frac{4 \, \mathrm{Det} \, \overset{\leftrightarrow}{G}{}^{(1)}(\mathbf{r}, t; \mathbf{r}, t)}{\left\{ \mathrm{Tr} \, \overset{\leftrightarrow}{G}{}^{(1)}(\mathbf{r}, t; \mathbf{r}, t) \right\}^2}}. \tag{85}$$

From Eqs. (73) and (85), one can readily show that for a beam of unpolarized photons $\mathscr{P} = 0$; and from Eqs. (82) and (85) it follows, at once, that for a completely polarized beam of photons $\mathscr{P}(\mathbf{r}; t) = 1$.

Although, formulas (28), and (85) look similar in form, one must bear in mind that the definition used in classical theory is not appropriate for light of low intensity, nor has it been shown that it is valid for fields that are not necessarily statistically stationary. On the other hand, the expression (85) for the degree of polarization applies also to low intensity light and for light whose statistical properties are characterized by non-stationary ensembles, such as, for example, fields associated with a non-stationary ensemble of pulses.

8. Wave-particle duality, partial coherence and partial polarization of light

Quantum systems (quantons[15]) possess properties of both particles and waves. Bohr's complementarity principle (Bohr, 1928) suggests that these two properties are mutually exclusive. In other words, depending on the experimental situation, a quanton will behave as a particle or as a wave. In the third volume of his famous lecture series (Feynman et al., 1966), Feynman emphasized that this *wave-particle duality* may be understood from Young's two-pinhole interference experiments (Young, 1802). In such an experiment, a quanton may arrive at the detector along two different paths. If one can determine which path the quanton traveled, then no interference fringe will be found (i.e., the quanton will show complete particle behavior). On the other hand, if one *cannot* obtain any information about the quanton's path, then interference fringes, with *unit visibility*, will be obtained (i.e., the quanton will show complete wave behavior), assuming that the intensities at the two pinholes are the same. In the intermediate case when one has partial "which-path information" (WPI), fringes with visibility smaller than unity are obtained, even if the intensities at the two pinholes are equal. For the sake of brevity, we will use the term "best circumstances" to refer to the situation when in an Young's interference experiment, the intensities at the two pinholes are equal, or to equivalent situations in other interferometric setups.

The relation between fringe visibility (degree of coherence) and WPI has been investigated by researchers [see, for example, (Englert, 1996; Jaeger et al., 1995; Mandel, 1991)]. It has been established that a quantitative measure of WPI and fringe-visibility obey a certain inequality. In this section, we will first recollect the results obtained by Mandel and then will show that it is not only the fringe-visibility, but also the polarization properties of the superposed light, which may depend on WPI in an interference experiment.

---

[15] This abbreviation is due to M. Bunge [see, for example, J.-M. Lévy-Leblond, Physica B **151**, 314 (1988)].

Suppose, that $|\psi_1\rangle$ and $|\psi_2\rangle$ represent two normalized single-photon states (eigenstates of the number operator), so that

$$\langle\psi_1|\psi_2\rangle = \langle\psi_2|\psi_1\rangle = 0, \tag{86a}$$
$$\langle\psi_1|\psi_1\rangle = \langle\psi_2|\psi_2\rangle = 1. \tag{86b}$$

We will consider the superposition of the two states in some interferometric arrangement, where a photon may travel along two different paths. Suppose that $|\psi_{ID}\rangle$ represents a state of light, which is formed by coherent superposition of the two states $|\psi_1\rangle$ and $|\psi_2\rangle$, i.e., that

$$|\psi_{ID}\rangle = \alpha_1|\psi_1\rangle + \alpha_2|\psi_1\rangle, \quad |\alpha_1|^2 + |\alpha_2|^2 = 1, \tag{87}$$

where α_1 and α_2 are, in general, two complex numbers. In this case, a photon may be in the state $|\psi_1\rangle$ with probability $|\alpha_1|^2$, or in the state $|\psi_2\rangle$ with probability $|\alpha_2|^2$, but the two possibilities are intrinsically *indistinguishable*. The density operator $\hat{\rho}_{ID}$ will then have the form

$$\hat{\rho}_{ID} = |\alpha_1|^2|\psi_1\rangle\langle\psi_1| + |\alpha_2|^2|\psi_2\rangle\langle\psi_2| + \alpha_1^*\alpha_2|\psi_2\rangle\langle\psi_1| + \alpha_2^*\alpha_1|\psi_1\rangle\langle\psi_2|. \tag{88}$$

In the other extreme case, when the state of light is due to incoherent superposition of the two states, the density operator $\hat{\rho}_D$ will be given by the expression

$$\hat{\rho}_D = |\alpha_1|^2|\psi_1\rangle\langle\psi_1| + |\alpha_2|^2|\psi_2\rangle\langle\psi_2|. \tag{89}$$

Here $|\alpha_1|^2$ and $|\alpha_2|^2$ again represent the probabilities that the photon will be in state $|\psi_1\rangle$ or in state $|\psi_2\rangle$, but now the two possibilities are intrinsically *distinguishable*. Mandel (Mandel, 1991) showed that in any intermediate case, the density operator

$$\hat{\rho} = \rho_{11}|\psi_1\rangle\langle\psi_1| + \rho_{12}|\psi_1\rangle\langle\psi_2| + \rho_{21}|\psi_2\rangle\langle\psi_1| + \rho_{22}|\psi_2\rangle\langle\psi_2| \tag{90}$$

can be uniquely expressed in the form

$$\hat{\rho} = \mathscr{I}\hat{\rho}_{ID} + (1 - \mathscr{I})\hat{\rho}_D, \quad 0 \le \mathscr{I} \le 1. \tag{91}$$

Mandel defined $\mathscr{I}$ as the *degree of indistinguishability*. If $\mathscr{I} = 0$, the two paths are completely distinguishable, i.e., one has complete WPI; and if $\mathscr{I} = 1$, they are completely indistinguishable, i.e., one has no WPI. In the intermediate case $0 < \mathscr{I} < 1$, the two possibilities may be said to be partially distinguishable. Clearly, $\mathscr{I}$ may be considered as a measure of WPI. According to Eqs. (90) and (91), one can always express $\hat{\rho}$ in the form

$$\hat{\rho} = |\alpha_1|^2|\psi_1\rangle\langle\psi_1| + |\alpha_2|^2|\psi_2\rangle\langle\psi_2| + \mathscr{I}(\alpha_1^*\alpha_2|\psi_2\rangle\langle\psi_1| + \alpha_2^*\alpha_1|\psi_1\rangle\langle\psi_2|). \tag{92}$$

Clearly, the condition of "best circumstances" requires that $|\alpha_1| = |\alpha_2|$.

8.1 WPI and partial coherence

Mandel considered a Young's interference experiment, in which the two pinholes (secondary sources) were labeled by 1 and 2. He assumed $|n\rangle_j$ to be a state representing n photons originated from pinhole j ($n = 0, 1$; $j = 1, 2$). Clearly, in this case

$$|\psi_1\rangle = |1\rangle_1|0\rangle_2, \qquad |\psi_2\rangle = |0\rangle_1|1\rangle_2. \tag{93}$$

Calculations show that [for details see (Mandel, 1991)] "visibility $\leq \mathscr{I}$", and the equality holds in the special case when $|\alpha_1| = |\alpha_2|$. Since fringe-visibility is a measure of coherence properties of light (modulus of degree of coherence is equal to the fringe visibility), Mandel's result displays an intimate relationship between wave-particle duality and partial coherence.

8.2 WPI and partial polarization

Let us now assume that $|\psi_1\rangle$ and $|\psi_2\rangle$ are of the form

$$|\psi_1\rangle = |1\rangle_x |0\rangle_y, \tag{94a}$$

$$|\psi_2\rangle = |0\rangle_x |1\rangle_y, \tag{94b}$$

where the two states are labeled by the same (vector) mode $\mathbf{k}$, and x, y are two mutually orthogonal directions, both perpendicular to the direction of $\mathbf{k}$. For the sake of brevity, $\mathbf{k}$ is not displayed in Eqs. (94). Clearly $|\psi_1\rangle$ represents the state of a photon polarized along the x direction, and $|\psi_2\rangle$ represents that along the y direction. In the present case, one may express $\widehat{E}_i^{(+)}(\mathbf{r};t)$ in the form

$$\widehat{E}_i^{(+)}(\mathbf{r};t) = Ce^{i(\mathbf{k}\cdot\mathbf{r}-\omega t)}\widehat{a}_i, \qquad (i = x, y), \tag{95}$$

where the operator $\widehat{a}_i$ represents annihilation of a photon in mode $\mathbf{k}$, polarized along the $i-$ axis, and C is a constant. From Eqs. (38), (92), and (95), one readily finds that the quantum polarization matrix $\overleftrightarrow{G}^{(1)}(\mathbf{r}, t; \mathbf{r}, t)$ has the form

$$\overleftrightarrow{G}^{(1)}(\mathbf{r}, t; \mathbf{r}, t) = |C|^2 \begin{pmatrix} |\alpha_1|^2 & \mathscr{I}\alpha_1^*\alpha_2 \\ \mathscr{I}\alpha_1\alpha_2^* & |\alpha_2|^2 \end{pmatrix}. \tag{96}$$

From Eqs. (84) and (96) and using the fact $|\alpha_1|^2 + |\alpha_2|^2 = 1$, one finds that, in this case, the degree of polarization is given by the expression

$$\mathscr{P} = \sqrt{(|\alpha_1|^2 - |\alpha_2|^2)^2 + 4|\alpha_1|^2|\alpha_2|^2\mathscr{I}^2}. \tag{97}$$

It follows from Eq. (97) by simple calculations that

$$\mathscr{P}^2 - \mathscr{I}^2 = (1 - \mathscr{I}^2)(2|\alpha_1|^2 - 1)^2. \tag{98}$$

Using the fact that $0 \leq \mathscr{I} \leq 1$, one readily finds that

$$\mathscr{P} \geq \mathscr{I}. \tag{99}$$

Thus, the degree of polarization of the out-put light in a single-photon interference experiment is always greater or equal to the degree of indistinguishability ($\mathscr{I}$) which a measure of "which-path information".

Let us now assume that the condition of "best circumstances" holds, i.e., one has $|\alpha_1| = |\alpha_2|$. It then readily follows from Eq. (97) that

$$\mathscr{P} = \mathscr{I}. \tag{100}$$

This formula shows that under the "best circumstances", the degree of indistinguishability (a measure of WPI) and the degree of polarization are equal. The physical interpretation of this result may be understood from the following consideration: If one has complete "which-path information" (i.e., $\mathscr{I} = 0$), then it follows from Eq. (100) that the degree of polarization of the light emerging from the interferometer is equal to zero. Complete "which-path information" in an single-photon interference experiment implies that a photon shows complete particle nature, and our analysis suggests that in such a case light is completely unpolarized. In the other extreme case, when one has no "which-path information", i.e., when a photon does *not* display any particle behavior, the output light will be completely polarized. Any intermediate case will produce partially polarized light. For details analysis see (Lahiri, 2011).

9. Conclusions

We conclude this chapter by saying that we have given a description of first-order coherence and polarization properties of light. The main aim of this chapter was to emphasize the fact that although, coherence and polarization seem to be two different optical phenomena, both of them can be described by analogous theoretical techniques. Our discussion also emphasizes some newly obtained results in quantum theory of optical coherence in the space-frequency domain, and in quantum theory of polarization of light beams.

10. Acknowledgements

The author sincerely thanks Professor Emil Wolf for many helpful discussions and insightful comments. This research was supported by the US Air Force Office of Scientific Research under grant No. FA9550-08-1-0417.

11. References

Agarwal, G. S. (1971). On the state of unpolarized radiation, *Lett. Al Nuovo Cimento* 1.

Arago, F. & Fresnel, A. (1819). L'action queles rayons de lumiere polarisee exercent les uns sur les autres, *Ann. Chim. Phys.* 2.

Berek, M. (1926a). *Z. Phys.* 36.

Berek, M. (1926b). *Z. Phys.* 36.

Berek, M. (1926c). *Z. Phys.* 37.

Berek, M. (1926d). *Z. Phys.* 40.

Berry, H. G., Gabrielse, G. & Livingston, A. E. (1977). Measurement of the stokes parameters of light, *App. Opt.* 16.

Bertolotti, M., Ferrari, A. & Sereda, L. (1995). Coherence properties of nonstationary polychromatic light sources, *J. Opt. Soc. Am. B* 12.

Bohr, N. (1928). Das quantenpostulat und die neuere entwicklung der atomistik, *Naturwissenschaften* 16.

Born, M. & Wolf, E. (1999). *Principles of Optics*, (Cambridge University Press, Cambridge, 7th Ed.).

Brosseau, C. (1995). *Fundamentals of Polarized Light: A Statistical Optics Approach*, (Wiley).

Brosseau, C. (2010). Polarization and coherence optics: Historical perspective, status, and future directions, *Prog. Opt.* 54.

Collett, E. (1993). *Polarized light : fundamentals and applications*, (New York, Marcel Dekker).

Dirac, P. A. M. (1957). *The Priciples of Quantum Mechanics, 4th Ed.*, Oxford University Press, Oxford, England.

Eberly, J. H. & Wódkievicz, K. (1977). The time dependent physical spectrum of light, *J. Opt. Soc. Am.* 67.

Englert, B.-G. (1996). Fringe vsibility and which-way information: an inequality, *Phys. Rev. Lett.* 77.

Feynman, R. P., Leighton, R. B. & Sands, M. (1966). *The Feynman Lectures on Physics, Vol. III*, Addison-Wesley Publishing Company, New York, USA.

Feynman, R. P., Leighton, R. & (Editor), E. H. (1985). *Surely you're joking Mr. Feynman!*, (W. W. Norton & Company, New York).

Friberg, A. T. & Wolf, E. (1995). Relationships between the complex degrees of coherence in the space-time and in the space-frequency domains, *Opt. Lett.* 20.

Gabor, D. (1946). Theory of communication, *J. Inst. Elec. Eng. (London), Pt. III* 93.

Gabor, D. (1955). Light and information, *Proceedings of a Symposium on Astronomical Optics and Related Subjects, Edited by Zdenek Kopal* .

Glauber, R. J. (1963). The quantum theory of optical coherence, *Phys. Rev.* 130.

Glauber, R. J. (2007). *Quantum Theory of Optical Coherence (selected papers and lectures)*, (Weinheim, WILEY-VCH Verlag Gmbh & Co. KGaA).

Hooke, R. (1665). *Micrographia: Or Some Physiological Descriptions of Minute Bodies Made by Magnifying Glasses with Observations and Inquities thereupon*, Martyn and Alleftry, Printers to the Royal Society, London, England.

Huygens, C. (1690). Traité de la lumière.

Jackson, J. D. (2004). *Classical electrodynamics*, (John Wiley, New York, 3rd Ed.).

Jaeger, G., Shimony, A. & Vaidman, L. (1995). Two interferometric complementarities, *Phys. Rev. A* 51.

James, D. F. V. (1994). Change in polarization of light beams on propagation in free space, *J. Opt. Soc. Am. A* 11.

Karczewski, B. (1963). Coherence theory of the electromagnetic field, *Nuovo Cimento* 30.

Lahiri, M. (2009). Polarization properties of stochastic light beams in the spaceŬtime and spaceŬfrequency domains, *Opt. Lett.* 34.

Lahiri, M. (2011). Wave-particle duality and partial polarization of light in single-photon interference experiments, *Phys. Rev. A* 83.

Lahiri, M. & Wolf, E. (2009). Cross-spectral density matrices of polarized light beams, *Opt. Lett.* 34.

Lahiri, M. & Wolf, E. (2010a). Does a light beam of very narrow bandwidth always behave as a monochromatic beam?, *Phys. Lett. A* 374.

Lahiri, M. & Wolf, E. (2010b). Quantum analysis of polarization properties of optical beams, *Phys. Rev. A* 82.

Lahiri, M. & Wolf, E. (2010c). Quantum theory of optical coherence of non-stationary light in the space-frequency domain, *Phys. Rev. A* 82.

Lahiri, M. & Wolf, E. (2010d). Relationship between complete coherence in the space-time and in the space-frequency domains, *Phys. Rev. Lett.* 105.

Lahiri, M., Wolf, E., Fischer, D. G. & Shirai, T. (2009). Determination of correlation functions of scattering potentials of stochastic media from scattering experiments, *Phys. Rev. Lett.* 102.

Lampard, D. G. (1954). Generalization of the wiener-khintchine theorem to nonstationary processes, *J. Appl. Phys.* 25.

Laue, M. V. (1906). Zur thermodynamik der interferenezerscheinungen, *Ann. d. Physik* 20.

Laue, M. V. (1907). *Ann. d. Physik* 23.

Mandel, L. (1991). Coherence and indistinguishability, *Opt. Lett.* 16.

Mandel, L. & Wolf, E. (1995). *Optical Coherence and Quantum Optics*, (Cambridge University Press, Cambridge).

Mark, W. D. (1970). Spectral analysis of the convolution and filtering of non-stationary stochastic processes, *J. Sound Vib.* 11.

Mehta, C. L. & Sharma, M.K. (1974). Diagonal coherent–state representation for polarized light, *Phys. Rev. A.* 10.

Michelson, A. A. (1890). Etude sur la constitution de la lumiere non polarisée et de la lumiere partiellement polarisée, *Phil. Mag.* 30.

Michelson, A. A. (1891a). *Phil. Mag.* 31.

Michelson, A. A. (1891b). *Phil. Mag.* 31.

Michelson, A. A. (1902). *Light Waves and Their Uses*, The University of Chicago Press, Chicago, USA.

Michelson, A. A. (1927). *Studies in Optics*, The University of Chicago Press, Chicago, USA.

Mujat, M., Dogariu, A. & Wolf, E. (2004). A law of interference of electromagnetic beams of any state of coherence and polarization and the fresnel–arago interference laws, *J. Opt. Soc. Am. A.* 21.

Page, C. H. (1952). Instantaneous power spectra, *J. Appl. Phys.* 23.

Poincaré, H. (1892). *Théorie Mathématique de la Lumiére, Vol 2*, Georges Carré, Paris, France.

Ponomarenko, S. A., Agrawal, G. P. & Wolf, E. (2004). Energy spectrum of a nonstationary ensemble of pulses, *Opt. Lett.* 29.

Prakash, H. & Chandra, N. (1971). Density operator of unpolarized radiation, *Phys. Rev. A* 4.

Roychowdhury, H. & Wolf, E. (2005). Statistical similarity and the physical significance of complete spatial coherence and complete polarization of random electromagnetic beams, *Opt. Commun.* 248.

Sereda, L., Bertolotti, M. & Ferrari, A. (1998). Coherence properties of nonstationary light wave fields, *J. Opt. Soc. Am. A* 15.

Setälä, T., Nunziata, F. & Friberg, A. T. (2009). Differences between partial polarizations in the spaceǓtime and spaceǓfrequency domains, *Opt. Lett.* 34.

Silverman, R. A. (1957). Locally stationary random processes, *Proc. I.R.E. (Trans. Inf. Th.)* 3.

Stokes, G. G. (1852). On the composition and resolution of streams of polarized light from different sources, *Trans. Cambr. Phil. Soc.* 9.

Titulaer, U. M. & Glauber, R. J. (1965). Correlation functions for coherent fields, *Phys. Rev.* 140.

van Cittert, P. H. (1934). *Physica* 1.

van Cittert, P. H. (1939). *Physica* 6.

Verdet, E. (1865). Etude sur la constitution de la lumiere non polarisée et de la lumiere partiellement polarisée, *Annales scientifiques de l'E.N.S.* 2.

Wiener, N. (1927). Coherency matrices and quantum theory, *J. Math. Phys. (M.I.T.)* 7.

Wiener, N. (1930). Generalized harmonic analysis, *Acta Math.* 55.

Wiener, N. (1966). *Generalized harmonic analysis and Tauberian theorems*, (The M.I.T. Press).

Wolf, E. (1954). Optics in terms of observable quantities, *Nuovo Cimento* 12.

Wolf, E. (1955). A macroscopic theory of interference and diffraction of light from finite sources ii. fields with a spectral range of arbitrary width, *Proc. Roy. Soc. (London) A* 230.

Wolf, E. (1959). Coherence properties of partially polarized electromagnetic radiation, *Nuovo Cimento* 13.

Wolf, E. (1981). New spectral representation of random sources and of the partially coherent fields that they generate, *Opt. Commun.* 38.

Wolf, E. (1982). New theory of partial coherence in the space-frequency domain. part i: spectra and cross spectra of steady-state sources, *J. Opt. Soc. Am.* 72.

Wolf, E. (1983). Young's interference fringes with narrow-band light, *Opt. Lett.* 8.

Wolf, E. (1986). New theory of partial coherence in the space-frequency domain. part ii: Steady-state fields and higher-order correlations, *J. Opt. Soc. Am. A* 3.

Wolf, E. (1987). Non-cosmological redshifts of spectral lines, *Nature* 326.

Wolf, E. (2003a). Correlation-induced changes in the degree of polarization, the degree of coherence, and the spectrum of random electromagnetic beams on propagation, *Opt. Lett.* 28.

Wolf, E. (2003b). Unified theory of coherence and polarization of random electromagnetic beams, *Phys. Lett. A* 312.

Wolf, E. (2007a). The influence of young's interference experiment on the development of statistical optics, *Progress in Optics* 50.

Wolf, E. (2007b). *Introduction to the Theory of Coherence and Polarization of Light*, (Cambridge University Press, Cambridge, 7th Ed.).

Wolf, E. (2009a). Solution to the phase problem in the theory of structure determination of crystals from x-ray diffraction experiments, *Phys. Rev. Lett.* 103.

Wolf, E. (2009b). Statistical similarity as a unifying concept in the theory of coherence and polarization of light, *Presented by M. Lahiri at the Koli Workshop on Partial Electromagnetic Coherence and 3D Polarization, Koli, Finland* .

Wolf, E. (2010a). Determination of phases of diffracted x-ray beams in investigations of structure of crystals, *Phys. Lett. A* 374.

Wolf, E. (2010b). Statistical similarity as a unifying concept of the theories of coherence and polarization of light, *Opt. Commun.* 283.

Wolf, E. (2011). History and solution of the phase problem in the theory of structure determination of crystals from x-ray diffraction measurements, *Adv. Imag. Elec. Phys.* 165.

Young, T. (1802). An account of some cases of the production of colours, not hitherto described, *Phil. Trans. Roy. Soc. Lond.* 92.

Zernike, F. (1938). The concept of degree of coherence and its application and its application to optical problems, *Physica* 5.

5

Hydrogen Bonds and Stacking Interactions on the DNA Structure: A Topological View of Quantum Computing

Boaz Galdino de Oliveira[1] and
Regiane de Cássia Maritan Ugulino de Araújo[2]
[1]Instituto de Ciências Ambientais e Desenvolvimento Sustentável,
Universidade Federal da Bahia, Barreiras – BA
[2]Departamento de Química, Universidade Federal
da Paraíba, João Pessoa, PB
Brazil

1. Introduction

Throughout these centuries, the origin of life is an intriguing question that still remains unanswerable (Nadeau & Subramaniam, 2011). Moreover, the growth and evolution of each living species in nature is governed by a life code, an acronym widely famous as DNA that means deoxyribonucleic acid (Rittscher, 2010). With a function of organic acid, the great DNA action consists into the organization and regulation of the genetic informations, which are simply so-called as genes (Gräslund et al, 2010). Thus, due to the immensity of genes to be taken into account, the DNA composition transformed it into a large olygomer formed by nucleotide subunits (Jung & Marx, 2005), and therefore DNA is considered one of the largest macromolecules ever known. Within the cellular environment, the DNA is organized into some very long structures so-named chromosomes, which are duplicated through the cell division process.

In the view of the recent history, however, in fact DNA is considered the cornerstone of the biological science, and therefore the appealing to investigate its structure was always an exciting task. Briefly, in 1868 the physicist Miescher (Dahm, 2008) has proposed the first DNA evidences. After that, precisely in 1878, the nucleic acid was isolated as primary nucleobases by Kossel (Jones, 1953). However, only in the beginning of the last century that these nucleobases were understood as being formed by phosphate groups linked by ester bonds of the 2-deoxyribose (see Fig. 1). Some time later, the discovering of the X-ray diffraction by Röntgen (Frankel, 1996) has aided Astbury (Astbury, 1947) to conclude that DNA had a regular structure. Furthermore, it was by the X-ray diffraction studies of Franklin and Gosling (Franklin & Gosling, 1953), in 1952, that Watson and Crick (Watson & Crick, 1953) informed the most modern structure of DNA as a double-helix form (Watson, 1980).

The DNA double-helix is stabilized by means of hydrogen bonds between nucleotides as well as stacking interactions among the aromatic nucleobases widely known as adenine (A),

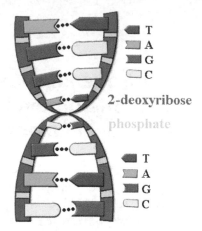

Fig. 1. Representation of the DNA structure.

cytosine (C), guanine (G) and thymine (T), which are tied to ester/phospate. In this context, it is widely established that these two types of base pairs form different hydrogen bonds. In other words, A and T form two hydrogen bonds N–H···O and N···H–N, whereas G and C form three hydrogen bonds O···H–N, N–H···N, and N–H···O, as can be seen in the Fig. 2. On the other hand, the stability of the DNA is also ruled by the interactions formed by G and C, which are recognized as intra-strand base type of stacking.

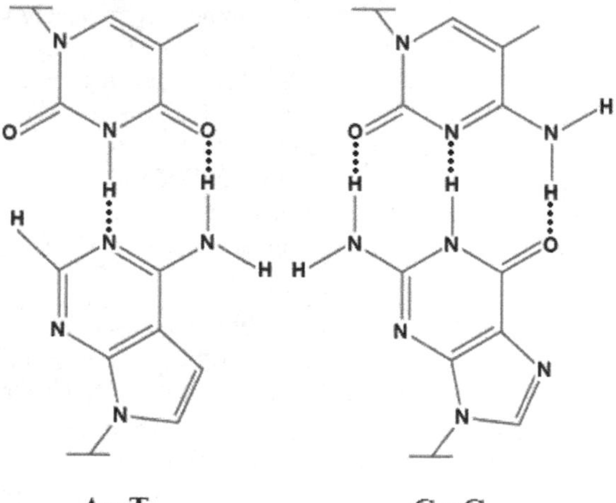

$$A \cdots T \qquad\qquad C \cdots G$$

Fig. 2. Illustration of the hydrogen bonds between nucleotides.

The biological science is well-known as one of the most interdisciplinary areas due to the large number of molecular processes occurring simultaneously within the living organisms (Cech & Rubin, 2004), in particular those related to the DNA functionality. Until nowadays, however, the discovery of the DNA structure is seen as one of the most important scientific conquests of all time. It was by this bioscientific scenery that an immense variety of contexts were grouped to congregate one unique idea: molecular recognition and its biochemical functionality (Iqbal *et al*, 2000; Laskowski, 1996). According to Hitaro (Hitaro, 2002), the biological understanding is closely related to the examination of structure and dynamics of the cellular functions in a cooperative way, not in isolated parts of the own organism. In corroborating to this, as a guide Stahl et al (Stahl et al, 2010) affirm that the molecular recognition is stated whether an attractive interaction provoked by the approaching of two molecules, which possess at least a slight difference of electronegativity between them. So, we would like to emphasize that a careful attention to the knowledge about the profile of the interaction types seems to be necessary.

Some time ago, a historical review signed by Martin and Derewenda put in proof a ransom of the concepts related to the hydrogen bonding (Martin & Derewenda, 1999). It was quoted some important researchers in this regard, such as Linski, Orgel, Nernst, Werner, and finally Lewis, that is considered one of the pioneers of the contemporary chemistry at work with systems formed via hydrogen bonds. However, Latimer and Rodebush have published the first report about hydrogen bond investigations in aqueous medium. Well, Astbury suggested a structure to the alfa-queratine caused by interactions of the NH and CO groups on the peptide bonds. Pauling and Mirsky revisited the protein structure and emphasized that peptide bonds were formed through the hydrogen bonds between the oxygen and peptide nitrogen. In meanwhile, Huggins has carefully analyzed the results reported by Astbury (Astbury, 1947), and noted that the amide hydrogen to behave out of the plane unless molecular resonance effects were enhanced, so that the single pair of electrons of nitrogen was also in the peptide chain.

Nevertheless, the theory cyclol of proteins in the peptide chain advocated by Wrinch have the form $-C(OH)\cdots N$ instead of $-(CO)\cdots(NH)-$, as it was known. In theory, no classical hydrogen bond could be formed. Thus, Pauling was quick to recognize the flaws in your publishing model in July 1939, in which he emphasized the planarity of the peptide bond. Pauling published his article weeks before the Nazis invaded Poland on September 1. However, the same year Pauling also released his classic book '*The Nature of the Chemical Bond*', who was one of the leading spokesmen for the dissemination and development of the history of chemical bond and hydrogen bond, so far. After several years of insights and discussions, Pauling affirmed in its theories that hydrogen bonds $(Y\cdots H)$ are formed by electronegative differences between proton donors (H) and their acceptor ones (Y), as already mentioned (Pauling, 1939). However, Pimentel and McClellan did not agree with this electronegative criterion, and they stated that hydrogen bonds can exist if the hydrogen is bound to any other atom (Pimentel & McClellan, 1960). Some years later, theoreticians established some physical conditions in order to unveil the nature of the hydrogen bond. For instance, when the electrostatic attractive is the dominant phenomenon undoubtedly the intermolecular system is stabilized by means of hydrogen bonds (Umeyama & Morokuma, 1977). In opposition to this, van der Waals systems are widely known as weakly bound because the London dispersion forces are the main contributions (Cukras & Sadlej, 2008).

Traditionally, besides oxygen, but fluorine and nitrogen are the most known proton acceptors in systems stabilized at light of the $H\cdots F$ and $H\cdots N$ hydrogen bonds. However, the proton character is a quite accepted parameter, and thereby, the hydrogen bond model leads to $X-H^{+\delta}\cdots Y^{-\delta}$. It can be perceived a charge separation interpreted as charge transference between HOMO and LUMO orbitals of the proton donor and acceptor, respectively. With this in mind, it was established that other proton acceptor types can be useful, such as the unsaturated hydrocarbon centers, by which the $X-H^{+\delta}\cdots\pi^{-\delta}$ hydrogen bonds emerged with great evidence. In this scenery, it become stated that a single element is not answerable for the formation of the hydrogen bond, but ideally the cornerstone of this interaction is site with high electronic density, which at this time is assumed as formed by electronegative elements or unsaturated bonds. The magnificence of the π centers becomes reliable upon the formation of the $\pi^{-\delta}\cdots\pi^{-\delta}$ sandwich stacking, whose profile is known as one of the weakest interaction with strength in range of 1-3 kcal/mol, being considered then as London's dispersion forces beyond the van der Waals contacts often devoted to weak hydrogen bonds.

The interpretation and forward comprehension of all kind of events and phenomena inherent to the DNA environment is not an easy task (Šponer *et al*, 2001-2002), but in recent years the applicability of the chemical methods, physical theories and spectroscopy analyses have been decisive in accurate investigations of the biological systems (Shogren-Knaak *et al*, 2001). On the other hand, this has yielded intense debates among the expert theoreticians, and a lot of computational approaches have been implemented with the purpose to decompose the total energy into the following terms: electrostatic, dispersion, charge transfer, polarizability, and exchange potential (Umeyama & Morokuma, 1977). Surely, other interaction types also occur, such as dihydrogen bonds, halogen bonds or stacking, but in practice the most important is the availability of appropriated methodologies to the examination of all properties of these interactions. In general, this requirement is displayed on the basis of theoretical calculations, such as those from *ab initio*, semi-empirical or DFT nature, where all of them are always implemented to seek and find the deeper potential energy surface.

On the other hand, a long time the scientific community would felt a necessity of a theoretical method by which the chemical bond content could be elucidated in its essence. Indeed, this theoretical method has emerged 40 years ago due to an insight of Bader based in catching information computed directly from the electronic density. Baptized as Quantum Theory of Atoms in Molecules (QTAIM) (Bader, 1990), this method models all points of molecular surface through the integration of the electronic density by taking into account the formalism of quantum mechanics for subspace. Thus, the principle adopted by Bader was purely based in quantum mechanics, but with the purpose to describe the atomic behaviour within the molecular environment. By revisiting the trajectory of the QTAIM development, Bader simply took into account the atomistic cooperative activity, by which atoms were defined in a molecule as open systems able to exchange charge and momentum with their neighbours.

The QTAIM benchmark is treat confined systems by means of boundary conditions, in which the molecular or atomic surfaces and their shapes are enable to transfer momentum.

$$\nabla\rho(r).n(r) = 0 \text{ for all points on surface } S(\Omega) \tag{1}$$

In a recent chapter, Bader (Bader, 2009) has discussed that proper open system are defined by equation of motion for an observable $\hat{G}$ as follows:

$$\left(\frac{N}{2}\right)\left\{(i/\hbar)\int\psi^*[\hat{H},\hat{G}]\psi d\tau + cc\right\} = \left(\frac{1}{2}\right)\left\{\oint dS(\Omega,r)J_G(r).n(r) + cc\right\} \tag{2}$$

Where the expressions for $J_G(r)$ and its property density $\rho_G(r)$ are given by:

$$J_G(r) = \left(\frac{\hbar}{2mi}\right) N \int d\tau \left\{\psi^* \nabla_r (\hat{G}(r)\psi) - \nabla_r \psi^* (\hat{G}(r)\psi)\right\} \tag{3}$$

$$\rho_G(r) = \left(\frac{N}{2}\right) \left\{\int d\tau \, \psi^* (\hat{G}(r)\psi) - \psi^* (\hat{G}(r)\psi)\right\} \tag{4}$$

The great goal here is transform each property into a particle density in a real space in according to the operation of $\int d\tau$, which sums the spins over all coordinates denoted by r in a surface space indicated by Ω. If we take into account that Schrödinger and Heisenberg's equations define the changes on state function and how these changes affect an average value of an observable. In this context, one of the most appropriated procedures to obtain a great relationship concerned to the observable, energy (E) for instance, is dedicated to the Ehrenfest's theorem, by which the time rate of change of the average values of an electronic position $\hat{r}$ and momentum $\hat{p} = i\hbar\nabla$ yields the following relation:

$$d\frac{<\hat{p}>}{dt} = < -\nabla\hat{U}(r) = < \hat{F}(r) > \tag{5}$$

$\hat{F}$ means a force exerted on an electron at position $\hat{r}$ by an average distribution of the remains electrons as well as by a nuclear framing yielding the force exerted on the electron density. In a real surface space, this kind of force is computed as:

$$\hat{F} = \int_\Omega \int d\tau \left\{\psi^* (-\nabla_r \hat{U})\psi\right\} = \int_\Omega dr \hat{F}(r) = -\oint dS(\Omega, r)\vec{\sigma}(r).n(r) \tag{6}$$

In this equation, the momentum of flux density of the QTAIM is distinguished by the stress tensor $\vec{\sigma}(r)$, whose physical nature indicates a dimension of energy density.

$$\vec{\sigma}(r) = \left(\frac{\hbar^2}{4m}\right) \left\{\psi^* \nabla(\nabla\psi) + \nabla(\nabla\psi^*)\psi - \nabla\psi^* \nabla\psi - \nabla\psi\nabla\psi^*\right\} \tag{7}$$

The stress tensor $\vec{\sigma}(r)$ is defined through the derivation of the Newton's equation of motion (Bader, 1991):

$$m\frac{d^2 <\hat{F}>}{dt^2} = < \hat{F}(r) \tag{8}$$

The Lagrangian formalism should be used to account the kinetic (K) and potential (U) energies, what results in the next equation:

$$L = \frac{m\dot{q}^2}{2} - U(q) \tag{9}$$

To set out the QTAIM formalism, it was used the principle of least action for particle motion in subspace conditions. Well, the principle of least action states that a quantity (q) derived from wave function is minimized in space and time (t₁ and t₂) and the atomic surface of a open system is modelled as a zero-flux surface, by which the time variations in end points is zero (see Fig. 3), as well as the surface also is zero in the extreme of functions, what can be summarized as:

$$\delta W_{12}(q) = \int_{t_1}^{t_2} \left\{ \left(\frac{\partial L}{\partial q} - \frac{d}{dt}\left(\frac{\partial L}{\partial \dot{q}}\right) \right) \right\} \delta q dt = 0 \tag{10}$$

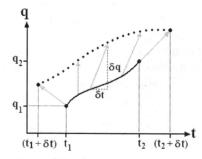

Fig. 3. Description of the principle of least action.

In this equation, L symbolizes the Lagrangian defined by the kinetic (K) and potential (U) energies. In surface, δW_{12} vanishes according to the Euler-Lagrange equation, and therefore, the Schrödinger's equation for normalized wave function can be determined as:

$$\hat{H}\Psi - E\Psi = 0 \quad \hat{H}\Psi^* - E\Psi^* = 0 \tag{11}$$

with $\hat{H} = -\frac{\hbar^2}{2m}\nabla^2 + \hat{U}$. In terms of quantum mechanics, the Lagrangian of the state functions is defined as: $L(\Psi, \nabla\Psi, \dot{\Psi}, t)$. In regards to the first-order variation in Ψ, it can be obtained the Schrödinger's equation $i\hbar\left(\frac{\partial \Psi}{\partial t}\right) = \hat{H}\Psi$ as the Euler-Lagrange equation $\Psi(q,t)$ dependent of time. In this context, by solving the Schrödinger's equation $\hat{H}\Psi = E\Psi$ on the basis of the classical Hamilton-Jacobi equation for a stationary state reduces the quantum Lagrangian to $J(\psi, \nabla\psi)$ in order to minimize the total energy. If the wavefunction is normalized, an undetermined multiplier in $J(\psi, \nabla\psi)$ is executed, thereby a new functional $G(\psi, \nabla\psi)$ is obtained. Moreover, it should be pointed out that $G(\psi, \nabla\psi)$ and $L(\Psi, \nabla\Psi, \dot{\Psi}, t)$ are functional of $\nabla\psi$ and $\nabla\Psi$ whose kinetic energy are respectively given as follows: $+\left(\frac{\hbar^2}{2m}\right) < \nabla\Psi . \nabla\Psi >$ and $-\left(\frac{\hbar^2}{2m}\right) < \Psi\nabla^2\Psi >$. Thus, it can be stated the difference between two forms of kinetic energy as proportional to the Laplacian (L) of the electronic density:

$$K - G = L - \left(\frac{\hbar^2}{2m}\right)\Psi^*\nabla^2\Psi - \left(\frac{\hbar^2}{2m}\right)\nabla\psi^* . \nabla\psi = -\left(\frac{\hbar^2}{4m}\right)\nabla^2\rho \tag{12}$$

Ruled by the Gauss's surface theorem over a spatial region $S(\Omega)$:

$$K - G = -\left(\frac{\hbar^2}{4m}\right)\int_\Omega \nabla^2\rho dr \tag{13}$$

$$K - G = -\left(\frac{\hbar^2}{4m}\right)\oint dS(\Omega)\,\nabla\rho(r).n \tag{14}$$

where $K(\Omega)$ and $G(\Omega)$ represent the kinetic energy densities, which are equivalent to the Laplacian of the charge density, $\nabla^2\rho_{(r)}$. If the surface $S(\Omega)$ is one of zero-flux at any point r where n is a normal vector, $K(\Omega) = G(\Omega)$ and becomes established the equation (1), whose meaning defines the surface by which the atom is delimited as zero-flux of charge density (Fig. 4). In other words, the value of the first electronic density derivative is zero, whereas the second derivatives go to a minimum or maximum of charge concentration.

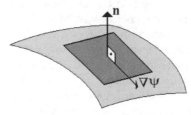

Fig. 4. Topology with representation of the zero-flux surface.

The relationship between surface conditions and high and low electronic density sites is ruled by the virial theorem. By assuming the contributions of the kinetic and potential energies, elevated and depressive charge density regions are modelled by the positive (kinetic energy density G is positive) and negative (electronic potential energy density U is negative) Laplacian values, as demonstrated by the equation (16):

$$2G + U = \left(\frac{\hbar^2}{4m}\right)\nabla^2\rho(r) \tag{15}$$

$$U = \left(\frac{\hbar^2}{4m}\right)\nabla^2\rho(r) - 2G \tag{16}$$

with $G = \dfrac{\hbar^2}{2m}N\int\nabla\psi^*.\nabla\psi d\tau$, in which G is the gradient kinetic energy density and Ψ is an antisymmetric many-electron wavefunction (Matta & Boyd, 2007). By the action of the kinetic (G) and potential (U) energy operators, QTAIM identifies maximum and minimum of electronic density in the molecular surface and the chemical bonds are classified as closed-shell whether $\nabla^2\rho_{(r)} > 0$ or shared interactions when $\nabla^2\rho_{(r)} < 0$. As aforesaid, the negative Laplacian indicates high concentration of charge density (uphill) whereas depletion of charge density is motivated by the positive Laplacian (downhill). The Laplacian $\nabla^2\rho_{(r)}$ is defined by the sum of the eigenvalues of the Hessian Matrix H ($\nabla^2\rho_{(r)} \equiv \lambda_1 + \lambda_2 + \lambda_3$) (see Equations 17), whereas the electronic density $\rho_{(r)}$ is described as a set of critical points, such as Cage Critical Points (CCP), Ring Critical Points (RCP), Bond Critical Points (BCP), and Nuclear Attractor (NA).

$$
H = \begin{pmatrix} \dfrac{\partial^2 \rho_{(r)}}{\partial x^2} & \dfrac{\partial^2 \rho_{(r)}}{\partial x \partial y} & \dfrac{\partial^2 \rho_{(r)}}{\partial x \partial z} \\[2mm] \dfrac{\partial^2 \rho_{(r)}}{\partial y \partial x} & \dfrac{\partial^2 \rho_{(r)}}{\partial y^2} & \dfrac{\partial^2 \rho_{(r)}}{\partial y \partial z} \\[2mm] \dfrac{\partial^2 \rho_{(r)}}{\partial z \partial x} & \dfrac{\partial^2 \rho_{(r)}}{\partial y \partial y} & \dfrac{\partial^2 \rho_{(r)}}{\partial z^2} \end{pmatrix} \quad H = \begin{pmatrix} \dfrac{\partial^2 \rho_{(r)}}{\partial x^2} & 0 & 0 \\[2mm] 0 & \dfrac{\partial^2 \rho_{(r)}}{\partial y^2} & 0 \\[2mm] 0 & 0 & \dfrac{\partial^2 \rho_{(r)}}{\partial z^2} \end{pmatrix} \quad H = \begin{pmatrix} \lambda_1 & 0 & 0 \\ 0 & \lambda_2 & 0 \\ 0 & 0 & \lambda_3 \end{pmatrix} \quad (17)
$$

All these critical points are specific, and their internal formalisms are ruled either by the sum of the eigenvalues signs $(\lambda_1 + \lambda_2 + \lambda_3)$ as well as by the number of non-zero eigenvalues. Therefore, it is obtained a coordinate pair (r, s), which can be used to classify the critical points above cited. For instance, the coordinates of CCP, RCP, BCP, and NA are (3,+3), (3,+1), (3,-1), and (3,-3). As explained above, r is a coordinate where a normal vector is aligned perpendicularly to molecular surface, but now r is interpreted as an intermediary point wherein two gradient paths of electronic density emerge from two bonded nuclei. Actually, this analysis is routinely applied in many investigations, in particular the application in systems formed by hydrogen bonds must be worthwhile. As widely established, these arguments have been also applied successfully to study π-systems (Oliveira & Araújo, 2011) and hydrogen-bonded complexes (Oliveira et al, 2009). As such, it can be seen critical points as extremes of electronic density, that is, maximum or minimum in each particular case. For instance, the BCP coordinates (3,-1) implies that the tridimensional (x, y, z) electronic density is extreme, whereas -1 is the summed result of two maximum (two -1 signs) and one minimum (one +1 sign) of electronic density. By the nature of the π··H hydrogen bonds, the proton donor is aligned perpendicularly to the π cloud, but in regards to QTAIM critical points, the BCP (3,-1) between the carbon atoms above mentioned is the attractor for the bond path linking the hydrogen to the C≡C, C=C, and C–C bonds. In this context, the coordinate (3,-1) is considered an able QTAIM source to accept protons along the CC bonds.

One of the most usual types of interactions existing in DNA structure is the hydrogen bond. As already mentioned, the formation of a hydrogen bond claims by one center with high electronic density, such as those containing lone electron pairs. In this context, a lot of proton acceptors possessing great electronic density have been exhaustively examined, e.g., hydrogen peroxides (HP). The great insight to investigate the capability of hydrogen peroxide in genetic environment is due to its presence in human blood as a metabolic bioproduct. It is widely reported the formation of several interaction complexes at the DNA level, of which the adenine base is one of the most used in this regard. Thus, the work elaborated by Dobado and Molina (Dobado & Molina, 1999) display great informations about the formation of hydrogen complexes on the DNA structure, in particular those composed by adenine and HP. As depicted in Fig. 5, there are multiple hydrogen bonds formed, in general, they are mutual once HP is functioning either as proton acceptor or proton donor, and due to this it is not easy to estimate the real strengths of these hydrogen bonds. From the structural point of view, the values of the hydrogen bond distances vary

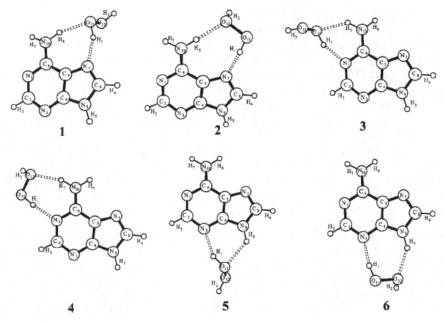

Fig. 5. The adenine ⋯ hydrogen peroxide complexes

between 1.8 Å and 2.1 Å. In terms of the QTAIM approach, by the topological contour plots of these geometries illustrated in the Fig. 6 became reliable to put in discussion the hydrogen bond profiles between adenine ⋯ HP.

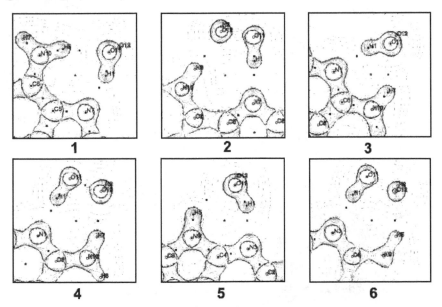

Fig. 6. $\nabla^2 \rho_{(r)}$ contour maps for the hydrogen bonds.

As can be seen, the structural nature of the hydrogen bond within these complexes is justly cyclic once two intermolecular BCP were located adenine and HP. According to QTAIM virial theorem of electronic energy, these BCP above mentioned are the source to obtain the Laplacian and electronic density quantities. It can be observed that the N–H···O (adenine···HP) and O–H···N (HP···adenine) hydrogen bonds were characterized not only in terms of the positive values of Laplacian fields and low amounts of electronic density, but also by the location of the RCP, what leads to the identification of large cyclic structures formed by seven or up to eight members. However, the charge concentration measurement on the RCP is valid to debate the hydrogen bond strength. As such, it was computed the higher $\rho_{(r)}$ value of 0.04 e/$a_o{}^3$ for O–H···N, whereas it was found 0.025 e/$a_o{}^3$ for N–H···O. In spite of this, the $\nabla^2\rho_{(r)}$ values of 0.1 e/$a_o{}^5$ and 0.06 e/$a_o{}^5$ also indicate that O–H···N is a pure closed-shell interaction albeit N–H···O cannot be one a typical one. In other words, the hydrogen bond is formed when HP is the strongest proton acceptor, what in this sense could be concluded that HP is a Lewis's base.

The formation of hydrogen bond is a quite diversified event and not occurs uniquely by means of independent species or isolated monomers, but also within the same structure whether the acceptor and donor of protons are located in appropriated molecular sites. This type of interaction is recognized as intramolecular, and its functionality on the DNA structure has been well examined. In this context, Hocquet (Hocquet, 2001) provided an explanation to the different conformations C3'-endo/anti and C2'-endo/anti of the deoxyribonucleosides, namely as 2'-deoxycytidine (dC), 2'-deoxythymidine (dT), 2'-deoxyadenosine (dA), and 2'-deoxyguanosine (dG) pictured in the Fig. 7 due to the formation of intramolecular hydrogen bonds between the purine base and the sugar group.

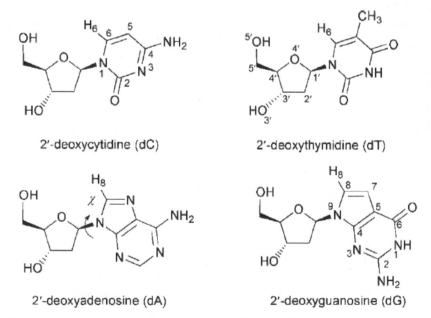

2'-deoxycytidine (dC)

2'-deoxythymidine (dT)

2'-deoxyadenosine (dA)

2'-deoxyguanosine (dG)

Fig. 7. Chemical structure and atom numbering of the four 2'-deoxyribonucleosides.

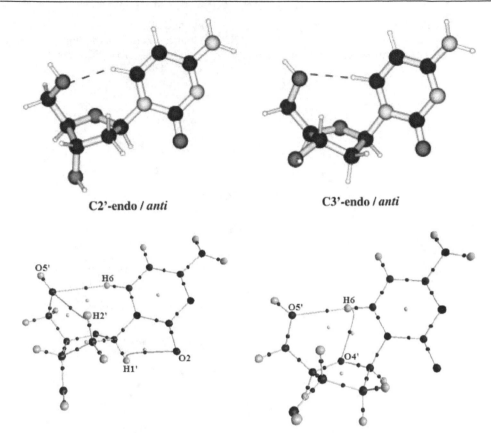

Fig. 8. Geometry optimized structures of 2'-deoxycytidine (dC) and the molecular graph showing all BCP and RCP.

The QTAIM calculations used to examine the conformations of these deoxyribonucleosides revealed the existence of BPs and an intermediate BCP along them. In according to the molecular graph (see Fig. 8), it is quoted the formation of the O5'···H6, O5'···H2' and O2···H1' in C2'-endo/anti, whereas O5'···H6 and O4'···H6 interactions in C3'-endo/anti. In comparison with other traditional works, the values of the electronic density and Laplacian correspond to median hydrogen bond strength, although it should be mentioned that low $\rho_{(r)}$ values followed by positive $\nabla^2\rho_{(r)}$ provide a closed-shell interaction. In this scenery, we would like to say that the proton donor feature of H6 diminish as follows dT > dC > dA > dG. Nevertheless, it was demonstrated that dT shows higher electronic density in comparison to the remaining deoxyribonucleosides. In exception, the C2'-endo/anti conformation of dC presents an O5'···H6 hydrogen bond weak, but in other hand, the C3'-endo/anti conformation has a normal electronic density value but its Laplacian is very high, what signify the existence of closed-shell interaction.

Subramanian et al (Parthasarathi et al, 2004) have used QTAIM topological parameters, such as electronic densities, Laplacian shapes, and chemical descriptors to investigate the formation of DNA base pairs and which hydrogen bond profiles are formed among them. In

Fig. 9 is illustrated the bond paths, BCPs as red dots, and RCPs as yellow dots of the Guanine··Cytosine Watson and Crick (GCWC) and 2amino-Adenine··Thymine (2aminoAT) DNA complexes, which are formed by means of three stable hydrogen bonds.

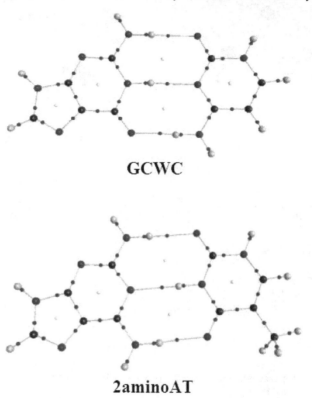

GCWC

2aminoAT

Fig. 9. Molecular graphs of DNA bases.

Initially, the QTAIM protocol indicates the existence of the hydrogen bonds N–H··O and N–H··N in conformity with their positive Laplacian values accompanied by low electronic density accounts, meaning the existence of a closed-shell interaction between these DNA entities. So, if we take into account the Koch and Popelier's criteria to ensure the characteristics of hydrogen bonds (Koch & Popelier, 1995), the alterations on the charge density of the proton donors are one of the most drastic events occurred after complexation. Of course that the QTAIM topological parameters are used in this insight, such as the appropriated values of electronic density and Laplacian values at the BCP, or even the mutual penetration between proton donors and acceptors.

Furthermore, one of the most important analyses in structures stabilized via hydrogen bonds is the measurement of its interaction strength, which can be obtained through the topological descriptors, e.g., electronic density, Laplacian, and electronic density energy, these in association with molecular parameters, such as interactions energies, structural distances, and vibrational stretching frequencies. In fact, these relationships have been very useful in studies of hydrogen-bonded complexes, but it was also used to investigate the interaction strength between the DNA base pairs. Well, in Fig. 10 is plotted a relationship between the interaction energies and the electronic densities computed in each intermolecular BCP not only in regards to GCWC and 2aminoAT, but other DNA types are also included in this analysis, of which we can cite Cytosine···Cytosine (CC), two Thymine···Thymine configurations (TT1 and TT2), two Adenine···Adenine configurations (AA1 and AA2), as well as other ones.

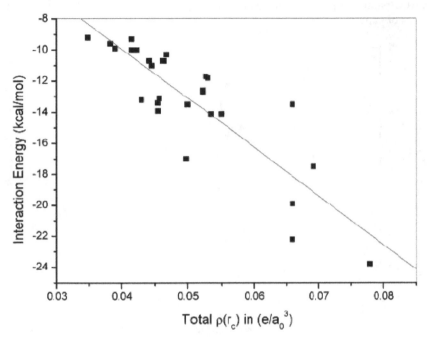

Fig. 10. Relationship between the interaction energy and total $\rho_{(r)}$ of the DNA base pairs.

Through the correlation coefficient value of 0.859 can be perceived a good and linear relationship between the electronic density in the range of 0.030 and 0.055 e/a$_o^3$ and the interaction energy between 9-15 kcal/mol. It can be seen that DNA pairing bases stabilized by three hydrogen bonds are most strongly bound, once the electronic densities of these systems are more than 0.05 e/a$_o^3$, and thereby they are not placed in the linear region. Notably, it is by the fact that the supermolecule approach is not accurate for determining the interaction energy in systems formed by three hydrogen bonds or higher, e.g. GCWC and 2aminoAT, we can assume that slight deviations in the linear adjustments should occur.

Nevertheless, additional hydrogen bonds beyond than two previously identified are possible, mainly in GG3 complex but in GG1 not. Ideally it could be possible to identify a bifurcate hydrogen bond O6···H(C8) and O6···H(N2) in GG1, although it was not possible to characterize any BCP or RCP for these two interactions, what makes QTAIM unfeasible to be used in this regard. However, the application of the Laplacian instead of the electronic density as topological descriptor to predict the interaction strength is very useful in many situations. To the best of our knowledge, the hydrogen bond strength on the DNA bases is also unveiled through the relationship between the interaction energy and the Laplacian computed in each intermolecular BCP, either those with two hydrogen bonds or even with three ones. This relationship is illustrated in Fig. 11, by which a correlation coefficient of 0.827 was obtained. We can observe that similar results were obtained in comparison to that presented for the electronic density. The low electronic density values as well as the depletion characteristic of the Laplacian corroborated themselves, and in this sense, these two QTAIM parameters show similar efficiency to predict the interactions strength of the DNA bases of pairs.

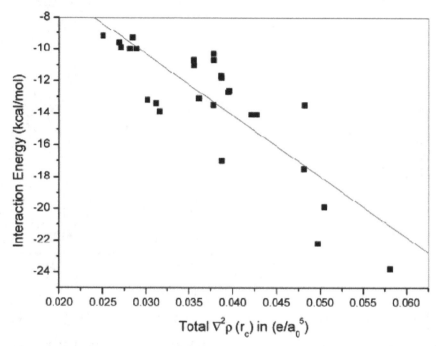

Fig. 11. Relationship between the interaction energy and total $\nabla^2\rho_{(r)}$ of the DNA base pairs.

As is widely known, the interaction strength is the cornerstone to preserve the molecular stabilization, and in the DNA scenery, it has been demonstrated that their nucleobases provide the molecular stability of the DNA chains due to the number of the hydrogen bonds to be formed, and indeed, their strengths are included in this context. Among the DNA structures well-known, it has been noticed that π stacking and hydrogen bonds are the most important types of interaction that retains the DNA helical structure with great influence of the guanine and cytosine nucleobases. In an overview, these nucleobases in olygonucleotides form are stabilized by distinct energies, i.e., 20 kcal/mol for hydrogen bonds whereas 2.40 kcal/mol for π stacking. In order to understand the connectivity between hydrogen bonds and π stacking, a symbolic model system was examined, in which the action of the benzene upon the formation of the $C_6H_6\cdots GC$ and $C_6H_6\cdots CG$ complex must be worthwhile (Robertazzi & Platts, 2006). In according to the Fig. 12, the bond paths and BCP of the $C_6H_6\cdots GC$ (a) and $C_6H_6\cdots CG$ (b) complexes can be analyzed. The QTAIM results show that no significant variation could be found between (a) and (b),

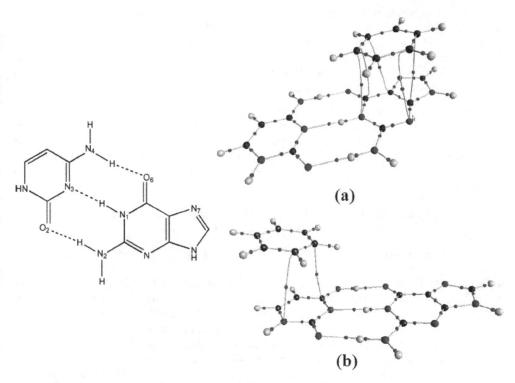

Fig. 12. Topologies of (**a**) benzene $\cdots$ GC and (**b**) benzene $\cdots$ CG.

i.e., the electronic density for the hydrogen bonds and π stacking are in the range of 0.001 e/a_o^3.

Likely, a decisive argument changes the conclusion highlighted above: inclusion of benzenic structures with the following substituents $-NO_2$, $-F$, $-CH_3$, $-CHO$, $-OH$, and $-NH_2$ into the ternary complexes (**a**) and (**b**). This action should be useful to demonstrate that the hydrogen bonds and π stacking can be affected by the hardness (n) of the substituted benzene, whose definition according to the Density Functional Theory (DFT) (Geerlings *et al*, 2003) is based on second derivative of electronic energy (E) with respect to the number of electrons (N) for a constant external potential $U_{(\vec{r})}$:

$$n = \frac{1}{2}\left(\frac{\partial^2 E}{\partial N^2}\right) U_{(\vec{r})} \tag{18}$$

The great goal of this insight is the reduction of the charge transfer from guanine (G) to cytosine (C) with stacked substituted groups on the benzene structure. For example, changing $-NO_2$ by $-NH_2$ cytosine is a better proton acceptor with increase of the electronic density at the BCP of the hydrogen bonds H1$\cdots$N3 and H2$\cdots$O2, but otherwise, a worse proton donor causes an increase in the electronic density, what could lead to confirm surely that π stacking does influence the formation of hydrogen bonds between G and C. As can be seen, hydrogen bonds and π stacking bring great deformations on the molecular sites of the DNA, but its ideal structure is preserved. In accord with Meggers *et al* (Meggers *et al,* 2005), DNA polymer analogous formed by π stacking interactions in agreement with Watson-Crick pairing scheme of bases produces α-double helix with absence of the backbone sugar residues. Definitively, hydrogen bonds and π stacking interactions are not independent, as already discussed the influence between them.

The nucleobases dimers are formed by π stacking interactions, which can be also subdivided into intrastrand (a) and interstrand (b), as illustrated below. As widely-known, the formation of stacking interactions is closely compromised with the formation of the gene codes. In this context is that, in addition to the hydrogen bonds, the π stacking contacts should be carefully reliable to interpret the DNA structure and the α-helix formation.

$$
\textbf{(a)} \;\downarrow \left| \begin{array}{ccc} \text{Base 1 - Base 4} \\ \updownarrow \quad \updownarrow \\ \text{Base 2 - Base 3} \end{array} \right| \uparrow
\qquad
\textbf{(b)} \;\downarrow \left| \begin{array}{c} \text{Base 1 - Base 4} \\ \searrow \\ \text{Base 2 - Base 3} \end{array} \right| \uparrow
\qquad
\downarrow \left| \begin{array}{c} \text{Base 1 - Base 4} \\ \nearrow \\ \text{Base 2 - Base 3} \end{array} \right| \uparrow
$$

Indeed, there is an intense discussion about the formation of hydrogen bonds on the nucleobases dimers. For instance, in recent years the triple hydrogen bonds occurrence on nucleobase has been evaluated through the application of high-level calculations, by which a slight difference in range of 3 kcal/mol between the dimer and the individual hydrogen bonds was discovered. Due to this, recently Matta *et al* (Matta *et al*, 2006) have developed a theoretical investigation of WC dimeric derived from the DNA fragments. It was quoted the existence of three types of hydrogen bonds, namely as N–H $\cdots$ O, C–H $\cdots$ O, and N–H $\cdots$ N. The first hydrogen bond type occurs between A and T as well as in G and C. The second hydrogen bond is recognized as triple between A and T. Finally, the last hydrogen bond model makes itself presents in a double format between A and T as well as G and C. Thus, it should be important to comment each one of these hydrogen bonds and their influence on the DNA structure.

It is observed a slight higher concentration of $\rho(r)$ in the CG complex in comparison to AT, in which the values are 0.028 e/a_o^3 and 0.025 e/a_o^3, respectively. Moreover, the H $\cdots$ O hydrogen bond is sensitively weak in AT once the value of $\rho_{(r)}$ is 0.006 e/a_o^3. Furthermore, the ellipticity curvature λ_3 is smaller in AT rather than in CG, what indicates a less charge density accumulation in the intermolecular region of the AT system. Only for mention, the remaining λ_1 and λ_2 are perpendicular to the BP of the hydrogen bonds, what makes their negative results and then are not taken into account. In according to the Equation 15, the virial potential operator U is negative over the entire molecule, whereas G is positive. If U is the dominant term, a high electronic density concentration is assumed, as can be seen in shared interactions such as covalent or unsaturated bonds. In other words, the electrons are placed on the BCP. The same reasoning can be dedicated to G, although the kinetic contribution diagnoses closed-shell interactions, or in this current work, the hydrogen bonds N–H $\cdots$ O and H $\cdots$ O. Thus, it was suggested an alternative approach to the virial expression in order to propose a novel term so-called as electronic energy density H:

$$
H = 2G + U \tag{19}
$$

Well, it is by the contributions of G and U that H is estimated. It was observed negative values of H not for N–H$\cdots$O, but actually to N–H$\cdots$H. The main feature of the N–H$\cdots$H hydrogen bond is its length, which is very short in AT than in CG. Thereby, the electronic density $\rho_{(r)}$ of 0.052 e/$a_o{}^3$ is AT is higher than CG, whose value is 0.038 e/$a_o{}^3$. Nevertheless, these hydrogen bonds exhibit positive value of $\nabla^2\rho_{(r)}$ as well as negative electronic energy density H, what is anomalous for closed-shell interactions, but it can be an indication of shared electronic density. However, Fig. 13 illustrates different π stacking interactions on DNA structure.

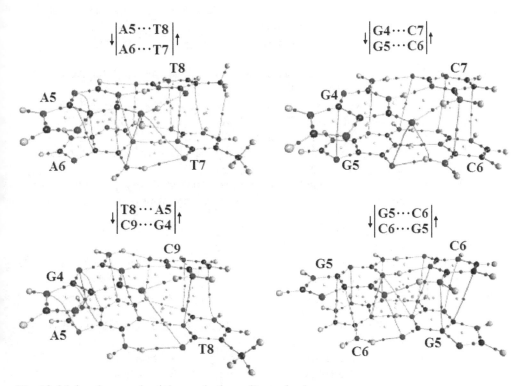

Fig. 13. Molecular graph of the nucleobase dimer duplexes.

It was discovered some diversity of π interactions formed by the N$\cdots$N, C$\cdots$C, C$\cdots$N, and O$\cdots$N contacts. These, some are intrastrand and other ones are interstrand. For the third structure, G4$\cdots$C7 and G5$\cdots$C6, in addition to the six hydrogen bonds, eight π stacking interactions are known, of which six are intrastrand whereas two are interstrand. As remarkably defined, the values of the electronic energy density H are positive due to the contribution of G accompanied by U with smaller negative amounts. By this relationship, the Laplacian fields are positive, and in this current analysis were obtained values of $\nabla^2\rho_{(r)}$ in range of 0.009-0.039 e/$a_o{}^5$, what no doubts in regards to the profile of closed-shell interactions remains about these interactions. In comparison with some typical hydrogen bonds formed, the $\rho_{(r)}$ values of the intrastrand and interstrand π-stacked contacts are very low, but the lowest charge concentration is found in intrastrand situations. In an overview, it was quoted that albeit these π stacking interactions are weak, surely they can influence the geometry and stabilization of the DNA structure.

2. Acknowledgment

CNPq and CAPES Brazilian funding agencies.

3. References

Astbury, W.T. (1947) X-ray studies of nucleic acids, Symposia of the Society for Experimental Biology, Vol. 1, pp. 66-76.

Bader, R.F.W. (1990) Atoms in Molecules: A Quantum Theory, USA: Oxford University Press.

Bader, R.F.W. (1991) A Quantum Theory of Molecular Structure and Its Applications Chemical Reviews, Vol. 91, pp. 893-928.

Bader, R.F.W. (2009) Confined atoms are open quantum systems, Vol. 57, pp. 285-318.

Bissantz, C.; Kuhn, B. & Stahl, M. (2010) A Medicinal Chemist's Guide to Molecular Interactions, Journal of Medicinal Chemistry, Vol. 53, pp. 5061-5084.

Cech, T.R. & Rubin, G.M. (2004) Nurturing interdisciplinary research, Nature Structural & Molecular Biology, Vol. 11, pp. 1166-1169.

Cukras, J. & Sadlej, J. (2008) Symmetry-adapted perturbation theory interaction energy decomposition for some noble gas complexes, Chemical Physics Letters, Vol. 459, pp. 44-48.

Dahm, R. (2008) Discovering DNA: Friedrich Miescher and the early years of nucleic acid research, Human Genetics, Vol. 122, pp. 565-581.

Dobado, J.A. & Molina, J. (1999) Adenine-Hydrogen Peroxide System: DFT and MP2 Investigation, The Journal of Physical Chemistry A, Vol. 103, pp. 4755-4761.

Frankel, R.I. (1996) Centennial of Röntgen's discovery of x-rays, Western Journal of Medicine, Vol. 164, pp. 497-501.

Geerlings, P.; De Proft, F. & Langenaeker, W. (2003) Conceptual Density Functional Theory, Chemical Reviews, Vol. 103, pp. 1793-1873.

Gräslund, A.; Rigler, R. & Widengren (2010), J. Single molecule spectroscopy in chemistry, physics and biology: nobel symposium, Springer Series in Chemical Physics, Vol. 96, pp. 3-559.

Hocquet, A. (2001) Intramolecular hydrogen bonding in 2'-deoxyribonucleosides: an AIM topological study of the electronic density, Physical Chemistry Chemical Physics, Vol. 3, pp. 3192-3199.

Iqbal, S.S.; Mayo, M.W.; Bruno, J.G.; Bronk, B.V.; Batt, C.A. & Chambers, J.P. (2000) A review of molecular recognition technologies for detection of biological threat agents, Biosens Bioelectronics, Vol. 15, pp. 549-578.

Jones, M.E. (1953) Albrecht Kossel: a biographical sketch, Yale Journal of Biology and Medicine, Vol. 26, pp. 80-97.

Franklin, R. & Gosling, R.G. (1953) Molecular configuration in sodium thymonucleate, Nature, Vol. 171, pp. 740-741.

Jung, K.-H. & Marx, A. (2005) Nucleotide analogues as probes for DNA polymerases, Cellular and Molecular Life Sciences, Vol. 62, pp. 2080-2091.

Kitano, H. (2002) Systems biology: a brief overview, Science, Vol. 295, pp. 1662-1664.

Koch, U. & Popelier, P.L.A. (1995) Characterization of C-H-O Hydrogen Bonds on the Basis of the Charge Density, The Journal of Physical Chemistry, Vol. 99, pp. 9747-9754.

Laskowski, R.A.; Luscombe, N.M. & Swindells, M.B. (1996) Protein clefts in molecular recognition and function, Protein Science, Vol. 5, pp. 2438-2452.

Martin, T.W. & Derewenda, Z.S. (1999) The name is bond — H bond, Nature Structural Biology, Vol. 6, pp. 403-406.

Matta, C.F. & Boyd, R.J. (2007) The Quantum Theory of Atoms in Molecules: From Solid State to DNA and Drug Design, Wiley-VCH, Weinheim.

Matta, C.F.; Castillo, N. & Boyd, R.J. (2006) Extended Weak Bonding Interactions in DNA: π-Stacking (Base-Base), Base-Backbone, and Backbone-Backbone Interactions, Journal of Physical Chemistry B, Vol. 110, pp. 563-578.

Nadeau, J.H. & Subramaniam, S. (2011) Wiley Interdisciplinary Reviews: Systems Biology and Medicine, all references herein cited.

Oliveira, B.G. & Araújo, R.C.M.U. (2011) The topology of π··H hydrogen bonds, Monatshefte fur Chemie, Vol. 142, pp. 861-873.

Oliveira, B.G.; Araújo, R.C.M.U.; Carvalho, A.B. & Ramos, M.N. (2009) The molecular properties of heterocyclic and homocyclic hydrogen-bonded complexes evaluated by DFT calculations and AIM densities, Journal of Molecular Modeling, Vol. 15, pp. 123-131.

Parthasarathi, R.; Amutha, R.; Subramanian, V.; Nair, B.U. & Ramasami, T. (2004) Bader's and Reactivity Descriptors'Analysis of DNA Base Pairs, The Journal of Physical Chemistry A, Vol. 108, pp. 3817-3828.

Pauling, L. (1939) The Nature of the Chemical Bond, Cornell University Press, Ithaca.

Pimentel, G.C. & McClellan, A.L. (1960) The Hydrogen Bond, W.H. Freeman, San Francisco.

Rittscher, J. (2010) Characterization of biological processes through automated image analysis, Annual Review of Biomedical Engineering, Vol. 12, pp. 315-344.

Robertazzi, A. & Platts, J.A. (2006) Gas-Phase DNA Oligonucleotide Structures. A QM/MM and Atoms in Molecules Study, The Journal of Physical Chemistry A, Vol. 110, pp. 3992-4000.

Umeyama, H. & Morokuma, K. (1977) The origin of hydrogen bonding. An energy decomposition study, The Journal of the American Chemical Society, Vol. 99, pp. 1316-1332.

Shogren-Knaak, M.A.; Alaimo, P.J. & Shokat, K.M. (2001) Recent advances in chemical approaches to the study of biological systems, Annual Review of Cell and Developmental Biology, Vol. 17, pp. 405-33.

Šponer, J.; Leszczynski, J. & Hobza, P. (2001-2002) Electronic Properties, Hydrogen Bonding, Stacking, and Cation Binding of DNA and RNA Bases, Biopolymers, Vol. 61, pp. 3-31.

Watson, J.D. & Crick, F.H.C. (1953) Genetical implications of the structure of deoxyribonucleic acid, Nature, Vol. 171, pp. 964-967.

Watson, J.D. (1980) The double helix: a personal account of the discovery of the structure of DNA, Atheneum.

Zhang, L.; Peritz, A. & Meggers, E. (2005) A Simple Glycol Nucleic Acid, The Journal of the American Chemical Society, Vol. 127, pp. 4174-4175.

6

The Role of Quantum Dynamics in Covalent Bonding – A Comparison of the Thomas-Fermi and Hückel Models

Sture Nordholm[1] and George B. Bacskay[2]

[1]*The University of Gothenburg*
[2]*School of Chemistry, The University of Sydney*
[1]*Sweden*
[2]*Australia*

1. Introduction

A brief history of thoughts on covalent bonding

The mechanism of covalent bonding is the basis of molecule formation and therefore of all of chemistry. It is a phenomenon which is extremely well characterized experimentally. The properties of covalently bonded molecules can also be reproduced and/or predicted by increasingly accurate calculations using the methods of quantum chemistry. These days, such calculations can be routinely performed using commercial program packages which are readily available to all chemists, who frequently find that molecular properties are simpler to calculate than to measure or track down in the literature. In the light of this success it is remarkable that the physical explanation of the origin of covalent bonding is still a subtle and contentious issue generating much discussion. The aim of this chapter is to reveal the origin of covalent bonding in simple terms and with illustrations allowing subtleties to be clarified and arguments to be settled. We shall propose that a deeper and more general analysis than hitherto widely known shows the merit and limitation of previous explanatory models and offers important clues for the understanding of the present computational methods of quantum chemistry and their future development. Our discussion will employ two very simple but very different theories of electronic structure – the Thomas-Fermi theory and the Hückel model of planar conjugated hydrocarbon molecules – which contain within them a full range of the dominant mechanisms needed to describe covalent bonding. The analysis presented here is in turn based on earlier articles [1-6] and a thesis which is not yet published in all its essential parts [7].

A brief historical review of the understanding of covalent bonding is in order. The first model of covalent bonding which remains very relevant today is that of Lewis who understood that the periodic variation in chemical stability of the elements was the origin of chemical bonding and of the covalent bonding mechanism in particular [8]. The model of Lewis suggested that atoms which do not have an inert gas like electronic structure seek to achieve such a structure and its particular stability by either electron transfer between atoms, forming ionic bonds, or by electron sharing in forming covalent bonds between atoms in molecules. The Lewis diagrams can be used to identify the bonds formed. Both mechanisms are in principle always present but more or less applicable and nature seems to confirm that molecules with very obvious Lewis structures according to ionic (valence electron transfer) or covalent (valence electron sharing) mechanisms are particularly stable. In our view the Lewis model – at its simple level without reference to any form of mechanics – is valid and useful. The difficulties start when one tries to interpret the bonding mechanisms identified by Lewis in more mechanical terms.

The ionic bonding mechanism lends itself readily to a mechanical interpretation. One needs only to note that ionic bonds form when the electron transfer to create inert gas like ions at infinite separation requires relatively little energy since electron affinities and ionization energies are unusually close. Once the ions have formed one can then estimate the binding energy released when they approach to form an ionic structure. Doing this for, e.g., NaCl, gives a good prediction of the ionic bond strength. One immediately also understands that such ionic molecules are very reactive due to their polar nature and form extended crystal structures rather than inert molecules. Thus the ionic bonding mechanism is very simple to understand in terms of well known properties of the atoms. The complicating quantum mechanical effects are hidden in the properties of the atoms and reflected in the periodically variable ionization energies and electron affinities.

The covalent bonding mechanism presents a far more difficult case. The physical explanation is still debated despite the apparent truth of the Lewis electron sharing model and the ability of modern quantum chemistry to account for the bond strengths to great accuracy. The question is how to physically interpret the calculations based on the quantum mechanical Schrödinger equation for the electronic structure of atoms and molecules. A first significant interpretation was offered by Hellman [9] already in the 1930-s soon after quantum mechanics had been developed. He suggested that covalent bonding should be understood as a quantum mechanical effect due to the lowering of ground state kinetic energy associated with the delocalization of valence electron motion between atoms in the covalently bound molecule. Although we shall find much which is correct in this proposal the picture is clouded by apparently contradictory facts. Most chemists at the time were reluctant to assign covalent bonding to a quantum effect. Instead the idea, seemingly supported by the Lewis diagrams with their electron pairs placed between bonded atoms, that electrostatic interaction between these pairs and the positively charged atomic centers in the molecule was the origin of the covalent bond became favored by most authors of introductory texts on chemistry. This view was supported by Coulson who in his famous book *Valence* [10] in 1961 pointed to the Virial Theorem of Coulombic systems which showed rigorously that when a molecule in its equilibrium (lowest energy) configuration was formed from infinitely separated atoms the kinetic energy rises by half as much as the potential energy drops. Thus it appeared that kinetic energy – in complete disagreement with Hellman's proposal – opposed bonding while the electrostatic interactions were the cause of bonding. It seemed therefore that the covalent bonding mechanism could be understood as an extension of the ionic bonding mechanism to include the effect of electron pairs located at bond midpoints between atoms. In more detailed discussions bond formation was seen to be associated with excess electron density placed around bond midpoints by constructive overlap of atomic orbitals in bonding molecular orbitals. For more recent elaboration of the electrostatic view of the mechanism of covalent bonding see, e.g. a book by Burdett [11] and an essay by Bader and coworkers [12].

This view of covalent bonding as an electrostatic interaction was generally accepted among chemists even though Ruedenberg [13] already in 1962 showed that the Virial Theorem was in small molecules like H_2^+ and H_2 a reflection of orbital contraction, i.e., the tightening of the electron density around the two nuclei rather than the displacement of electron density into the region around the bond midpoint. Ruedenberg suggested that this contraction of atomic orbital or electron density around the atomic nucleus could be thought of as a "promotion" of the atom to a smaller size appropriate in the molecule. If this promotion was performed at a certain cost in increased energy at infinite separation then the subsequent formation of the molecule would follow the proposal of Hellman and show a decrease in kinetic energy due to valence electron delocalization which stabilizes the bond. An alternative procedure followed by most early applications of quantum chemistry to the calculation of covalent bond energies is to use frozen atomic orbitals in a minimal basis set calculation in which case the orbital contraction identified by Ruedenberg is completely absent. In this case the Virial Theorem is completely violated in the bond formation which is again – as predicted by Hellman – due to the lowering of the kinetic energy upon electron delocalization. One may then – as suggested by Kutzelnigg [14]– at the end of such a calculation allow the orbital exponents reflecting degree of contraction around the nuclei to be optimized. The Virial Theorem now is satisfied but only a small decrease in both total energy and bond length is found. Thus one can see that the ability to resolve the covalent bonding mechanism is hardly dependent on whether the Virial Theorem is satisfied or only weakly so. Recent discussion of this issue can be found in articles by Ruedenberg [15] and Kutzelnigg [16] and work cited therein.

2. The current impasse

The problem we are currently faced with in the understanding of covalent bonding is that we have a well formed empirical concept of a covalent bond which is arguably the very foundation of chemistry and we have within modern quantum chemistry many approximate theories which allow us to compute the structures, energies and other properties of covalently bonded molecules with high and ever increasing accuracy. Remarkably we still have not agreed upon how to provide an appropriate physical explanation of the observed and calculated covalent bonding. Clearly the arguments of Hellman and Ruedenberg in terms of kinetic energy lowering and that of Coulson and many authors of chemistry textbooks in terms of electrostatic interactions as the source of covalent bonding have their "grains of truth" but they also contain apparent contradictions which may confuse students or make textbook authors avoid the issue of physical understanding of covalent bonding altogether.

Most quantum chemists appear to be pragmatic about this state of affairs and are not greatly concerned with the lack of physical understanding as long as their methods can quantitatively resolve covalent bonding. Bader [12] and others strongly maintain that chemical bonding must be thought of as due to an electrostatic interaction connected with the redistribution of electron density to yield a build up in the interatomic (~ bond midpoint) region. Others, e.g. Frenking, [17] seek insight into chemical bonding from an energy decomposition analysis (EDA) using the results of modern quantum chemistry for molecules with various types of chemical bonds. This can be thought of as a refinement of the discussion about the roles of kinetic or potential energies and whether one or the other is the key to the covalent stabilization. There are of course different types of kinetic and potential energies. While contributing insight, these interpretations of modern quantum chemistry have not provided a convincing and fully accepted resolution of the dispute concerning the nature of covalent bonding. Our own work over a period of 25 years [1-6] has pointed to a deeper quantum dynamical origin of covalent bonding. This work has departed significantly from earlier mechanistic interpretations of covalent bonding, but thus far it has had little impact on the earlier and long-held views of the mainstream chemical community. It is this both general and more personal impasse which forms the starting point of our present contribution.

3. The present focus – The nonbonding of Thomas-Fermi Theory contrasted with the successful bonding model of Hückel Theory

Our view proposed here in the light of a long program of research is that i) there is a need for a deeper physical understanding of covalent bonding, ii) such an understanding will show that covalent bonding is a dynamical phenomenon relating to the motion of valence electrons in the molecule and iii) this insight casts helpful new light on the range of chemical bonds in molecules and materials and the applicability of quantum chemical methods in the calculation of atomic reactivity and covalent bonding. The aim here is to expose the main points of our analysis in their simplest context. Thus we focus on two old and simple methods, i.e., the original electronic density functional theory of Thomas and Fermi [18, 19] and the Hückel model of π-electron states in planar conjugated hydrocarbon molecules [20]. TF theory is known [21, 22] to fail completely to describe covalent bonding while the Hückel model is the simplest approach to the calculation of covalent bonds and provides within it a pedagogical illustration of the generality of its underlying mechanism [23].

The complete failure of Thomas-Fermi (TF) density functional theory (DFT) with respect to covalent bonding is also a very significant advantage in the search for the mechanism of covalent bonding. The TF DFT is in many ways very successful and it contains representation of all the major electrostatic forces proposed as the origin of covalent bonding. Its variable electron density can clearly be made to reflect excess density at bond midpoints and in bonding regions between nuclei in molecules. Yet it returns no covalent bonding whatsoever. It would appear there must be a fundamental reason for this which in turn may tell us what covalent bonding is really about.

The Hückel theory of π-electron resonance stabilization is – in stark contrast – almost inexplicably successful at capturing covalent bonding. It is fundamentally a quantum mechanical theory of one-electron wave functions, the π-molecular orbitals. The empirical form of its one-electron Hamiltonian is exceedingly simple. The bonding is essentially captured by one parameter β, which is placed off-diagonally, so as to allow the formation of band-like molecular orbitals delocalized over the planar hydrocarbon structure. There would not appear to be any room for the intricacies of virial theorems or electron density rearrangement to bond midpoints or bonding regions. Yet the π-bonding of two carbon atoms in ethylene or the resonance in benzene seem well resolved by this simple theory.

It will be our premise here that the mechanism of covalent bonding can be identified by noting first that it is contained in the error made by the Thomas-Fermi theory in representing quantum mechanics and secondly noting that whatever this error is it is not affecting the Hückel model of π-bonding. Covalent bonding, whether normal or resonant in character, is clearly contained in this model with extensions to σ-electrons and its nature should therefore be deducible from the simple ansatz of the Hückel theory. In mathematical terms the mechanism has been captured in the intersection of (1) the complement to the TF theory and (2) Hückel theory.

4. Thomas-Fermi Theory – Quantization by the semiclassical Correspondence Principle

Thomas-Fermi (TF) theory is the original density functional theory (DFT) of electronic structure. It proposes that the ground state energy of electrons in an external potential $V_{ext}(\mathbf{r})$ can be found by minimizing the functional –

$$E(\rho) = C\int d\mathbf{r}\left[\rho(\mathbf{r})\right]^{5/3} + \tfrac{1}{2}\int d\mathbf{r}\rho(\mathbf{r})\int d\mathbf{r}'\,\rho(\mathbf{r}')/r_{12} + \int d\mathbf{r}\rho(\mathbf{r})V_{ext}(\mathbf{r}) \qquad (1)$$

where atomic units have been used such that C is $3(3\pi^2)^{2/3}/10$ and the Coulomb interaction between two electrons appears as $1/r_{12}$ with r_{12} being the separation between the electrons. The first term on the right is the kinetic energy functional, the second is the electrostatic repulsion between electrons and the third is the potential energy of interaction with the external field $V_{ext}(\mathbf{r})$. It should immediately be noted that this – in distinction to most modern forms of DFT – is a true density functional in that no reference is made to any wave function. Thus solving for ground state electronic structure and energy is far simpler in TF theory than in modern DFT where Kohn-Sham or Hartree-Fock expressions are used for the kinetic energy with the introduction of one-electron orbital wave functions. Many features of electronic structures are well reproduced by solutions obtained by TF theory but the failures are also dramatic. Given that covalent bonding is completely absent TF theory eliminates chemical reactivity and thereby nearly all of chemistry. Beyond that singular failure lie quantitative errors – particularly for small systems.

A full derivation of TF theory can be carried out by semi-classical arguments mapping quantum mechanics into classical phase space by the following assumptions:

1. Each quantum state of an electron corresponds precisely to a phase space volume of h^3.
2. Each one-electron energy eigenstate can be found by sequentially slicing up classical phase space by energy surfaces separated by a phase space volume h^3. The phase space slices so formed are uniformly occupied and the energy of the corresponding quantum energy eigenstate is the uniform slice average of the classical energy.
3. The formation of classical quantum states is carried out ergodically, i.e., without considering any possible dynamical constraints that may decompose phase space. The electronic phase space is extended to have a spin up and a spin down half-space which are both included in the summation to find the state volume h^3.
4. Many-electron systems are assumed to satisfy the mean-field approximation where the individual electrons are moving independently in an average field created by all electrons.

The details can be found in a thesis [7] and will be published independently. Here we want to point out how the known errors and limitations of TF theory enters through the construction of the functional on this semiclassical basis. A good summary of the performance of Thomas-Fermi theory for atoms and molecules can be found in the book by Parr and Yang on "The Density-Functional Theory of Atoms and Molecules" [24] where also many correction schemes have been discussed. It is our purpose here to reveal as clearly as possible the underlying causes of the failure of the original TF theory – particularly the reason why atomic reactivity and covalent bonding are absent. It then becomes possible to not only understand how to correct the theory but also to identify the origin of atomic reactivity and the covalent bonding mechanism in its general form.

4.1 Thomas-Fermi Theory for one-dimensional motion

We begin by considering the familiar particle-in-the-box system where a single electron is moving in a potential $V(x) = 0$ for $0 < x < L$, and $= \infty$ elsewhere. Given the classical kinetic energy $T = p^2/2m$ it is clear that the classical phase space volume below an energy E is $\Omega(E) = 2p(E)L = 2(2mE)^{1/2}L$ and the density of states is $\rho(E) = (2m/E)^{1/2}L$. The energies separating states are obtained from $\Omega(E_{F,n}) = nh$, which yields $E_{F,n} = (nh/2L)^2/2m$. Finally the TF energy eigenvalues are obtained as –

$$E_n^{(TF)} = h^{-1} \int_{E_{F,n-1}}^{E_{F,n}} dE\rho(E) = \frac{h^2\left(n^2 - n + \frac{1}{3}\right)}{8mL^2} \tag{2}$$

This result strongly resembles the exact relation $E_n = h^2n^2/8mL^2$ as shown in Figure 1, but the TF result is always slightly lower.

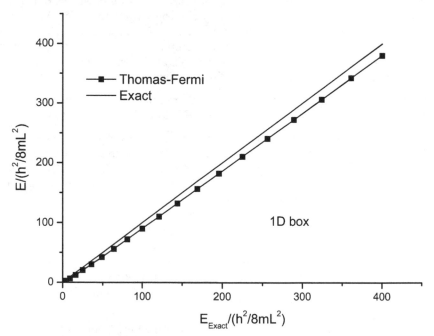

Fig. 1. Comparison of Thomas-Fermi (filled square) with exact (diagonal line) 1D particle-in-the-box energy eigenvalues

In fact we note that $E_{F,n} = E_n$ so that the exact result is the upper bound on the slice of phase space making up the TF energy eigenstate. The reason for this deviation is that the correct quantum energy has a component of gradient kinetic energy which sets in as the wave function is brought to zero at the hard walls defined by the rise of the potential to infinity. The TF energy lacks this type of gradient kinetic energy and allows the density to be constant $1/L$ in the interval $[0, L]$ rather than of the form $2\sin^2(\pi x)/L$ which vanishes at both endpoints with a peak in the middle.

The gradient kinetic energy is related to the change in motion and thereby to the dynamical mechanism that TF theory completely neglects. It is easily shown that only this type of gradient kinetic energy arises for electrons in ground state orbitals and it can be obtained in terms of the electron density by a functional first noted by von Weizsäcker [25]. For many-electron systems where more than the ground state orbital is filled there is another type of kinetic energy which we might call "orthogonality kinetic energy". It is directly connected with the Pauli principle and the Pauli repulsion between molecules. In the semi-classical derivation of TF theory one can see that the TF kinetic energy functional is estimating orthogonality kinetic energy but in a way that extends it also to one or two electron ground states. Interestingly the TF estimate of the energy levels of a harmonic oscillator in one dimension is in perfect agreement with the exact result (see Figure 2). This is due to a cancellation of the gradient kinetic energy rise and the lowering of energy that follows from the tunneling of orbital wave functions into the potential wall.

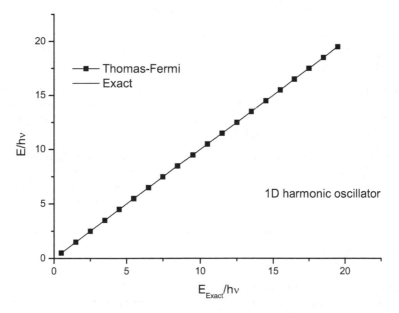

Fig. 2. Comparison of Thomas-Fermi (filled squares) with exact (diagonal line) energy eigenvalues for the 1D harmonic oscillator.

If we consider electron motion in a three dimensional cubic box we find two types of errors in the TF results compared with exact results as shown in Figure 3. The TF energies are

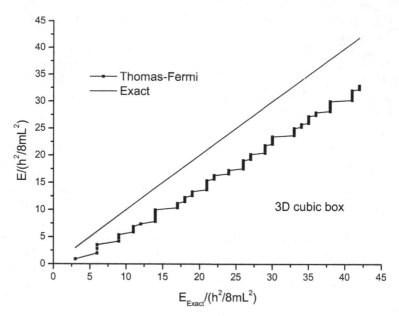

Fig. 3. Comparison of Thomas-Fermi (filled squares forming jagged line) and exact (diagonal line) energy eigenvalues for the cubic box potential in 3D.

generally too low due to the lack of gradient kinetic energy but they are also nondegenerate and smoothly progressing with decreasing energy gaps while the exact results show larger and very variable energy gaps but strong degeneracies of the energy levels.

The same comparison of exact and TF energies for the isotropic 3D harmonic oscillator in Figure 4 show only this latter error where the TF energy levels are nondegenerate with smoothly decreasing energy gaps while the exact spectrum shows constant energy gaps as for the 1D harmonic oscillator but rapidly rising degeneracy.

The "degeneracy deficiency" seen in the TF eigenvalues of both the particle-in-the-cube and the isotropic 3D harmonic oscillator is due to the assumption of rapid ergodic electron dynamics which ignores the separability of the x, y and z-directed motion of the electron in these two potential wells. If we were to account for this separability and apply the TF theory to each 1D motion in the axial directions independently then the degeneracies would be recovered and we would have perfect agreement between 3x1D TF results and exact results for the harmonic oscillation and only the gradient error remaining in the case of the motion in a cube as shown in Figure 3.

The identification of the "dynamical error" in standard TF theory above is of the utmost importance since the degeneracies in the spectra of the electron motion in a cube or in an isotropic 3D harmonic oscillation are a reflection of what we call "shell structure" in the case of atoms. Thus we can see already that the TF theory ignores shell structure and we shall show below that it thereby also ignores atomic reactivity. In order to correct TF theory for this error we need to explicitly account for the dynamical constraints, i.e. non-ergodicity and hindered electron dynamics, which is present in real electronic structures. This is easy to do

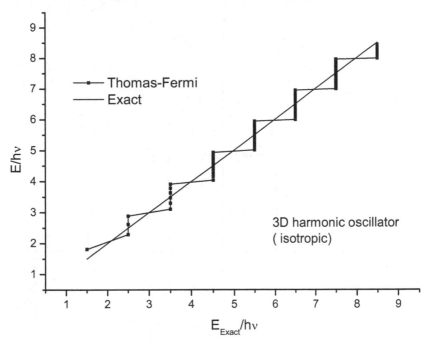

Fig. 4. Comparison of Thomas-Fermi (filled squares in jagged line) and exact (diagonal line) energy eigenvalues for the isotropic 3D harmonic oscillator.

in the case of separable electronic motion as above but will be more subtle below as we approach realistic atomic and molecular systems.

The dynamical error illustrated above can be seen to be related to the loss of shell structure in the TF theory. The molecular dynamical error which eliminates covalent bonding can be illustrated for motion in a simple one-dimensional double well potential:

$$
\begin{aligned}
V(x) &= \infty, \ x < -(L+a)/2 \quad \text{or} \quad x > (L+a)/2 \\
&= 0, \ -(L+a)/2 \le x \le -a/2 \quad or \quad a/2 \le x \le (L+a)/2 \\
&= V_0, \ -a/2 < x < a/2
\end{aligned} \tag{3}
$$

This potential is shown in Figure 5.

With a barrier height of 2.0 E_h (5.25 MJ/mol) and L = 8.0 a_0 (4.23 Å) the energy splitting between the ground and first excited state energies according to TF theory, is compared with the exact result ΔE_{21} (= $3h^2/8mL^2$ for a = 0) in Figure 6.

We see that while the exact result – as is well known – is very sensitive to the barrier width a, the TF result is always the same irrespective of barrier width. Thus near degeneracies in molecular spectra of orbital energies are not accounted for in TF theory. Such near degeneracies are related to hindered electron motion between the two wells and thus to the reactivity of the system. In the case of interacting atoms this type of near orbital degeneracy is relaxed as the atoms bond by the covalent mechanism. Thus the formation of the covalent

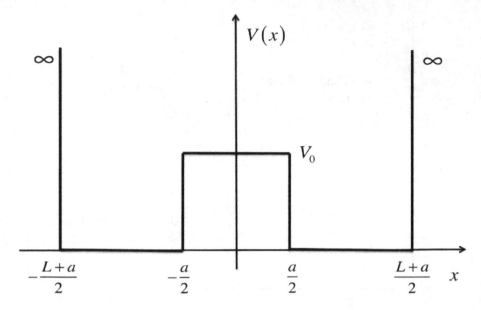

Fig. 5. The 1D double square well potential.

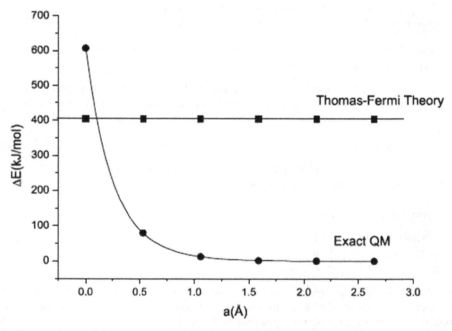

Fig. 6. Comparison of the energy splitting between ground and first excited states of a double square well motion in one dimension. The TF theory shows constant and large splitting while the exact result decreases rapidly towards zero with increasing barrier width.

bond will be found to be directly related to the facilitation of interatomic electron transfer which is reflected in the transition of the energy spectrum of the separated atoms from degenerate or near degenerate form towards a nondegenerate form for molecules as in the TF theory. The failure of TF theory to describe covalent bonding is then clearly due to the ergodic quantization procedure which produces smooth progression of electronic energy levels without degeneracies or near degeneracies at all atomic separations.

4.2 The hydrogen atom

In the case of the hydrogen atom (in atomic units) the 3D potential is –

$$V(r) = -1/r. \tag{4}$$

The phase space volume below the energy E can be worked out analytically as

$$\Omega(E) = 2^{3/2}\pi^3/3(-E)^{3/2}. \tag{5}$$

The energy eigenvalues according to Thomas-Fermi theory readily follow and are shown compared with the exact energy eigenvalues of the hydrogen atom in Figure 7. We have

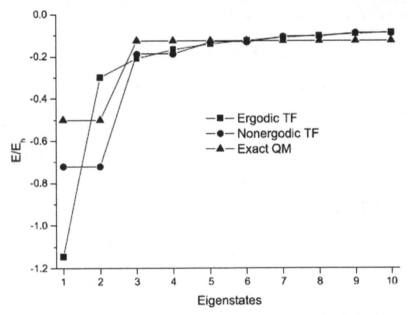

Fig. 7. Comparison of Thomas-Fermi and exact energy eigenvalues for the hydrogen atom. The original TF theory is called "ergodic" while the spin-conserving TF theory is denoted "nonergodic".

$$E_n^{(\text{TF})} = -2^{1/2}(2^{1/2}/12)^{-1/3}/4 \text{ for } n = 1, \tag{6}$$

$$= [-2^{1/2}(2^{1/2}/12)^{-1/3}/4] \times [n^{1/3} - (n-1)^{1/3}] \text{ for } n = 2,3,..... \tag{7}$$

as compared to the exact values –

$$E_{n,l,m} = -1/2n^2 \,,\ n = 1,2,\ldots,\ l = 0,1,..\ n-1,\ m = -l,\ -l+1,\ \ldots l-1,\ l. \tag{8}$$

However, the electron has spin – either up or down. We have so far considered spinless particles. Allowing for spin while maintaining the ergodic quantization of the original TF theory will double the phase space as the electron is allowed to rapidly flip from up to down or vice versa. The results above therefore describe doubly degenerate energy levels which correspond to a form of "nonergodic" TF theory which conserves spin. The standard TF theory, on the other hand, will populate both spin up and spin down phase space with one-electron states which are spin-depolarized giving nondegenerate energy levels. Figure 7 shows both the fully ergodic (original) TF energy eigenvalues as well as those obtained by the spin-nonergodic TF theory described above. The clear picture seen is that the Thomas-Fermi energies behave reasonably apart from the inability to resolve the highly degenerate shell structure of the correct quantum results. This will be a key fact in the resolution of reactivity and covalent bonding proposed below.

4.3 The hydrogen molecule ion

The simplest molecule, the hydrogen molecule ion H_2^+, has an electronic potential composed of two displaced hydrogen atom potentials, hence the total potential energy including the internuclear repulsion (all in atomic units) is -

$$V(r) = -\frac{1}{r_a} - \frac{1}{r_b} + \frac{1}{R} \tag{9}$$

where r_a and r_b represent the distance of the electron from nuclei a and b respectively and R is the internuclear separation.

The first two terms in equation (9) correspond to a double well potential of two superposed hydrogen Coulomb wells while the third term represents the repulsion between the two protons. The integral determining the phase space volume below an energy E, $\Omega(E)$, is now harder to find analytically but still easy to determine numerically. We leave those details aside and consider the results obtained by the "nonergodic" (i.e., spin conserving) form of Thomas-Fermi theory. In Figure 8 we show the electronic, internuclear and total energy of the ground state as a function of the internuclear separation R.

It is clear from the figure that while the electronic energy is weakly attractive the internuclear repulsion is stronger and the total energy is weakly repulsive. The phase space integrations are readily extended to generate excited state energies and the corresponding bond energy curves for the first ten states of H_2^+ in the spin conserving TF theory are shown in Figure 9 below.

The non-bonding character of the Thomas-Fermi results is not related to the choice of spin conserving or original spin-unpolarized theory. As will be even more clear after consideration of the successfully bonding Hückel theory below the problem of the Thomas-Fermi theory is related to the inability to resolve the constraints on the electron dynamics. In the case of the calculations for H_2^+ above the loss of bonding is related to the inability to resolve left-right electron transfer between the two coulombic potential wells located on the protons. As seen very clearly in the case of the two non-overlapping square wells in 1D above the TF theory will effectively quantize the motion as if the dynamical hindrance of the

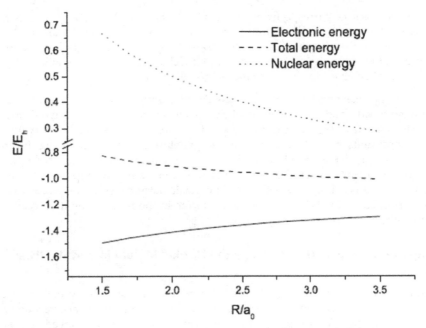

Fig. 8. The Thomas-Fermi energies for H_2^+ obtained in the "nonergodic" (spin conserving) theory are shown as a function of bond length R. From the top we have the internuclear, total and electronic energies.

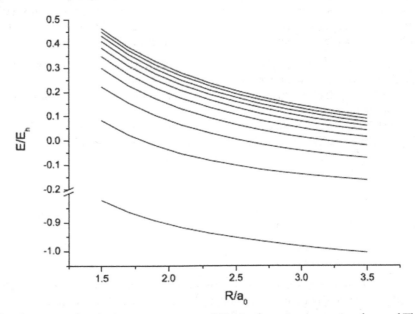

Fig. 9. Total energies for the lowest ten states of H_2^+ in the spin conserving form of Thomas-Fermi theory.

barrier was not present – irrespective of its width. In the same way the TF results for H_2^+ above have a character as if the electron transfer is rapidly delocalized for all bond lengths in stark contrast to the slowing down of this transfer by an increasingly broad barrier for increasing bond length seen in the exact quantum mechanical results. This will be discussed in greater detail in connection with the Hückel model below.

We conclude this examination of the Thomas-Fermi theory with the observation that it is able to follow broad trends in the density of energy eigenstates quite well but fails to resolve degeneracies and near-degeneracies in the spectrum due to dynamical conservation laws and other constraints such as potential barriers penetrable only by tunneling. In order to confirm our notion that this failure is also the cause of its inability to predict covalent bonding we now turn to the Hückel theory which despite its simple empirical nature is able to resolve covalent bonding very well. We shall particularly consider the representation of dynamics afforded by the Hückel theory to find therein the reason why it succeeds where the Thomas-Fermi theory fails.

5. Covalent bonding as described by the Hückel Molecular Orbital (HMO) model

Hartree-Fock and modern DFT based on the Kohn-Sham kinetic energy both successfully describe covalent bonding – albeit with some quantitative errors due to incorrect description of electron correlation effects. Common to these theories is the key role played by molecular orbitals formed from modified atomic orbitals as the covalently bound molecule is created. It is clearly the nature of these molecular orbitals and their orbital energies which carries the information about the covalent bonding mechanism. In order to most graphically illustrate the covalent bonding mechanism we shall therefore reexamine perhaps the simplest form of molecular orbital quantum chemistry, the Hückel molecular orbital model, which successfully captures the covalent bonding mechanism. In particular we shall

- claim that Hückel theory, appropriately interpreted and extended, is the minimal and precisely focussed explanation of all forms of covalent bonding,
- use the Hückel model to illustrate the concept of quantum ergodicity in the context of electron dynamics,
- show that Hückel theory displays the direct relation between electron delocalization and quantum ergodicity which in turn is the key to covalent bonding, and
- argue that all main types of bonding, including resonance stabilization, can be systematically organized according to the degree of delocalization employed in the stabilization.

The molecular orbitals (MO) of a molecule are, by definition, solutions of the Hartree-Fock equations, i.e. eigenfunctions of the Fock operator

$$\hat{F}\psi_i = \varepsilon_i\psi_i \tag{10}$$

where $\hat{F}$ is an effective one-electron Hamiltonian operator for a single electron within the molecule which interacts with the nuclei and all other electrons. Thus we have

$$\hat{F} = \hat{T} + \hat{V}_N + \hat{J} - \hat{K}, \tag{11}$$

where $\hat{T}$ is the kinetic energy operator of an electron, $\hat{V}_N$ and $\hat{J}$ represent the Coulomb interaction of an electron with the nuclei and all other electrons respectively, and $\hat{K}$ is the exchange operator which accounts for the fermion nature of the electrons, i.e. the anti-symmetry requirement on the total many-electron wave function. In the majority of quantum chemical applications the MO-s are constructed from atom centred basis functions $\{\chi_k\}$,

$$\psi_i = \sum_k C_{ki}\chi_k . \tag{12}$$

The Hartree-Fock equations, when projected onto the space spanned by the above atomic orbital (AO) basis, become matrix eigenvalue equations

$$\mathbf{FC} = \mathbf{SC}\varepsilon , \tag{13}$$

where $\mathbf{S}$ and $\mathbf{F}$ are the overlap and Fock matrices,

$$S_{kl} = \langle \chi_k | \chi_l \rangle , \tag{14}$$

$$F_{kl} = \langle \chi_k | \hat{F} | \chi_l \rangle . \tag{15}$$

The Hückel MO method is developed from Hartree-Fock theory through a series of assumptions and approximations.

1. As Hückel theory was formulated primarily for conjugated hydrocarbons, i.e. planar molecules, the $\sigma - \pi$ separability applies and hence the Fock equations are block-diagonal:

$$\begin{pmatrix} \mathbf{F}_\sigma & 0 \\ 0 & \mathbf{F}_\pi \end{pmatrix}\begin{pmatrix} \mathbf{C}_\sigma & 0 \\ 0 & \mathbf{C}_\pi \end{pmatrix} = \begin{pmatrix} \mathbf{S}_\sigma & 0 \\ 0 & \mathbf{S}_\pi \end{pmatrix}\begin{pmatrix} \mathbf{C}_\sigma & 0 \\ 0 & \mathbf{C}_\pi \end{pmatrix}\begin{pmatrix} \varepsilon_\sigma & 0 \\ 0 & \varepsilon_\pi \end{pmatrix} \tag{16}$$

Hückel theory is then concerned with the calculation of the π-MO-s only, by solving

$$\mathbf{F}_\pi\mathbf{C}_\pi = \mathbf{S}_\pi\mathbf{C}_\pi\varepsilon_\pi . \tag{17}$$

2. The π basis set is assumed to be a minimal basis, consisting of atomic $2p$ orbitals, one on each carbon atom. Assuming that the molecule is in the xy plane, its π MO-s are then constructed in terms of the set of $2p_z$ AO-s of the carbon atoms.

3. All diagonal elements of the π Fock matrix are represented by a single parameter, α, which represents the energy of an electron in a $2p_z$ atomic orbital that includes its interaction with the rest of the molecule, according to the definition of a diagonal Fock matrix element. All off-diagonal elements between AO-s on neighbour, i.e. bonded atoms, are assigned a single parameter, β, which is assumed to be a negative quantity. All other off-diagonal Fock matrix elements are assumed to be negligible. The neglect of integrals between non-neighbour atoms is known as the *tight-binding approximation*, and, as Ruedenberg [26] pointed out, it is a justifiable first approximation, whereas "the outright neglect of neighbour overlap is not." The Hückel (Fock) matrix is thus

$$\mathbf{F}_\pi = \alpha \mathbf{I} + \beta \mathbf{M} \; , \tag{18}$$

where $\mathbf{I}$ is the unit matrix and $\mathbf{M}$ is the topological or connectivity matrix which specifies the presence or absence of bonds between all pairs of atoms. For butadiene, for example, we have

$$\mathbf{M} = \begin{pmatrix} 0 & 1 & 0 & 0 \\ 1 & 0 & 1 & 0 \\ 0 & 1 & 0 & 1 \\ 0 & 0 & 1 & 0 \end{pmatrix} , \tag{19}$$

where the numbering of carbon atoms and hence of the $2p_z$ atomic orbitals is sequential, starting at one end of the molecule.

4.   In the simplest formulation of Hückel theory the $2p_z$ atomic orbitals are assumed to be orthonormal. The Hückel π-MO equations are then simply

$$\mathbf{F}_\pi \mathbf{C}_\pi = \mathbf{C}_\pi \boldsymbol{\varepsilon}_\pi \; . \tag{20}$$

The corresponding eigenvalues are

$$\varepsilon_i = \alpha + \lambda_i \beta \; , \tag{21}$$

where the coefficients $\{\lambda_i\}$ are eigenvalues of the topological matrix $\mathbf{M}$. We will refer to this approach as the *standard Hückel model*. We recall here, however, that the neglect of overlap between neighbour atoms is not a justifiable approximation according to Ruedenberg's analysis [26].

Since β is negative, the bonding, non-bonding and antibonding MO-s correspond to positive, zero and negative eigenvalues of $\mathbf{M}$, respectively. The energies of the MO-s directly correlate with the number of nodal planes. Thus, for butadiene the normalized MO-s and their energies are

$$\begin{pmatrix} \psi_1 & \psi_2 & \psi_3 & \psi_4 \end{pmatrix} = \begin{pmatrix} \chi_1 & \chi_2 & \chi_3 & \chi_4 \end{pmatrix} \begin{pmatrix} 0.37 & 0.60 & 0.60 & 0.37 \\ 0.60 & 0.37 & -0.37 & -0.60 \\ 0.60 & -0.37 & -0.37 & 0.60 \\ 0.37 & -0.60 & 0.60 & -0.37 \end{pmatrix} \tag{22}$$

$$\varepsilon_1 = \alpha + 1.62\beta, \quad \varepsilon_2 = \alpha + 0.62\beta, \quad \varepsilon_3 = \alpha - 0.62\beta, \quad \varepsilon_4 = \alpha - 1.62\beta \; . \tag{23}$$

For any given MO the presence of a nodal plane between two neighbour atoms is indicative of antibonding character between those particular atoms, as the node signifies destructive interference of the AO-s. For butadiene thus, occupancy of the second MO would weaken the π-bond between atoms 2 and 3, while strengthening those between atoms 1 and 2, as well as between 3 and 4. Such effects are given a quantitative measure by the first order reduced density matrix in the AO representation, $\mathbf{D}$, which in Hückel theory is generally known as the charge and bond order matrix. It is defined as

$$D = C_\pi n C_\pi^\dagger , \tag{24}$$

where n is the diagonal MO occupancy matrix with elements 2, 1 or 0. The off-diagonal elements of D are the bond orders, with values ranging from 0 to 1, where $D_{pq} = 1$ signifies a conventional two-centre π-bond, as in ethylene. In butadiene the π-bond orders are 0.89, 0.45 and 0.89. Thus, Hückel theory predicts in a qualitatively correct manner that there is an alternation of π-bond strengths in conjugated polyenes which accounts for the observed variation of CC bond lengths. In fact, the Hückel-SCF model [27], in which the β-parameters are defined to be bond length dependent, while the bond lengths in turn are determined on the basis of the corresponding bond orders in a self-consistent way, has been found to be a useful and reliable computational approach in a number of applications [28-32].

As noted above, the most obvious discrepancy of the standard Hückel model is the neglect of the overlap integrals between the $2p_z$ AO-s on neighbour atoms, i.e. atoms which are directly bonded to each other. This approximation actually introduces an inconsistency in that if such an approximation were justified, then β would also be negligible since the latter is overlap dependent. The justification, in the words of Coulson, is that "Historically, people have tended not to include it (overlap) because it turns out not to make much numerical difference" [33]. However, in the spirit of the Hückel parametrization, this problem is easily remedied, by approximating the overlap matrix S as

$$S = I + \gamma M , \tag{25}$$

where γ, a third Hückel parameter, is a measure of the average overlap integral between the $2p_z$ AO-s on bonded atoms. Since F_π, M and S have a common set of eigenvectors, U, i.e.,

$$SU = (I + \gamma M)U = U(I + \gamma \varepsilon),$$
$$F_\pi U = (\alpha I + \beta M)U = U(\alpha I + \beta \varepsilon), \tag{26}$$

where λ is the diagonal matrix of eigenvalues of M. The eigenvalues, ε, and eigenvectors, C_π, of the generalized eigenvalue equations (17) are simply

$$\varepsilon_i = \frac{\alpha + \beta \lambda_i}{1 + \gamma \lambda_i} \tag{27}$$

$$C_\pi = U(I + \gamma \lambda)^{-1/2} . \tag{28}$$

This formalism, originally proposed by Wheland [34], will be referred to as the *standard Hückel model with overlap*. In comparison with the two-parameter standard model, the inclusion of differential overlap integrals results in higher energies, except for non-bonding MO-s whose energies would be unaffected. Thus, the bonding MO-s become less bonding, while the antibonding ones become more antibonding. The MO energy levels for butadiene obtained by the application of these two Hückel models are compared in Figure 10. An important consequence of the above parametrization is that the only change in the MO-s brought about by the introduction of explicit overlap is their renormalization, whereby all elements of the i-th column vector of U are multiplied by a (re)-normalization constant of $(1 + \gamma \lambda_i)^{-1/2}$.

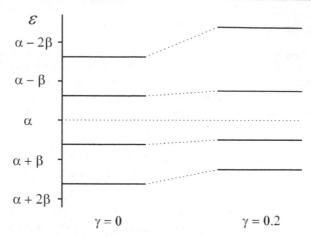

Fig. 10. Butadiene: Comparison of Hückel π-MO energy levels calculated by standard model ($\gamma = 0$) and with the inclusion of overlap ($\gamma = 0.2$), with $\alpha = 0$.

An alternative, implicit way of incorporating overlap in the Hückel model is to assume that the $2p_z$ AO basis is a rigorously orthonormal basis, $\{\phi_i\}$, consisting of Löwdin orthogonalized AO-s [35]. The Löwdin process [36], known also as symmetric orthogonalization, is defined by the linear transformation

$$\phi = \chi S^{-1/2} . \tag{29}$$

This formalism, termed *Hückel model in OAO basis*, where the α and β parameters of Hückel theory are defined with respect to the Löwdin orthogonalized $2p_z$ AO basis, is formally equivalent to the standard (two-parameter) model developed above, but the interpretation of the parameters, especially from the point of view of chemical bonding is quite different, as discussed below.

5.1 The hydrogen molecule ion

The hydrogen molecule ion, H_2^+, is the textbook example of covalent chemical bonding as well as of the basic principle of molecular orbital (MO) theory. It also represents a natural starting point for Hückel theory. The simplest representation of the (normalized) bonding and antibonding wave functions, $\psi_\pm$, are in- and out-of-phase combinations of the $1s$ atomic orbitals (AO) of hydrogen, χ_a, χ_b, centred on the nuclei a and b (with a fixed orbital exponent $\zeta = 1$):

$$\psi_\pm = \left[2(1 \pm S_{ab})\right]^{-1/2} (\chi_a \pm \chi_b), \tag{30}$$

where S_{ab} is the overlap matrix element of the two AO-s. The corresponding energies (in atomic units) in terms of overlap (S), kinetic (T) and nuclear attraction (V) integrals are

$$E_\pm = \frac{(T_{aa} \pm T_{ab}) + (V_{aa} \pm V_{ab})}{(1 \pm S_{ab})} + \frac{1}{R} \tag{31}$$

where the last term is the nuclear repulsion energy at an arbitrary internuclear distance R. Thus the interaction between the AO-s is quantified by the three coupling matrix elements

$$S_{ab} = \langle \chi_a | \chi_b \rangle, \quad T_{ab} = \langle \chi_a | \hat{T} | \chi_b \rangle, \quad V_{ab} = \langle \chi_a | \hat{V} | \chi_b \rangle, \tag{32}$$

where $\hat{V}$ represents the attractive Coulomb potential of both nuclei. The resulting total energies $E_{\pm}$ as a function of the bond length are shown in Figure 11.

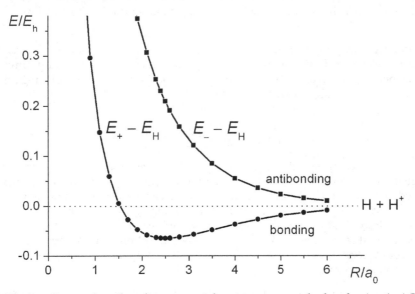

Fig. 11. H_2^+ bonding and antibonding potential energy curves calculated using $1s$ AO-s with a fixed orbital exponent of 1.

The calculation predicts the existence of a stable molecule with an equilibrium bond length of 2.5 a_0 and a dissociation (D_e) or bond energy of 0.0648 E_h, i.e. 170 kJ mol^{-1}. The experimental values are 2.0 a_0 and 270 kJ mol^{-1}. Thus this simplest of MO calculations yields results that are qualitatively correct.

While non-zero overlap is manifested in non-zero kinetic and potential interactions between the AO-s, it also gives rise to a direct energetic effect known as Pauli repulsion. So as to separate the effect of the latter from the direct kinetic and potential components of the bond energy we orthogonalize the AO-s by Löwdin's symmetric orthogonalization method, as discussed in the previous section. The resulting OAO-s have the property that they resemble the original atom-centred AO-s to the maximum degree, i.e. the dominant contribution to each OAO is from a specific AO. In the present case of H_2^+, the normalized OAO-s are

$$\phi_1 = N(\chi_1 - \mu \chi_2),$$

$$\phi_2 = N(\chi_2 - \mu \chi_1), \tag{33}$$

where the constants μ (<1) and N can be shown to be

$$\mu = \frac{1-\left(1-S_{12}^2\right)^{1/2}}{S_{12}}, \tag{34}$$

$$N = \left(1-2\mu S_{12}+\mu^2\right)^{-1/2}. \tag{35}$$

Since $N \geq 1$, the orthogonalization process results in increased density close to the centre of the dominant AO which is compensated by a reduction in density in the nodal region of the OAO.

An important aspect of the orthogonalization procedure is the behaviour of the kinetic energy matrix elements. In the AO basis the diagonal elements are large and positive, while the off-diagonal elements are at least an order of magnitude smaller. With regard to the latter, it is instructive to rewrite such a quantity as

$$\left\langle f\left|\hat{T}\right|g\right\rangle = \frac{\hbar^2}{2m}\int_{-\infty}^{\infty} d\mathbf{r}\nabla f(\mathbf{r})\cdot\nabla g(\mathbf{r}) \tag{36}$$

where $\hbar = h/2\pi$, h being Planck's constant, and m is the electronic mass. According to the above equation (obtained via integration by parts) the sign of the kinetic matrix element is determined by the relative signs of the gradients of the wave functions f and g in the overlap region. In the case of two AO-s on different atoms the signs of the bond-perpendicular gradients of the two AO-s are the same, but would be different for the bond-parallel components for all reasonable bond distances. Hence an off-diagonal kinetic energy matrix element would consist of a negative bond-parallel component and a positive bond-perpendicular component.

The kinetic energy matrix elements obtained from the above H_2^+ calculations are shown in Figure 12. We note that as the molecule forms, i.e. the internuclear distance decreases, the effect of Löwdin orthogonalization is to increase the diagonal kinetic energy elements but decrease the off-diagonal elements to negative values. The former effect is the increasing Pauli repulsion with increasing overlap, while the latter represents the energetic stabilization associated with electron delocalization, i.e. covalent bond formation.

The effect of Löwdin orthogonalization on the potential energy (nuclear attraction) matrix elements is quite different, as shown in Figure 13. In the OAO basis the coupling matrix element V_{ab} is near-zero at all distances, while V_{aa} varies inversely with the kinetic term T_{aa}. In fact the sum $T_{aa} + V_{aa}$ is largely constant throughout the range of bond lengths: it varies between $-0.698\ E_h$ at $R = 5\ a_0$ and $-0.871\ E_h$ at $R = 1\ a_0$.

The total energy, when expressed in terms of the OAO matrix elements, is simply

$$E_+ = \left[T_{aa}+V_{aa}+V_{ab}+1/R\right]+T_{ab}. \tag{37}$$

Noting that the sum of the bracketed terms monotonically increases with decreasing bond distance, it follows that the equilibrium binding, which occurs in the region of 2.5 a_0, is entirely due to the contribution of the coupling kinetic energy matrix element T_{ab} to the energy. The conclusion that follows is well known, inasmuch as an MO calculation using a fixed exponent minimal basis predicts that covalent bonding is due to a drop in the kinetic

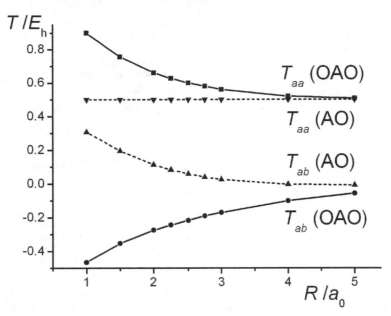

Fig. 12. Kinetic energy matrix elements for H_2^+ with respect to (non-orthogonal) 1s AO-s and Löwdin OAO-s.

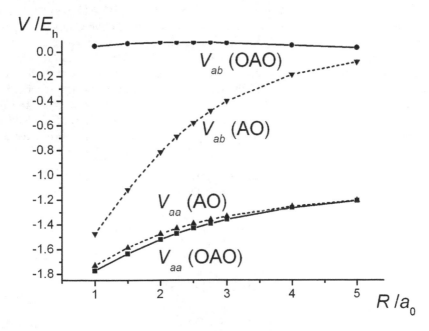

Fig. 13. Potential energy (nuclear attraction) matrix elements for H_2^+ with respect to (non-orthogonal) 1s AO-s and Löwdin OAO-s.

energy as the molecule's constituent atoms come together. The same conclusions apply in the more general case of optimised exponents, provided we allow for orbital contraction, as pointed out by Ruedenberg.

To provide more insight into the mathematical reasons for these changes consider the diagonal and off-diagonal matrix elements of the (Hermitean) kinetic energy operator $\hat{T}$ that arise when two real orbitals are Löwdin orthogonalized, as described above. The transformed matrix elements are

$$
\begin{aligned}
\left\langle \phi_1 \middle| \hat{T} \middle| \phi_1 \right\rangle &= N^2 \left(\left\langle \chi_1 \middle| \hat{T} \middle| \chi_1 \right\rangle + \mu^2 \left\langle \chi_2 \middle| \hat{T} \middle| \chi_2 \right\rangle - 2\mu \left\langle \chi_1 \middle| \hat{T} \middle| \chi_2 \right\rangle \right) \\
&\approx N^2 \left\langle \chi_1 \middle| \hat{T} \middle| \chi_1 \right\rangle > \left\langle \chi_1 \middle| \hat{T} \middle| \chi_1 \right\rangle
\end{aligned}
\tag{38}
$$

$$
\begin{aligned}
\left\langle \phi_1 \middle| \hat{T} \middle| \phi_2 \right\rangle &= N^2 \left(-\mu \left\langle \chi_1 \middle| \hat{T} \middle| \chi_1 \right\rangle - \mu \left\langle \chi_2 \middle| \hat{T} \middle| \chi_2 \right\rangle + \left(1 + \mu^2\right) \left\langle \chi_1 \middle| \hat{T} \middle| \chi_2 \right\rangle \right) \\
&\approx -2N^2 \mu \left\langle \chi_1 \middle| \hat{T} \middle| \chi_1 \right\rangle < 0.
\end{aligned}
\tag{39}
$$

Using the overlap and kinetic energy matrix elements computed for H_2^+ at a range of distances we find that in both expressions the terms involving the diagonal matrix elements, i.e $\left\langle \chi_1 \middle| \hat{T} \middle| \chi_1 \right\rangle$ and $\left\langle \chi_2 \middle| \hat{T} \middle| \chi_2 \right\rangle$, are dominant. (The same pattern has been found for the p_π orbitals of ethylene and benzene, computed at the SCF/STO-3G level of theory, to be discussed later.) The net result is that the kinetic energy of an OAO is larger than that for the AO, and the off-diagonal kinetic energy matrix element between two OAO-s is negative, as noted above. The physical explanation for the increased kinetic energy content, in the words of Weber and Thiel [37], is that "the orthogonalization accounts for the Pauli repulsion of the electrons at other centres and prevents penetration into these regions, thus effectively reducing the volume for an electron in the OAO and increasing its kinetic energy". In the case of Löwdin orthogonalized AO-s the dominant contributions arise from the regions where the dominant component of a given OAO interacts with the "orthogonalization tail" of the other, and where the gradients of the OAO-s with respect to all coordinates differ in sign.

In the context of Hückel MO theory, we note that the energy of H_2^+ at any distance can be written in terms of distance dependent α and β parameters:

$$
E_+(R) = \alpha(R) + \beta(R) + 1/R,
\tag{40}
$$

where

$$
\alpha(R) = T_{aa}(R) + V_{aa}(R)
\tag{41}
$$

and

$$
\beta(R) = T_{ab}(R) + V_{ab}(R),
\tag{42}
$$

where the matrix elements are evaluated in the OAO basis. While similar, but more complex expressions can be easily derived for the *standard Hückel model with overlap*, the inherent simplicity of the former makes it more attractive. More importantly, however, in the *Hückel model in OAO basis*, the Pauli repulsion is rigorously separated from the effects of delocalization and covalent bonding. In other words, use of OAO bases represents a natural starting point for the analysis and ultimately the understanding of covalent bonding.

5.2 The ethylene molecule

Ethylene is the classic π-bonded molecule which often serves as the starting point for the parametrization of Hückel MO theory. The bonding and antibonding π MO-s of ethylene are formally the same as the σ MO-s of H_2^+, but there are two important differences: (1) π MO-s are constructed from p_z-type MO-s which are parallel to each other, hence their interaction matrix elements are in general smaller than those of s and p AO-s in head to head arrangements. (2) In a many-electron molecule such as ethylene the potential energy of a given electron includes its interaction with all other electrons, as well as the field due to the nuclei, as implemented in the SCF process.

Computed values of the kinetic and potential matrix elements of the $2p_z$ AO-s and OAO-s are shown for a range of C-C distances in Figures 14 and 15. The CH distances and bond angles are fixed at their experimental values. The matrix elements were calculated from SCF/STO-3G kinetic energy and Fock matrix elements of ethylene. The potential energy matrix elements are simply

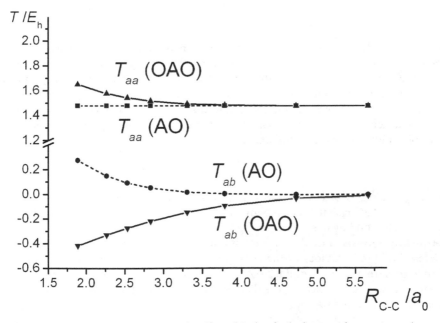

Fig. 14. Kinetic energy matrix elements for $2p_z$ orbitals of ethylene with respect to (non-orthogonal) AO-s and Löwdin OAO-s.

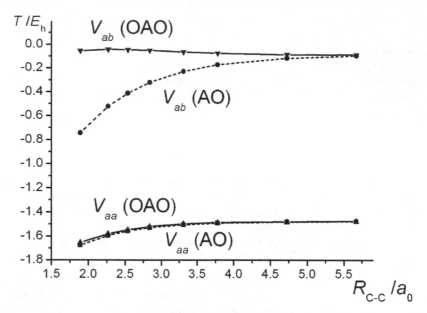

Fig. 15. Potential energy matrix elements for $2p_z$ orbitals of ethylene with respect to (non-orthogonal) AO-s and Löwdin OAO-s.

$$V_{ij} = F_{ij} - T_{ij}, \quad i,j = a,b. \tag{43}$$

where F_{ij} is the Fock matrix element between AO-s i, j.

The C-C distance dependence of these matrix elements is qualitatively the same as what was obtained for H_2^+ (Figures 12 and 13). Consequently the same observations and conclusions apply for the formation of a simple π-bond in ethylene as for the σ-bond in H_2^+. Thus, in the OAO representation π-bonding is the result of the negative kinetic interaction between the two OAO-s. As the π MO-s are eigenfunctions of the π block of the Fock matrix (see equation (10)), the Hückel parameters can readily be identified as the appropriate Fock matrix elements of a minimal basis computation, i.e.

$$\alpha = F_{aa}, \quad \beta = F_{ab}. \tag{44}$$

The π orbital energies of ethylene as a function of the C-C distance, as obtained in the current SCF/STO-3G calculations, are given in Figure 16. As expected, the splitting of the bonding and antibonding energies is strongly dependent on the interatomic distance. This is a consequence of the distance dependence of the kinetic energy matrix element T_{ab}, which in turn determines the behaviour of the Fock matrix element F_{ab}, ie. β of Hückel theory. The kinetic energies (relative to that of a $2p_z$ AO) of the bonding π and antibonding π^* MO-s are also shown in Figure 16. Clearly, the distance dependence of the MO-s is determined by their kinetic energy content.

We conclude this section with the observation that the Hückel description of π- bonding in ethylene is exactly the same as that of the covalent bond in H_2^+. In both molecules the

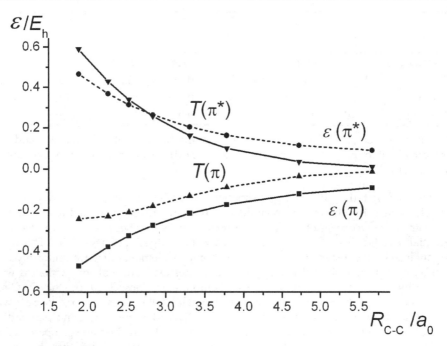

Fig. 16. Total orbital and kinetic energies of the bonding π and antibonding π^* MO-s of ethylene computed at the SCF/STO-3G level of theory. The kinetic energies are relative to that of a $2p_z$ AO (1.4777 E_h).

bonding and antibonding energy levels can be determined in terms of the Hückel α and β parameters, where α represents the energy of an electron in an OAO and β is the coupling matrix element between the OAO-s on the bonded atoms of the molecule. β is negative and predominantly kinetic in character. Thus for the bonding MO, which is the in-phase combination of the OAO-s, the energy stabilization relative to the OAO-s is due to delocalization and the concomitant drop in kinetic energy. The opposite holds for the antibonding MO, the out-of-phase combination of OAO-s, where the energetic destabilization is due to the presence of the extra node in the MO.

5.3 Benzene

In conjugated molecules there is an extra degree of stabilization of the bonding MO-s due to the extra degree of possible π delocalization. To further test the performance of the above Hückel models with regard to describing and interpreting π-bonding in planar conjugated hydrocarbons, we decided to parametrize the Hückel models on the basis of an ab initio minimal basis (STO-3G) Hartree-Fock calculation on benzene at its experimental equilibrium geometry. The high (D_{6h}) molecular symmetry is evident in the overlap and Fock matrices; there are only four distinct type of elements in each, as indicated in Table 1, which lists the converged Fock, overlap and kinetic energy matrix elements in the $2p_z$ AO basis and also in the Löwdin orthogonalized (OAO) basis. The in-plane (x,y) components of the kinetic energy matrix elements are also shown (as $T_{\parallel}$).

i,j	AO				OAO		
	S	F	T	$T_\parallel$	F	T	$T_\parallel$
1,1; 2,2; 3,3;...	1.000	−0.103	1.478	0.296	0.004	1.590	0.362
1,2; 2,3; 3,4;...	0.214	−0.254	0.075	−0.081	−0.251	−0.263	−0.157
1,3; 2,4; 3,5;...	0.026	−0.041	0.000	−0.018	0.010	0.017	0.010
1,4; 2,5; 3,6;...	0.010	0.014	−0.004	−0.009	0.023	−0.013	−0.008

Table 1. Benzene: Overlap, Fock and kinetic energy matrix elements (in E_h) in AO and OAO bases from ab initio (HF/STO-3G) calculations. (The orbitals are numbered sequentially around the ring.)

First, we note that the Hückel approximation, whereby off-diagonal overlap and Fock matrix elements between orbitals on atoms which are not directly bonded to each other (as listed in the last two rows of Table 1) are neglected, is justified. Secondly, however, the differential overlaps, i.e. S_{12}, S_{23},...can hardly be judged as negligible. Thirdly, the orthogonalization has resulted in diagonal Fock matrix elements that are uniformly higher in energy, as expected, since the orthogonalization results in orbital contraction as well as delocalization by virtue of the "orthogonalization tails" possessed by the OAO-s. As discussed earlier in the section on H_2^+, an off-diagonal kinetic energy matrix element would consist of a negative bond-parallel component, $T_\parallel$, and a positive bond-perpendicular component, $T_\perp$. In the case of benzene the former is actually the in-plane component. In the case of non-orthogonal AO-s the two components are positive and negative. As expected, Löwdin orthogonalization increases the diagonal elements of the kinetic energy matrix by ~0.11 E_h, but decreases both bond-parallel and bond-perpendicular components of the off-diagonal elements to large negative values.

As the data in Table 1 indicate, in the (non-orthogonal) AO basis the Fock matrix elements, in particular the off-diagonal elements, F_{12}, F_{23},...are dominated by potential energy contributions. However, the large and definitely non-negligible overlap integrals, S_{12}, S_{13},...are indicative of substantial Pauli repulsion, so that in the OAO basis the off-diagonal Fock matrix elements are almost entirely kinetic in character. Moreover, the increase in the diagonal Fock matrix elements brought about by the orthogonalization process is largely due to the analogous increase in the diagonal kinetic energy matrix elements.

Another notable consequence of the orthogonalization procedure is the decrease, by approximately a factor of 4, in the magnitude of the Fock matrix elements between AO-s on next nearest neighbouring atoms, viz. 1,3; 2,4; 3,5;... elements.

Utilizing the appropriate ab initio Fock and overlap matrix elements of benzene we can parametrize the three Hückel models as follows:

1.   Standard Hückel model:

$$\alpha = F_{11} = F_{22} = ... = -0.103\, E_h \tag{45}$$

$$\beta = F_{12} = F_{23} = ... = -0.254\, E_h$$

$$\gamma = 0\,.$$

2.   Standard Hückel model with overlap: In addition to α and β, as chosen for the standard model,

$$\gamma = S_{12} = S_{23} = ... = 0.214 \qquad (46)$$

3.   Hückel model in OAO basis:

$$\alpha = F_{11}^{OAO} = F_{22}^{OAO} = ... = 0.004\ E_h \qquad (47)$$

$$\beta = F_{12}^{OAO} = F_{23}^{OAO} = ... = -0.251\ E_h$$

Applying these three models to benzene, the energy levels obtained are shown in Figure 17, along with the *ab initio* SCF energies for comparison.

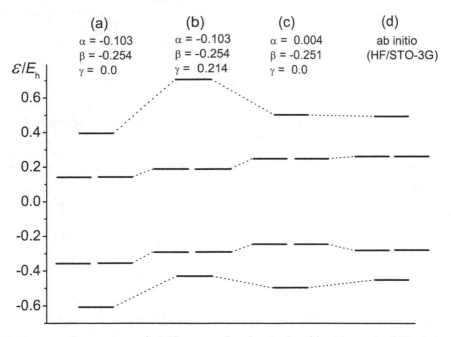

Fig. 17. Benzene: Comparison of π MO energy levels calculated by (a) standard Hückel model (b) standard Hückel model with overlap (c) Hückel model in OAO basis and (d) ab initio (HF/STO-3G) method.

The energy levels calculated by the standard model are uniformly too low in comparison with the *ab initio* values. This is attributable largely to the neglect of the differential overlap integrals. Their inclusion, according to the standard model with overlap, results in significant improvements in the energies of the bonding MO-s, but also an overcorrection to the energy of the highest antibonding MO. The best overall agreement with the *ab initio* results is obtained by the Hückel model in the OAO basis. The energies calculated by the latter approach are essentially the same as those from the standard model calculations plus a positive shift of ~0.1 E_h, since the respective β parameters are nearly the same.

Despite its simplicity, the standard model with overlap has not gained much popularity, largely because most workers in the field, especially prior to the development of more sophisticated semi-empirical and indeed *ab initio* methods, found that the standard model works well without the explicit allowance for overlap, as noted for example by Coulson [33]. It was also appreciated that the standard model can be regarded as being formally equivalent to the Hückel model in OAO basis, i.e. the neglect of overlap is rigorously justifiable [35]. Depending on the problem at hand, be it the estimation and/or rationalisation of stabilities in terms of π-delocalization, or the interpretation of electronic spectra, the α and β parameters of Hückel theory are treated as adjustable empirical parameters, chosen so as to best fit experimental data. As remarked above, choosing α and β to best model ab initio minimal basis SCF MO-s and their energies is a pedagogical exercise - we do not suggest that such a procedure is superior or preferable to the accepted practice when it comes to "real" applications. The difference between the two models arises, of course, in the computation of the MO coefficients. Use of the OAO basis implies that the MO-s are

$$\psi = \chi S^{-1/2} U \tag{48}$$

where U is the matrix of eigenvectors of the of the topology matrix, M.

As a further test of the three Hückel models for benzene, the kinetic and potential energy components of the computed orbital energies are listed in Table 2, where each MO is labelled by the number of its nodal planes.

Number of nodes	ε				$\langle T \rangle$				$-\langle V \rangle$			
	1	2	3	4	1	2	3	4	1	2	3	4
Standard Hückel	−0.61	−0.36	0.15	0.41	1.63	1.55	1.40	1.33	2.24	1.91	1.25	0.92
Hückel with overlap	−0.43	−0.29	0.19	0.71	1.14	1.28	1.79	2.33	1.57	1.57	1.59	1.61
Hückel in OAO basis	−0.50	−0.25	0.26	0.51	1.06	1.33	1.85	2.12	1.56	1.57	1.60	1.61
ab initio (HF/STO-3G)	−0.46	−0.28	0.27	0.50	1.08	1.32	1.82	2.16	1.54	1.60	1.55	1.66
ab initio $\langle T_{\parallel} \rangle$					0.06	0.20	0.50	0.70				

Table 2. Benzene: Total orbital energies of Hückel and ab initio MO-s and their kinetic and potential energy components (in E_h).

The two Hückel models, which include overlap, reproduce the ab initio kinetic and potential orbital energies quite accurately, providing further support for the view that these models capture the essential features of the more sophisticated ab initio approach to MO theory. Neglecting differential overlap integrals, as invoked in the standard model, yields qualitatively different results: the ordering of the MO-s in the standard model is determined by the potential energy, as the apparent kinetic energies of the MO-s depend *inversely* on the number of nodes, due to the neglect of Pauli repulsion. To help with our understanding of these effects consider the kinetic energy associated with a given MO, as obtained from equation (27):

$$\langle T \rangle = \frac{\left\langle \psi_i \left| \hat{T} \right| \psi_i \right\rangle}{\left\langle \psi_i \middle| \psi_i \right\rangle} = \frac{\alpha_T + \lambda_i \beta_T}{1 + \lambda_i \gamma} \tag{49}$$

where α_T and β_T are the kinetic energy components of α and β respectively. As noted already, in the standard model the kinetic energy contributions to the β parameters, although effectively negligible, are actually positive. The variation in kinetic energy among the MO-s is largely determined by the variation in energy denominators which depend inversely on the eigenvalues, $\{\lambda_i\}$, which in turn are in a direct relationship with the total energies, $\{\varepsilon_i\}$. However, in the absence of Pauli repulsion, i.e. overlap effects, the kinetic energy of the i-th MO would be determined by the energy numerator $\alpha_T + \lambda_i \beta_T$ alone, which gives rise to the incorrect nodal dependence of the kinetic energy.

The apparent non-physical behaviour of the MO energies is a basic flaw of the standard Hückel model. It is caused by the inconsistency which is implicit in the formalism, i.e. the neglect of differential overlap in the parametrization of the overlap matrix, but *not* in the parametrization of the Fock matrix. We believe that from a pedagogical point of view we should not advocate this model as a qualitatively correct representation of the electronic structure of conjugated π-electron molecules, but rather as the first step in the development of an acceptable, i.e. physically correct, theory.

The above results graphically illustrate the importance of overlap integrals in the calculation of molecular wave functions and energies as well as in the description and interpretation of chemical bonding. Overlap integrals make an important contribution to the quantum mechanical coupling of (non-orthogonal) AO-s and, more generally, of atom-centred basis functions. Their neglect, as in the standard Hückel model, results in MO-s and energetics which would provide an incorrect interpretation of covalent bonding in terms of kinetic and potential energies. With the help of judiciously chosen empirical or semi-empirical parameters such overlap effects can be "folded" into those parameters in a range of popular semi-empirical methods which neglect differential overlap, although ultimately, as shown by the work of Weber and Thiel [37], significant improvements in accuracy and reliability can be achieved by the direct incorporation of overlap into these models. In the case of Hückel MO theory this can be achieved simply by adopting the OAO representation.

The interpretation of covalent bonding via the Hückel theory is particularly transparent by utilizing the model in OAO representation, as the role of kinetic energy and its origins in the context of bonding can be clearly seen. Löwdin orthogonalization of the AO-s accounts for the presence of Pauli repulsion between electrons in the AO-s and results, most significantly, in the contraction of the AO-s. The coupling between the resulting OAO-s is, however, very strong, negative and predominantly kinetic in character. Electronic delocalization, as described by the formation of bonding MO-s, leads to stabilization, largely because of the concomitant drop in the kinetic energy. This is graphically illustrated by the results for benzene which are summarized in Figure 18, where the kinetic and potential energies of the MO-s are plotted against their total energies.

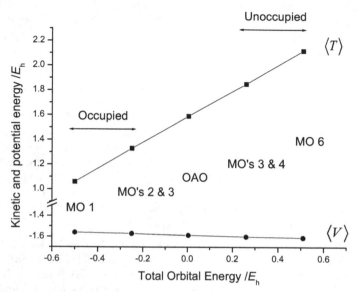

Fig. 18. Kinetic and potential energies π-MO-s and OAO-s of benzene as obtained by Hückel model in OAO basis.

The energetic stabilization and destabilization of the occupied and unoccupied MO-s (relative to an OAO) are clearly a consequence of the kinetic energy content of the MO-s. There is negligible variation in the potential energies of the MO-s. This picture is essentially the same as the description of Ruedenberg and co-workers [13, 38-41].

In summary, we note that, despite its apparent simplicity of application, the Hückel model clearly contains all the necessary quantum mechanical elements to describe π-bonding in polyenes. It enables us to interpret covalent bonding in a manner that is physically correct as well as transparent. It relies on just three parameters: α, representing the binding of a π-electron occupying a $2p_z$ AO to an atom; β, representing the quantum mechanical interaction between AO-s on neighbour atoms; and γ, which is the overlap between the coupled AO-s. β and γ can be thought of as two sides of the same coin, as both are crucial in the correct description of the coupling between two AO-s. By reformulating the Hückel model in terms of a Löwdin orthogonalized AO basis, the number of parameters is reduced, as only α and β need to be defined. Solution of the Hückel equation results in delocalized MO-s with a range of energies which are lower as well as higher than α, depending on the number of nodes in the MO-s and on the magnitudes of β and γ.

The degree of π-bonding, i.e. π stabilization, that occurs in a given molecule is determined by the overall accumulated contributions of the β coupling terms resulting from the appropriate occupancy of the lowest energy MO-s. It will be argued in a later section, that

Hückel MO-s are maximally delocalized, and therefore they naturally provide an optimal description of π-bonding, since the degree of delocalization is a key factor that effectively governs the degree of stabilization. An important point that emerges from this analysis is that the mechanism of binding an electron within an atom, as represented by α is of no consequence. All that is needed is an attractive atomic potential. In other words, the particular form of the Coulomb attraction that binds electrons does not play a crucial role in the covalent bonding mechanism. The latter depends on delocalization and quantum mechanical coupling of (localized) AO-s and is not affected by the exact nature of the inter-particle forces.

5.4 Extensions of Hückel Theory

Hückel MO theory in its standard form describes the π-electron structure of conjugated polyenes. It can be, as indeed it has been, generalized in a number of ways so as to be applicable e.g. to π systems with heteroatoms, the treatment of σ-bonding – as we have done above – and the description of solids.

The generalized Hückel method, more widely known as the Pariser-Parr-Pople (PPP) method [42-45] is a semi-empirical π-MO theory, where electron repulsion effects are explicitly, although approximately, allowed for. PPP utilizes the tight binding approximation and it is parametrized in terms of one- and two-electron integrals chosen largely on the basis of experimental data.

The extended Hückel MO (EHMO) method, developed by Hoffmann [46], is a valence MO theory, i.e. σ- and π-electrons are treated on equal footing. The diagonal Fock matrix elements, $\{F_{ii}\}$ (in a minimal basis) are parametrized on the basis of atomic ionization energies, while the off-diagonal elements, $\{F_{ij}\}$, following the recommendation of Wolfsberg and Helmholtz [47], are specified simply as

$$F_{ij} = KS_{ij}\left(F_{ii} + F_{jj}\right)\big/2 , \qquad (50)$$

where S_{ij} is an overlap integral computed in a minimal Slater orbital basis and K is an adjustable parameter, usually ~1.75. The MO-s and their energies are obtained by solving the generalized eigenvalue equations (13); the total molecular energy is simply defined as the sum of the occupied MO energies. As the overlap integrals are computed at any given molecular geometry (rather than parametrized), the EHMO method allows geometry optimizations to be carried out.

Basic Hückel theory lies at the heart of the widely used tight binding method of material modelling [48]. The formalism was first developed by Slater and Koster [49] in an attempt to generalize the LCAO method to infinite periodic crystals by means of Bloch sums. The AO basis in a tight binding calculation is understood to be a Löwdin orthonormalized basis. The Hamiltonian, i.e. Fock matrix elements, are formally defined with respect to the orthogonal AO basis, although it is generally parametrized, such that, e.g. the two-centre integrals are represented by parameters that depend on the distance between the atomic centres, rather like in Hoffmann's EHMO method. The popularity of the method, as explained by Goringe et al. [48], stems from the view that the tight binding method combines physical transparency and quantum mechanical sophistication, as well as being a method that combines computational speed with surprising accuracy. The same applies, in our view, to standard π Hückel MO theory, although in most practical molecular applications today the level of accuracy that's required cannot be achieved by such a simple model. It can, however, help us understand covalent bonding.

6. The quantum dynamical content of Hückel Theory

We have seen above that Hückel theory resolves covalent bonding in terms of a kinetic coupling allowing π-electrons to form delocalized molecular orbitals. Now we shall show how this in fact resolves the interatomic flow of electrons which can vary in extent (degree of delocalization) and rate. Thus the Hückel model can be seen as a simple theory of molecular electron dynamics. In turn we shall be able to deduce that covalent bonding is most generally understood as due to a quantum dynamical mechanism relating to the facilitation of interatomic electron transfer.

As suggested by the examples above, the Hückel π-MO-s of polyenes are expected to be delocalized. A simple, yet convincing way to demonstrate this is to utilize the analogy between Hückel theory and the finite element method (FEM) representation of wave functions of a "particle in a box", which is a familiar example to chemistry students. It is also known as the Free Electron Molecular Orbital (FEMO) model for conjugated π systems [42,50]. It represents the simplest quantum mechanical treatment of conjugated molecules, whereby the π-electrons are assumed to be fully delocalized within a one-dimensional box, the length of which is chosen on the basis of the total distance between the end-atoms of the conjugated system. The system, used also in our discussion of Thomas-Fermi theory above, is defined by the square-well potential:

$$V(x) = 0, \quad 0 < x < L$$
$$= \infty, \text{ elsewhere,} \tag{51}$$

and solution of the one-dimensional Schrödinger equation yields the following energy eigenvalues and (delocalized) eigenfunctions:

$$E_n = \frac{n^2 h^2}{8mL^2} , \tag{52}$$

$$\psi_n(x) = \sqrt{\frac{2}{L}} \sin \frac{n\pi x}{L}, \quad n = 1, 2, \ldots \infty, \tag{53}$$

where h is Planck's constant and m is the mass of the particle, i.e. here the electron.

An unusual but, in the context of this work, instructive way to solve this problem is to use the finite element method. This is a basis set method where, in the simplest implementation, the basis functions are *roof functions*, centred at equidistant grid-points $\{x_k\}$:

$$\phi_k(x) = \sqrt{\frac{3}{2\Delta}} \frac{(x - x_{k-1})}{\Delta}, \quad x_{k-1} < x < x_k , \tag{54}$$

$$= \sqrt{\frac{3}{2\Delta}} \frac{(x_{k+1} - x)}{\Delta}, \quad x_k < x < x_{k+1}$$
$$= 0, \quad \text{elsewhere.}$$

The N basis functions (for $N = 7$) are illustrated in Figure 19, and the grid-spacing Δ is of course equal to $L/(N + 1)$.

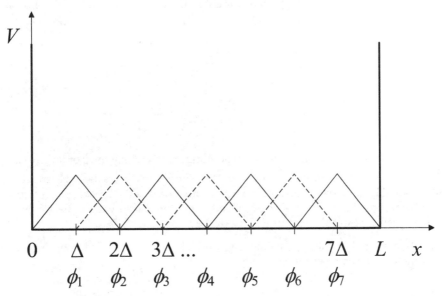

Fig. 19. Square well potential of particle in a box system showing grid-points and the FEM roof functions $\{\phi_k\}$ for $N = 7$.

Note that these basis functions are normalized but not orthogonal. The overlap matrix $\mathbf{S}$ is obtained by simple integration:

$$S_{kl} = 1, \quad \text{if } k = l \tag{55}$$

$$= 1/4, \quad \text{if } k = l \pm 1,$$
$$= 0, \quad \text{otherwise.}$$

The Hamiltonian consists of only the kinetic energy operator:

$$\hat{H} = -\frac{\hbar^2}{2m} \frac{d^2}{dx^2}, \tag{56}$$

and we obtain the Hamiltonian matrix elements via integration by parts (c.f. equation 36), whereby only the first derivatives of the roof functions are required. Thus,

$$H_{kl} = \frac{3}{2} \frac{\hbar^2}{m\Delta^2}, \quad \text{if } k = l, \tag{57}$$

$$= -\frac{3}{4}\frac{\hbar^2}{m\Delta^2}, \quad \text{if } k = l \pm 1,$$

$$= 0, \qquad \text{otherwise.}$$

Now, we note that the structures of these translational Hamiltonian and overlap matrices in the FEM roof-function basis are precisely those of the *standard Hückel model with overlap* which we discussed above. Thus we merely identify the Hückel parameters as follows:

$$\alpha = \frac{3}{2}\frac{\hbar^2}{m\Delta^2}, \tag{58}$$

$$\beta = -\frac{3}{4}\frac{\hbar^2}{m\Delta^2} = -\frac{\alpha}{2}$$

$$\gamma = \frac{1}{4}.$$

The energy eigenvalues, $\{\varepsilon_i\}$, and the corresponding matrix of eigenvectors, $\mathbf{C}$, are given by equations (27) and (28). Since the basic form of the eigenvectors is independent of the magnitudes of the parameters α, β and γ, we can conclude that the Hückel eigenvectors for a linear polyene will be completely delocalized over the entire carbon back-bone of the molecule the same way as the eigenfunctions of the particle in a box.

It is worth noting that the eigenvalues, λ, and eigenvectors, $\mathbf{U}$, of the corresponding topology matrix, $\mathbf{M}$, can be written in analytic form [42]:

$$\lambda_n = -2\cos\frac{n\pi}{N+1}, \tag{59}$$

$$U_{kn} = \sqrt{\frac{2}{N}}\sin\frac{nk\pi}{N+1} \tag{60}$$

$$= \sqrt{\frac{L}{N}}\psi_n(k\Delta),$$

where ψ_n is the n-th (exact) eigenstate of the particle in a box, as given by equation (53). The coefficients of the roof functions in a given global wave function ψ_n are thus the same as the coefficients for the $2p_z$ AO-s at corresponding carbon sites in the Hückel scheme, all given by the numerical value of ψ_n evaluated at the grid-point in question. The analytic formulae in equations (59) and (60) actually found their way into the chemical literature as solutions of the Hückel equations for linear polyenes [42]. Analogous analytical formulae exist for (mono-) cyclic polyenes, where complete delocalization of the Hückel MO-s can be similarly demonstrated by analogy with the particle on a ring problem.

Since the eigenvectors of the particle in a box correspond to wave functions that are fully delocalized within the square well, the model describes free translation of the particle inside the box. In other words, if at some time $t = 0$ the particle is represented by a localized wave-packet $\xi(x,0)$, the time-evolution of the latter is given by

$$\xi(x,t) = \sum_n \langle \psi_n | \xi(x,0) \rangle \exp(-iE_n t/\hbar) \psi_n(x) \tag{61}$$

and therefore the probability density, $P(x, t)$, of this wave packet evolves in time according to

$$P(x,t) = |\xi(x,t)|^2 = \sum_{n,l} \langle \psi_n | \xi(x,0) \rangle \langle \xi(x,0) | \psi_l \rangle \cos \frac{(E_n - E_l)t}{\hbar} \psi_n(x) \psi_l(x), \tag{62}$$

where the eigenfunctions, $\{\psi_n\}$, of the (time-independent) Hamiltonian, are assumed to be real.

As an illustration of the movement of a wave-packet, let it initially correspond to the first roof function, ϕ_1, i.e.,

$$\xi(x,0) = \phi_1(x). \tag{63}$$

The probability density at a given grid-point x_k at time t is then

$$P(x_k,t) = \sum_{n,l} (C_{1n} + \gamma C_{2n})(C_{1l} + \gamma C_{2l}) \cos \frac{(E_n - E_l)t}{\hbar} C_{kn} C_{kl}. \tag{64}$$

For $N = 4$ the application of this equation (utilising equations (27), (28), (59) and (60) for the coefficients $\{C_{kn}\}$ and energies $\{E_n\}$ yields the results shown in Figure 20.

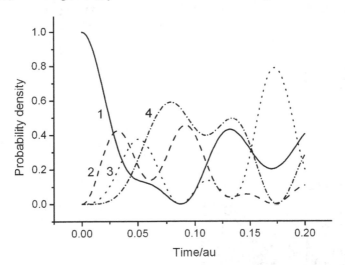

Fig. 20. Time evolution of probability density $P(x_k,t)$ at a given grid-point x_k for a wave packet defined as the roof function ϕ_1 at $t = 0$, in a basis of 4 roof functions ($N = 4$), with the grid-point index k shown in the figure.

The time-evolution of the densities at the four grid points within the box demonstrates the movement of density from one end of the box to the other including reflections from the ends and the broadening of the initial wave packet. Since the time-independent eigenstates of the Hamiltonian are completely delocalized, an electron initially localized at one end of the box is able to freely traverse the box, i.e. fully access the available phase space. The time average of the probability density at a given grid-point is simply

$$\left\langle P(x_k) \right\rangle = \lim_{\tau \to \infty} \frac{1}{\tau} \int_0^\tau dt\, P(x_k, t) . \tag{65}$$

Approximating the integral by discrete summation, we define $\left\langle P(x_k) \right\rangle_\tau$ which represents the time average of the probability density calculated for a finite value of τ. A plot of the computed values of $\left\langle P(x_k) \right\rangle_\tau$ as a function of τ for the current problem is shown in Figure 21. It provides a convincing illustration of the rapid convergence and uniformization of the probability densities at the various grid points. The fully converged values are $\left\langle P(x_1) \right\rangle = \left\langle P(x_4) \right\rangle = 0.3$ and $\left\langle P(x_2) \right\rangle = \left\langle P(x_3) \right\rangle = 0.2$.

As shown above, the Hückel type Hamiltonian reflects translation which is free within the region spanned by the delocalized eigenfunctions. Hückel MO theory therefore describes the free translation of electrons among the atoms of the linear structures established by the carbon atoms of a planar conjugated hydrocarbon molecule, which form a "molecular wire". The values of β and γ will give the moving electron an effective mass and a velocity. The shifts in orbital energies, downward for bonding MO-s, which brings about the energetic stabilization of the molecule, are clearly associated with the "delocalization of the electronic motion." The fundamental mechanism of covalent bonding is therefore electron delocalization.

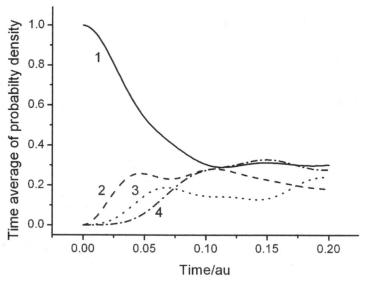

Fig. 21. Time averages of probability densities $<P(x_k)>_\tau$ as a function of the elapsed time τ, with the grid-point index k shown in the figure.

7. Quantum Ergodicity and the Hückel Model

The Hückel model is about the coupling of carbon atoms, formation of delocalized molecular orbitals and electron translation along "molecular wires" but most of all it is about stabilization of π-bonded conjugated hydrocarbon molecules. Why does delocalization lead to stabilization? The answer can be given at different levels of explanation. The simplest answer may be to appeal to the Uncertainty Principle, as it clearly defines the connection between spatial localization and zero-point energy of a quantum particle. This relationship is particularly clearly demonstrated by the behaviour of a particle in a box, as summarized by equation (52), whereby the (kinetic) energy of the particle is inversely dependent on the size of the box, L. The more we localize particle motion, by reducing L, the higher the zero-point energy becomes along with increased spacing between energy levels, since

$$E_n - E_{n-1} = \frac{(2n+1)h^2}{8mL^2} . \tag{66}$$

Thus, it is clear that if the interaction of atoms results in delocalization of the valence electrons, then this must, in general, decrease the occupied orbital energies of the system and thereby its total energy, resulting in the formation of a molecule. This was discussed by Hellmann [9], who, although aware of the Virial Theorem, emphasised the drop in kinetic energy due to delocalization upon molecule formation. Reality is however a little more complex. As indicated in the Introduction, the accurate energetics of molecule formation (which satisfy the requirements of the Virial Theorem) appear to contradict Hellmann's ideas, because they also account for orbital contraction. As shown by Ruedenberg and co-workers [13, 38-41], the correct analysis of the energetics from the point of view of covalent bonding needs to separate the atomic orbital contraction effects from the molecular interference and delocalization effects.

At another, and possibly deeper level of explanation, we point out that the reactivity of atoms is related to the presence of nonergodic constraints which are relaxed as the molecule forms, thereby lowering the energy. The concept of quantum ergodicity is not widely known nor without subtlety. A classical (microcanonical) system is said to be ergodic if its trajectory dynamics uniformly accesses, or *covers*, all parts of the (momentum-position) phase space which are consistent with a given total energy. Following the original suggestion of von Neumann, it was proposed [51] that a quantum system is ergodic if and only if the spectrum of energy eigenstates is non-degenerate. Thus a system with a degenerate energy specrum would be classified nonergodic. Subsequently it has been recognized that a *degree* of ergodicity could be discerned from the corresponding degree of regularity of the spectrum of energy eigenvalues, with an ergodic system displaying a smooth variation of energy gaps which translate into a corresponding smooth variation in the total density of states of the system [51]. By contrast, a nonergodic system would display random variations in the energy gaps. While there is no doubt about the validity of such a correlation between ergodic properties of a system and its spectral properties, we believe that quantum ergodicity is more directly related to the character of the energy eigenfunctions. This suggestion has been made and explored in earlier work on anharmonically coupled harmonic oscillators [52]. Consequently, the definition we adopt here is that a quantum system is ergodic if the energy eigenfunctions are maximally delocalized in any local basis set. The full implications of this definition may not be immediately obvious. There are many conceivable local basis sets to examine and the meaning of "maximal delocalization" is not readily apparent either. However, in some cases the application of the above definition is straightforward. Such is the case with the Hückel model of conjugated polyenes, because the relevant local basis set is obviously the set of carbon $2p_z$ AO-s which are used to construct the Hückel MO-s, as illustrated for benzene in Figure 22.

In the absence of the coupling the localized basis functions form degenerate sets of zeroth order energy eigenfunctions. The coupling transforms them into π-MO-s with a band of energies. The MO-s may be maximally delocalized or not, depending on the underlying molecular geometry. The larger the coupling and the greater the degree of delocalization, the broader the energy band becomes, with energy levels which are more evenly spaced. As the occupancy of the MO-s is in Aufbau manner, in order of increasing energy, the broader the band of energies, the greater the energetic stabilization. Thus, the ergodic character of the π-MO-s, being indicative of the degree of delocalization and rate of electron transfer, leads to maximal binding.

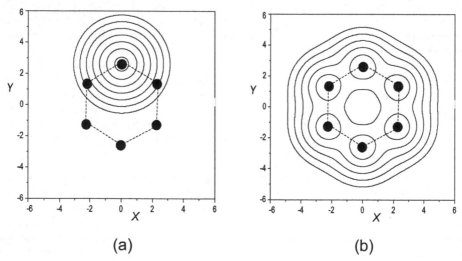

(a)   (b)

Fig. 22. Benzene: (a) Contour maps of $2p_z$ AO, and (b) lowest energy Hückel π-MO displaying delocalization and maximal spread in terms of the localized basis. (Plane of contour map: $0.32\ a_0$ above molecular plane. Contour levels are 0.002×2^n, $n = 0,1,2,...$)

A remarkable feature of the Hückel model is that the actual nature of the eigenstates is unaffected by the strength of the coupling. As indicated by equation (28), the relative contributions of the AO-s to a given MO is determined entirely by the topology of the molecule. The presence of overlap coupling between the AO-s only affects the normalization of the MO. Thus the slightest coupling leads to the formation of delocalized MO-s, as dictated by the topology of the molecule, which reflects a degree of ergodicity that is a geometrical feature, being independent of the coupling matrix element β and/or overlap γ. Only the rate of electron transfer among the atoms reflects the magnitude of the coupling. (Indeed, in the absence of any coupling, a polyene with n C-atoms would be n-fold degenerate which would be unaffected by *any* unitary transformation of the AO basis, including the one that diagonalizes the topology matrix **M** of the molecule.)

It should be clear then that from a fundamental quantum mechanical point of view covalent bond formation can be interpreted as the relaxation of dynamical constraints which make the non-bonded atoms or molecular fragments nonergodic. In H_2^+ for example, the important constraint is the spatial confinement of the electron to the vicinity of one of the protons when the internuclear distance is large. As the distance decreases towards the equilibrium value the localization constraint is relaxed and the electron is able to more rapidly delocalize over the two protons. The same mechanism operates in the formation of the 2-electron σ-bond in H_2 or the analogous π-bond in ethylene, but the process now involves two electrons of opposite spin. In the case of butadiene there is an additional stabilizing delocalization as the π-electrons are able to move along the whole chain of four carbon atoms. Similar additional delocalization is present in longer polyene chains and in cyclic molecules such as benzene. The covalent bonding mechanism in all these molecules is consistent with a more ergodic form of electron dynamics in the molecules than in the corresponding H atoms or in a hypothetical ethylene system without a π-bond ($\beta = \gamma = 0$).

It is instructive to examine the character of the energy eigenstates of a polyene chain with reference to the ergodic hypothesis. The conventional valence-bond picture of a conjugated polyene is in terms of alternating single and double bonds between the carbon atoms, although the importance of delocalization is recognised by the introduction of *resonance* involving several canonical structures. In the Hückel MO model each carbon atom "donates" a $2p_z$ AO towards the formation of delocalized π-MO-s, which are then occupied by the π-electrons of the molecule, one from each atom. Thus, π-bonding is delocalized right from the start, although it is customary to describe the individual CC bonds in terms of bond orders, which are readily obtained from the MO coefficients and which for a polyene typically range from ~0.5 to ~0.8. As noted already, correction for non-zero overlap will renormalize the MO-s but will not affect the relative magnitudes of the coefficients. We claim that the Hückel eigenstates exhibit maximal delocalization in the $2p_z$ AO basis. The support for this claim comes from the fact that the Hückel MO-s are formally equivalent to the particle-in-a-box eigenfunctions expanded in terms of roof functions, as discussed in the previous section. As this analogy has general applicability, we conclude that Hückel models of polyene chains exhibit full ergodicity and free translation of all electrons within the bounds established by the molecular chain.

In the case of cyclic polyenes, exemplified by benzene, the rotational symmetry and the cyclic boundary conditions impose restrictions which, as shown in Figure 17, result in two pairs of doubly degenerate (highest occupied and lowest unoccupied) MO-s. Such a pattern is typical of cyclic polyenes: Systems with an even number of carbon atoms have just two non-degenerate π-MO-s and only one, if the number of carbons is odd [42]. Application of the von Neumann definition to the latter would imply that the electron dynamics in cyclic polyenes is not fully ergodic. The degeneracies arise, however, as a result of the rotational character of electronic motion in systems with only cyclic boundary conditions and they correspond to the classical nonergodicity associated with the absence of reflection. Thus, the direction of motion, forward or backward, is an additional constant of motion. The motion is nonergodic for the trivial reason that direction cannot be reversed, but the rotation is still "free" in the sense that there are no barriers of any kind.

The analogous free electron MO model of cyclic systems is "the particle on a ring" whose energy eigenvalues and eigenfuctions are [42]

$$E_n = \frac{n^2 h^2}{8\pi I},$$ (67)

$$\psi_n = \frac{1}{\sqrt{2\pi}} \exp(in\theta)$$ (68)

where n = 0, ±1, ±2,..., I is the moment of inertia of the particle and θ is the angular coordinate. With the exception of the ground state (n = 0) all states are doubly degenerate, forming forward and backward rotating pairs of eigenstates at each energy.

8. Chemical bonding and delocalization

Recognizing the central role of delocalization in the quantum mechanical description of molecular electron structure, chemical bonding can be given a unified and physically correct description. We recognise that there are two separate sources of stabilization:

i.   Transfer of electrons due to differences in electronegativity, and
ii.   Delocalization of electronic motion.

In both cases the process of bonding brings the system towards ergodic and rapid electron dynamics. Due to the inherent asymmetry between bonded atoms in the former case, the ergodic dynamics in such systems is largely local, resulting in the type of interaction which, in the limiting case of complete charge transfer, we describe as ionic bonding. The covalent bonding mechanism, illustrated so clearly by the application of Hückel theory to polyenes, which implicitly accounts for resonance stabilization that is associated with conjugation and aromaticity, is a direct consequence of the relaxation of constraints which releases the bonding electrons to move freely among the bonded atoms of the molecule.

It is widely appreciated that ionic and covalent bonding represent two limiting bonding scenarios, with a continuous spectrum in-between, including bonds described as ionic bonds with covalent character, polar covalent bonds or covalent bonds with ionic character. Molecular orbital theory naturally covers all possible cases. This is most easily demonstrated by using perturbation theory to deduce the eigenvectors and eigenvalues (correct to first and second order respectively) of the following model Hamiltonian for a hypothetical heteronuclear diatomic molecule with two valence electrons, in a basis of two valence orbitals ϕ_a, ϕ_b, which are localized on the two atoms, a and b:

$$\mathbf{H} = \begin{pmatrix} \alpha & \beta \\ \beta & \alpha + \Delta\alpha \end{pmatrix}.$$ (69)

In this equation $\Delta\alpha$ (>0) is nominally the difference in ionization energies of the two atoms, i.e. an energetic measure of their electronegativity difference. Assuming that ϕ_a and ϕ_b are orthogonal, the MO-s (in non-normalized form) and their energies are readily shown to be

$$\psi_a = \phi_a - \frac{\beta}{\Delta\alpha}\phi_b, \quad \psi_b = \phi_b + \frac{\beta}{\Delta\alpha}\phi_a,$$ (70)

$$\varepsilon_a = \alpha - \frac{\beta^2}{\Delta\alpha}, \qquad \varepsilon_b = (\alpha + \Delta\alpha) + \frac{\beta^2}{\Delta\alpha}, \tag{71}$$

provided $\beta/\Delta\alpha \ll 1$. In the limiting case of very large electronegativity difference and sufficiently small coupling, i.e. $\beta/\Delta\alpha \approx 0$, the model predicts ionic bonding, whereby ψ_a, the lower energy MO, accommodating both electrons, is entirely localized on atom a. The energetic stabilization of the ion pair is of course due to the Coulomb attraction of the oppositely charged ions. For small but finite values of $\beta/\Delta\alpha$, ψ_a contains some contribution from ϕ_b, i.e. there is some delocalization, signalling a degree of covalent character to the predominantly ionic bond. These effects increase with decreasing $\Delta\alpha$, corresponding to increasing covalency (through the regime when low order perturbation theory is no longer applicable). In the limit of $\Delta\alpha = 0$ we have a homonuclear diatomic with fully delocalized MO-s

$$\psi_\pm = \phi_a \pm \phi_b \tag{72}$$

and a fully covalent bond with the two electrons equally shared by the two atoms.

In extended systems, ranging from just triatomics such as NO_2 to polyenes like benzene to crystalline solids such as graphite there is an increased degree of delocalization that can extend over the whole molecule, culminating in macroscopic delocalization that takes place in metallic systems such as solid sodium or graphite. Thus, metallic bonding is just a natural extension of the very basic covalent bonding mechanism that occurs in H_2^+! The success and reliability of computational models based on MO theory to systems ranging from diatomics to proteins and to solids (in the form of the tight binding method briefly discussed above) provides convincing practical support for our belief that the MO model contains all the essential physics needed to correctly account for the wide range of bonding interactions of atoms and molecules that account for the chemistry of our world.

Depending on the system of interest and the accuracy we require of a specific calculation, the basic MO model may need to be extended so that the computed molecular wave function better approximates the exact wave function. This requires the use of large orbital basis sets as well the explicit inclusion of dynamical electron correlation, predominantly between electrons of opposite spins, which, by definition, is neglected in the Hartree-Fock formalism. For example, the quantitative prediction of a covalent bond dissociation energy requires the inclusion of correlation in the quantum chemical model since correlation contributions to bond dissociation energies are typically 25 - 50%. Another well-known deficiency of the (single configuration) Hartree-Fock method is its inability to correctly describe dissociation processes which involve the breaking of one or more two-electron covalent bonds, as occurs e.g. in the dissociation of H_2 (from its ground state), which requires a multi-configuration treatment, i.e. the inclusion of non-dynamical (static) electron correlation. Irrespective, however, of the nature of the correlated wave function, whether it is multi-configuration SCF, configuration interaction or coupled cluster, it implicitly includes electron delocalization. This is obvious when a given correlated wave function is constructed in terms of delocalized MO-s, such as canonical SCF MO-s, or even localized SCF MO-s, e.g. two- or three-centre bond orbitals and lone pair orbitals, which, relative to the atomic orbitals, are of course considerably delocalized. Delocalization is less obvious when the total wave function is constructed directly from atomic orbitals, as in Valence

Bond Theory. For example, the Heitler-London wave function of H_2 is built in terms of atomic configurations $1s_a(1)1s_b(2)$ and $1s_b(1)1s_a(2)$. While these correspond to localized electrons (on atom a or b), their (in-phase) combination, representing the ground state of H_2, is clearly a delocalized two-electron wave function.

9. Implications for Thomas-Fermi and modern Density Functional theories

We have seen above the direct relation between covalent bonding and valence electron dynamics which in turn is related to canonical orbital delocalization and energy splitting. The original Thomas-Fermi theory effectively assumes ergodicity of electronic motion and maximal transfer rates which in molecular orbital representation would mean maximally delocalized and energy separated canonical orbitals. Thus the atoms are according to the TF theory already relaxed into inert gas form and molecules are unstable since the effects on stability due to dynamical constraints posed by atomic shell structure and potential barrier hindrance of interatomic electron transfer cannot be resolved to favor molecule formation. Here we have emphasized the value of the understanding this failure provides concerning the mechanism of covalent bonding. The obvious follow up question is whether it is – given this understanding – possible to correct Thomas-Fermi theory and obtain a more useful but still "orbital free" form of TF-theory.

It is clearly possible to remove the assumed ergodicity with respect to spin-flipping as we have done here in our analysis of the hydrogen atom and the hydrogen molecular ion. It is also possible to recognize that quantization must be done separately and individually for the atoms of an infinitely expanded molecule. Thus we could immediately see that the energy of H_2^+ would be lower at equilibrium separation than at infinite bond length where the energy should – with local quantization – be that of a hydrogen atom. In fact, we have in earlier work [4-7] explored the possibility of comparing traditional ergodic TF energies with nonergodic locally quantized TF energies to estimate the reactivity of atoms and the maximal stabilization possible by assuming that there were no dynamical constraints on the electronic motion for a molecule at its equilibrium geometry. This work was done using simplified exponential electron densities and empirical von Weiszäcker correction to allow atomic TF energies to agree with experiment or accurate quantum chemical results. The results have clearly shown that Thomas-Fermi theory, with or without spin and space localization imposed on the electronic motion and therefore also on the quantization, is capable of providing a measure of atomic reactivity and bonding efficiency in the formation of molecules. The picture one finds is in good agreement with chemical intuition. First row atoms are more reactive than those in the second row. Oxygen is the most reactive atom and – due to Pauli and internuclear repulsion and remaining constraints on electronic motion – only a small fraction (typically a quarter) of the stabilization energy available in principle is actually realized in the molecule at equilibrium.

An "orbital free" true density functional theory that can accurately predict reactivity and covalent bonding at all stages of molecule formation has still not been found. This is in itself a signal that covalent bonding is not – despite the Hohenberg-Kohn theorem [53] – readily described in terms of electron density. Wave functions, here in terms of one-electron orbitals, provide the best route to covalent bonding. The reason is, we hope, very clear from our analysis above: the canonical orbitals of Hartree-Fock, Hückel or modern density functional theory capture the character of the electron dynamics in the most direct manner and thereby also reactivity and bonding. It has been noticed in connection with DFT applications to solids [54] that Thomas-Fermi theory can be corrected for the "nonbonding flaw" by a zeroth order orbital calculation where the one-electron Hamiltonian is generated from the Thomas-Fermi electron density and no iterations to consistency are performed. We have independently investigated such orbital corrections and verified [7] that the Thomas-Fermi results for electron densities of molecules can similarly be used to generate approximate one-electron Hamiltonians of the Fock or Kohn-Sham type which then yield energies of a quality nearly as good as that obtained by converging the corresponding Hartree-Fock or DFT calculations. This again illustrates the power of the molecular orbital analysis and the fact that it so effectively comprises and corrects the flaw of the Thomas-Fermi theory.

A final reflection on the evolution of modern DFT suggests that it is – perhaps surprisingly – more beholden to the Kohn-Sham method of estimating kinetic energy by an introduction of orbitals than it is to the Hohenberg-Kohn theorem which gave the density functional theory of electronic structure its legitimacy. Without a solution to the "nonbonding flaw" DFT would have remained of marginal interest in the great computational advance on molecular and solid state electronic structures over recent decades. The Kohn-Sham scheme inserted the molecular orbital analysis which – precisely as in the Hückel theory above or in the orbital correction of Thomas-Fermi theory – effectively resolved the electron dynamics and related mechanisms of reactivity and bonding. This suggests that improved functionals of either higher accuracy or computational efficiency must retain the ability to resolve the electron dynamics – if not by orbitals then by some other method no less able to describe the intimate coupling between electronic motion and stability.

10. Conclusions

The contrast between the total failure of the Thomas-Fermi theory and the success of the Hückel molecular orbital model in describing covalent bonding has shown very clearly that the mechanism of bonding is fundamentally of dynamical origin. Interatomic transfer of valence electrons, present to a degree in the molecule but absent between the separated atoms, relaxes the energy penalty of the constrained quantization in a dissociated molecule. The molecular orbitals obtained either by the application of Hückel theory, or by ab initio Hartree-Fock methods, capture the delocalization and orbital energy splitting which represent the spatial degree as well as the rate of valence electron transfer which underlies covalent stabilization. Thomas-Fermi theory employs an ergodic quantization scheme as if the delocalization and the rate of electron transfer were always maximal irrespective of bond lengths and potential and other barriers restraining such a flow. As soon as it is recognized that quantization must be done independently for separated atoms, Thomas-Fermi theory becomes capable of addressing reactivity and bonding. The non-bonding flaw is then identified, but without the use of molecular orbitals the quantitative prediction of covalent stabilization as a function of bond length is still too demanding for Thomas-Fermi theory. The obvious remedy to the problem was proposed by Kohn and Sham [55] in their famous prescription for the evaluation of kinetic energy. Our view is that the reintroduction of molecular orbitals, as in the Hückel model, provides the key to the quantitative representation of the valence electron dynamics and its effect on covalent stability.

The dynamical mechanism of covalent bonding as described above allows us to understand why earlier explanations of covalent bonding as due either to kinetic energy lowering or electrostatic potential stabilization through electron density redistribution and build up of density around bond midpoints have led to persistent disputes. Both these mechanisms play a vital but subsidiary role to the dynamical relaxation which is the fundamental mechanism of covalent stabilization. Thus there are molecules with particular electronic structures and/or wave functions obtained by specific quantum chemical methods for which the description of bonding may agree or disagree with either of these earlier explanations of covalent bonding. This is because the fundamental valence electron transfer between atomic centres in the molecule may result in either potential or kinetic energy lowering which in themselves represent stabilization processes, but which are nevertheless less general than the underlying dynamical mechanism. Only the most general description of the mechanism will give us the full benefit of understanding the bonding seen in all types of molecules and computational methods.

Our analysis above has focused on the qualitative understanding of covalent bonding but we are convinced that a gain in such understanding will also bring advantages in the quantitative prediction of bonding. There are good reasons to believe that such benefits will first be found in the development of new density functional methods with improved performance with respect to computational efficiency and accuracy. Our work so far indeed contains a number of suggestions for the improvement of DFT and we look forward to much further progress in the future.

11. Acknowledgements

We thank our former graduate students and colleagues William Eek and Andreas Bäck for their many contributions to our earlier work some of which has specifically been drawn upon in the analysis above. The support of the Swedish Science Research Council over a long period, while this work was carried out, is also gratefully acknowledged.

12. References

[1] Nordholm, S. J. Chem. Phys. 1987, 86, 363.
[2] Nordholm, S. J. Chem. Ed. 1988, 65, 581.
[3] Bacskay, G. B.; Reimers, J. R.; Nordholm, S. J. Chem. Ed. 1997, 74, 1494.
[4] Eek, W.; Nordholm, S. Theor. Chem. Acc. 2006, 115, 266.
[5] Nordholm, S.; Bäck, A.; Bacskay, G.B. J. Chem. Ed. 2007, 84, 1201 and Supplementary Information.
[6] Nordholm, S.; Eek, W. Int. J. Quant. Chem. 2011, 111, 2072.
[7] Eek, W. Understanding Atoms and Covalent Bonds , PhD thesis, The University of Gothenburg, 2008.
[8] Lewis, G. N. J. Am. Chem. Soc. 1916, 38, 762.
[9] Hellmann, H. Z. Phys. 1933, 85, 180.
[10] Coulson, C. A. Valence, 2nd ed.; Oxford University Press: London, 1961.
[11] Burdett, J. K. Chemical Bonds - A Dialog; Wiley: Chichester, 1997; Chapter 1.
[12] Bader, R. F. W.; Hernandez-Trujillo, J.; Cortés-Guzman, F. J. Comput. Chem. 2007, 28, 4.
[13] Ruedenberg, K. Rev. Mod. Phys. 1962, 34, 326.

[14] Kutzelnigg, W. In *Theoretical Models of Chemical Bonding*; Maksič, Z. B., Ed.; Springer: Berlin, 1990; p.1.

[15] Bitter, T; Ruedenberg, K.; Schwarz, W. H. E. J. Comput. Chem. 2007, *28*, 411.

[16] Kutzelnigg, W. Angew. Chem. Internat. Ed. 2007, *28*, 25.

[17] Esterhuysen, C.; Frenking, G. Theor. Chem. Acc. 2004, *111*, 381.

[18] Thomas, L. H. *Proc. Cambridge Philos. Soc.* 1926, *23*, 542.

[19] Fermi, E. *Rend. Accad. Lincei* 1927, *6*, 602.

[20] Hückel, E. Z. *Phys.* 1930, *60*, 423; *ibid.* 1931, *70*, 204; *ibid.* 1931, *72*, 310; *ibid.* 1932, *76*, 628.

[21] Teller, E. *Rev. Mod. Phys.* 1962, *34*, 627.

[22] Balazs, N. L. Phys. Rev. 1967, *156*, 42.

[23] Kutzelnigg, W. *J. Comput. Chem.* 2007, *28*, 25.

[24] Parr, R. G.; Yang, W. *Density Functional Theory of Atoms and Molecules* , Oxford University Press, New York, 1994.

[25] von Weizsäcker, C. F. Z. *Phys.* 1935, *96*, 431.

[26] Ruedenberg, K. *J. Chem. Phys.* 1961, *34*, 1861.

[27] Longuet-Higgins, H. C.; Salem, L. *Proc. Roy. Soc. (London) A*, 1959, *251*, 172.

[28] Pople, J. A.; Walmsley, S. H. *Mol. Phys.* 1962, *5*, 15.

[29] Su, W. P.; Schriffer, J. R.; Heeger, A. J. *Phys. Rev. Lett.* 1979, *42*, 1698.

[30] Klimkāns, A.; Larsson, S. *Chem. Phys.* 1994, *189*, 25.

[31] Larsson, S.; Rodriguez-Monge, L. *Int. J. Quant. Chem.* 1998, *67*, 107.

[32] Blomgren, F.; Larsson, S. *Theor. Chem. Acc.* 2003, *110*, 165.

[33] Coulson, C. A.; O'Leary, B.; Mallion, R. B. *Hückel Theory for Organic Chemists*; Academic Press: London, 1978.

[34] Wheland, G. W. *J. Am. Chem. Soc.* 1942, *64*, 900.

[35] Parr, R. G. *J. Chem. Phys.* 1960, *33*, 1184.

[36] Löwdin, P.-O. *J. Chem. Phys.* 1950, *18*, 365.

[37] Weber, W.; Thiel, W. *Theor. Chem. Acc.* 2000, *103*, 495.

[38] Feinberg, M. J.; Ruedenberg, K.; Mehler, E. L. In *Advances in Quantum Chemistry*, Vol. 5; Löwdin, P. O., Ed.; Academic Press: New York, 1970; p. 28.

[39] Feinberg, M. J.; Ruedenberg, K. *J. Chem. Phys.* 1971, *54*, 1495.

[40] Feinberg, M. J.; Ruedenberg, K. *J. Chem. Phys.* 1971, *55*, 5804.

[41] Ruedenberg, K. in *Localization and Delocalization in Quantum Chemistry*; Chalvet, O.; Daudel, R.; Diner, S.; Malrieu, J. P. Eds.; Reidel: Dordrecht, 1975; p. 223.

[42] Pilar, F. L. *Elementary Quantum Chemistry*; McGraw-Hill: New York, 1968.

[43] Pariser, R.; Parr, R. G. *J. Chem. Phys.* 1953, *21*, 466; *ibid.* 1953, *21*, 767.

[44] Pople, J. A.; *Trans. Farad. Soc.* 1953, *49*, 1375;

[45] Pople, J. A. *J. Phys. Chem.* 1957, *61*, 6.

[46] Hoffmann, R. *J. Chem. Phys.* 1963, *39*, 1397.

[47] Wolfsberg, M.; Helmholz, L. *J. Chem. Phys.* 1959, *20*, 837.

[48] Goringe, C. M.; Bowler, D. R.; Hernández, E. *Rep. Prog. Phys.* 1997, *60*, 1447.

[49] Slater, J. C.; Koster, G. F. *Phys. Rev.* 1954, *94*, 1498.

[50] Bayliss, N. S. *Quart. Rev.* 1952, *6*, 319.

[51] Jancel, R. *Foundations of Classical and Quantum Statistical Mechanics*; Pergamon Press: London, 1969.

[52] Nordholm, S.; Rice, S. A. *J. Chem. Phys.* 1974, *61*, 203.

[53] Hohenberg, P.; Kohn, W. *Phys. Rev.* 1964, *136*, B864.

[54] Zhou, B.; Wang, Y. A. J. Chem. Phys. 2006, *124*, 081107.

[55] Kohn, W.; Sham, L. J. Phys. Rev. 1965, *140*, A1133.

Part 3

Quantum Gravity

7

Quantum Fields on the de Sitter Expanding Universe

Ion I. Cotăescu
West University of Timişoara
Romania

1. Introduction

The quantum theory of fields on curved space-times, which is of actual interest in astrophysics and cosmology, is faced with serious difficulties arising from the fact that some global properties of the theory on flat space-times become local when the background is curved. The principal impediment on curved manifolds is the absence of the Poincaré symmetry which gives rise to the principal invariants of special relativity (i.e. the mass and spin) and assures the stability of the vacuum state. This drawback encouraged many authors to avoid the principal steps of the quantum theory based on canonical quantization looking for new effective methods able to be used in theories on curved manifolds. Thus a particular attention was payed to the construction of the two-point functions of the axiomatic quantum field theory (Allen & Jacobson, 1986) or to some specific local effects as, for example, the Unruh's one (Unruh, 1976). However, in this manner some delicate problems related to the fields with spin may remain obscure.

For this reason we believe that turning back to the traditional method of canonical quantization we may open new perspectives in developing the quantum field theory on curved space-times. According to the standard interpretation of the quantum mechanics, our basic assumption is that the quantum states are *prepared* by a *global* classical apparatus which includes the natural and local frames. This means that the quantum observables must be defined globally as being *conserved* operators which commute with that of the field equation. Therefore, the quantum modes have to be determined as common eigenfunctions of several complete systems of commuting operators which include the operator of the field equation. These mode functions must be orthonormalized with respect to a suitable relativistic scalar product such that the subspaces of functions of positive and respectively negative frequencies remain orthogonal to each other in any frame. This conjecture leads to a stable vacuum state when we perform the canonical quantization which enables us to derive the propagators and especially the one-particle operators. Moreover, the theory of the interacting fields may be developed then as in the flat case using the S-matrix and the perturbation theory since the normal ordering of the operator products do make sense thanks to the stability of the vacuum state.

The theory of quantum fields with spin on curved backgrounds has a specific structure since the spin half can be defined only in orthogonal local (non-holonomic) frames. Therefore, in general, the Lagrangian theory of the matter fields has to be written in local frames assuming that this is tetrad-gauge covariant. Thus the gauge group $L_+^\uparrow \subset SO(1,3)$ (of the principal fiber bundle) and its universal covering group, $SL(2,\mathbb{C})$, (of the spin fiber bundle) become crucial since their finite-dimensional (non-unitary) representations *induce* the *covariant* representations of the universal covering group of the isometry one according to which the matter fields transform under isometries (Cotăescu, 2000). The generators of these covariant representations are the differential operators given by the Killing vectors associated to isometries according to the generalized Carter and McLenagan formula (Carter & McLenaghan, 1979). The theory of the fields with integer spin can be written either in local frames or exclusively in natural ones where we have shown how the spin must be defined in order to recover the Carter and McLenagan formula (Cotăescu, 2009). Thus the external symmetry offers us the conserved operators among them we may select different sets of commuting operators which have to define the free quantum modes. A convenient relativistic scalar product and a stable vacuum state have to complete the framework we need for performing the canonical quantization.

Our method is helpful on the de Sitter space-time where all the free field equations can be analytically solved while the $SO(1,4)$ isometries provide us with a large collection of conserved operators. A particular feature of this symmetry is that the energy and momentum operators do not commute to each other. Consequently, there are no mass-shells and the energy and momentum are diagonal in different bases, called the energy and respectively momentum bases (or representations). In spite of this new behavior, we pointed out that the principal invariants of the covariant representations can be expressed in terms of rest energy and spin as in special relativity (Cotăescu, 2011a). Moreover, we have derived the principal sets of quantum modes of the free scalar (Cotăescu et al., 2008), Dirac (Cotăescu, 2002; Cotăescu & Crucean, 2008), Proca (Cotăescu, 2010) and Maxwell (Cotăescu & Crucean, 2010) fields applying the same procedure of canonical quantization. In what follows we would like to present these results for the scalar and Dirac fields which are the typical examples of the field theory formulated exclusively either in natural or in local frames.

In the second section we briefly review our theory of external symmetry focusing on the de Sitter isometries and their invariants. The third section is devoted to the de Sitter space-time and its isometries. The next two sections are devoted to the second quantization of the scalar and Dirac fields. For these fields we derive the quantum modes in momentum representation using the free field equations in the (co)moving frames of the de Sitter expanding universe and the local frames defined by the diagonal gauge in the case of the Dirac equation. The polarization of the Dirac field is given in the helicity basis which can be easily defined in the momentum representation we use. Considering appropriate scalar products in each particular case, a special attention is paid to the orthogonality and completeness properties as well as to the choice of the vacuum state. We argue that the vacuum of Bunch-Davies type (Bunch & Davies, 1978) we define here is stable as long as the particle and antiparticle sets of mode functions remain orthogonal among themselves in any frame. Under such circumstances the method of canonical quantization is working well allowing us to obtain important pieces as propagators and one-particle operators. Finally, we present our conclusions marking out what is new in interpreting the global quantum modes studied here.

2. External symmetry in general relativity

In general relativity the space-time symmetries are the isometries of the background associated to the Killing vectors. The physical fields minimally coupled to gravity take over this symmetry transforming according to appropriate representations of the isometry group.

In the case of the scalar vector or tensor fields these representations are completely defined by the well-known rules of the general coordinate transformations since the isometries are in fact particular automorphisms. However, the theory of spinor fields is formulated in orthogonal local frames where the basis-generators of the spinor representation were discovered by Carter and McLenaghan (Carter & McLenaghan, 1979).

For this reason we proposed a new theory of external symmetry in the context of the gauge-covariant theories in local orthogonal frames (Cotăescu, 2000). We introduced there new transformations which combine isometries and gauge transformations such that the tetrad fields should remain invariant. In this way we obtained the external symmetry group and we derived the general form of the basis-generators of the representations of this group shoving that these are given by a formula which generalizes the Carter and McLenaghan one. Moreover, we pointed out that this theory can be formulated exclusively in natural frames if there are only scalar, vector and tensor fields (Cotăescu, 2009).

2.1 Tetrad gauge covariance

Let us consider the pseudo-Riemannian space-time (M, g) and a local chart (or natural frame) of coordinates x^μ (labeled by natural indices, $\mu, \nu, \ldots = 0, 1, 2, 3$) (Wald, 1984). Given a gauge, we denote by $e_{\hat\mu}$ the tetrad fields that define the local frames and by $\hat{e}^{\hat\mu}$ those of the corresponding coframes. These have the usual duality, $\hat{e}^{\hat\mu}_\alpha e^\alpha_{\hat\nu} = \delta^{\hat\mu}_{\hat\nu}$, $\hat{e}^{\hat\mu}_\alpha e^\beta_{\hat\mu} = \delta^\beta_\alpha$, and orthonormalization, $e_{\hat\mu} \cdot e_{\hat\nu} = \eta_{\hat\mu\hat\nu}$, $\hat{e}^{\hat\mu} \cdot \hat{e}^{\hat\nu} = \eta^{\hat\mu\hat\nu}$, properties. The metric tensor $g_{\mu\nu} = \eta_{\hat\alpha\hat\beta} \hat{e}^{\hat\alpha}_\mu \hat{e}^{\hat\beta}_\nu$ raises or lowers the natural indices while for the local indices ($\hat\mu, \hat\nu, \ldots = 0, 1, 2, 3$) we have to use the flat Minkowski metric $\eta = \text{diag}(1, -1, -1, -1)$. The derivatives in local frames are the vector fields $\hat\partial_{\hat\nu} = e^\mu_{\hat\nu} \partial_\mu$ which satisfy the commutation rules $[\hat\partial_{\hat\mu}, \hat\partial_{\hat\nu}] = C^{\cdot\cdot\hat\sigma}_{\hat\mu\hat\nu\cdot} \hat\partial_{\hat\sigma}$ defining the Cartan coefficients.

The metric η remains invariant under the transformations of its gauge group, $O(1,3)$. This has as subgroup the Lorentz group, $L^\uparrow_+$, of the transformations $\Lambda[A(\omega)]$ corresponding to the transformations $A(\omega) \in SL(2, \mathbb{C})$ through the canonical homomorphism (Tung, 1984). In the standard covariant parametrization, with the real parameters $\omega^{\hat\alpha\hat\beta} = -\omega^{\hat\beta\hat\alpha}$, the $SL(2, \mathbb{C})$ transformations $A(\omega) = \exp(-\frac{i}{2} \omega^{\hat\alpha\hat\beta} S_{\hat\alpha\hat\beta})$ are generated by the covariant basis-generators of the $sl(2, \mathbb{C})$ algebra, denoted by $S_{\hat\alpha\hat\beta}$. For small values of $\omega^{\hat\alpha\hat\beta}$ the matrix elements of the transformations Λ in the local basis can be expanded as $\Lambda^{\hat\mu}_{\cdot\hat\nu}[A(\omega)] = \delta^{\hat\mu}_{\hat\nu} + \omega^{\hat\mu}_{\cdot\hat\nu} + \cdots$.

Assuming now that (M, g) is orientable and time-orientable we can consider $G(\eta) = L^\uparrow_+$ as the gauge group of the Minkowski metric η (Wald, 1984). This is the structure group of the principal fiber bundle whose basis is M. The group $\text{Spin}(\eta) = SL(2, \mathbb{C})$ is the universal covering group of $G(\eta)$ and represents the structure group of the spin fiber bundle (Lawson & Michaelson, 1989). In general, a matter field $\psi_{(\rho)} : M \to \mathcal{V}_{(\rho)}$ is locally defined over M with values in the vector space $\mathcal{V}_{(\rho)}$ of a representation ρ, generally reducible, of the group $\text{Spin}(\eta)$. The covariant derivatives of the field $\psi_{(\rho)}$,

$$D^{(\rho)}_{\hat\alpha} = e^\mu_{\hat\alpha} D^{(\rho)}_\mu = \hat\partial_{\hat\alpha} + \frac{i}{2} \rho(S^{\hat\beta}_{\cdot\hat\gamma}) \hat\Gamma^{\hat\gamma}_{\hat\alpha\hat\beta}, \tag{1}$$

depend on the connection coefficients in local frames,

$$\hat{\Gamma}^{\hat{\sigma}}_{\hat{\mu}\hat{\nu}} = e^{\alpha}_{\hat{\mu}} e^{\beta}_{\hat{\nu}} (\hat{e}^{\hat{\sigma}}_{\gamma} \Gamma^{\gamma}_{\alpha\beta} - \hat{e}^{\hat{\sigma}}_{\beta,\alpha})$$

$$= \frac{1}{2} \eta^{\hat{\sigma}\hat{\lambda}} (C_{\hat{\mu}\hat{\nu}\hat{\lambda}} - C_{\hat{\mu}\hat{\lambda}\hat{\nu}} - C_{\hat{\nu}\hat{\lambda}\hat{\mu}}), \tag{2}$$

which assure the covariance of the whole theory under tetrad gauge transformations produced by automorphisms A of the spin fiber bundle. This is the general framework of the theories involving fields with half integer spin which can not be treated in natural frames.

2.2 External symmetries in local frames

A special difficulty in local frames is that the theory is no longer covariant under isometries since these can change the tetrad fields that carry natural indices. For this reason we proposed a theory of external symmetry in which each isometry transformation is coupled to a gauge one able to correct the position of the local frames such that the whole transformation should preserve not only the metric but the tetrad gauge too (Cotăescu, 2000). Thus, for any isometry transformation $x \to x' = \phi_{\xi}(x) = x + \xi^a k_a + ...$, depending on the parameters ξ^a $(a, b, ... = 1, 2...N)$ of the isometry group $I(M)$, one must perform the gauge transformation A_{ξ} defined as

$$\Lambda^{\hat{\alpha}}_{\cdot\hat{\beta}}[A_{\xi}(x)] = \hat{e}^{\hat{\alpha}}_{\mu}[\phi_{\xi}(x)] \frac{\partial \phi^{\mu}_{\xi}(x)}{\partial x^{\nu}} e^{\nu}_{\hat{\beta}}(x) \tag{3}$$

with the supplementary condition $A_{\xi=0}(x) = 1 \in SL(2, \mathbb{C})$. Then the transformation laws of our fields are

$$(A_{\xi}, \phi_{\xi}): \quad \begin{aligned} e(x) &\to & e'(x') = e[\phi_{\xi}(x)] \\ \hat{e}(x) &\to & \hat{e}'(x') = \hat{e}[\phi_{\xi}(x)] \\ \psi_{(\rho)}(x) &\to & \psi'_{(\rho)}(x') = \rho[A_{\xi}(x)]\psi_{(\rho)}(x). \end{aligned} \tag{4}$$

We have shown that the pairs (A_{ξ}, ϕ_{ξ}) constitute a well-defined Lie group we called the external symmetry group, $S(M)$, pointing out that this is just the universal covering group of $I(M)$ (Cotăescu, 2000). For small values of ξ^a, the $SL(2, \mathbb{C})$ parameters of $A_{\xi}(x) \equiv A[\omega_{\xi}(x)]$ can be expanded as $\omega^{\hat{\alpha}\hat{\beta}}_{\xi}(x) = \xi^a \Omega^{\hat{\alpha}\hat{\beta}}_a(x) + \cdots$, in terms of the functions

$$\Omega^{\hat{\alpha}\hat{\beta}}_a \equiv \frac{\partial \omega^{\hat{\alpha}\hat{\beta}}_{\xi}}{\partial \xi^a}\Big|_{\xi=0} = \left(\hat{e}^{\hat{\alpha}}_{\mu} k^{\mu}_{a,\nu} + \hat{e}^{\hat{\alpha}}_{\nu,\mu} k^{\mu}_a \right) e^{\nu}_{\hat{\lambda}} \eta^{\hat{\lambda}\hat{\beta}} \tag{5}$$

which depend on the Killing vectors $k_a = \partial_{\xi_a} \phi_{\xi}|_{\xi=0}$ associated to ξ^a.

The last of Eqs. (4) defines the operator-valued representations $T^{(\rho)} : (A_{\xi}, \phi_{\xi}) \to T^{(\rho)}_{\xi}$ of the group $S(M)$ which are called the *covariant* representations (CR) *induced* by the finite-dimensional representations ρ of the group $SL(2, \mathbb{C})$. The covariant transformations,

$$(T^{(\rho)}_{\xi} \psi_{(\rho)})[\phi_{\xi}(x)] = \rho[A_{\xi}(x)]\psi_{(\rho)}(x), \tag{6}$$

leave the field equation invariant since their basis-generators (Cotăescu, 2000),

$$X_a^{(\rho)} = i\partial_{\xi^a} T_\xi^{(\rho)} \Big|_{\xi=0}$$

$$= -ik_a^\mu \partial_\mu + \frac{1}{2}\Omega_a^{\hat\alpha\hat\beta}\rho(S_{\hat\alpha\hat\beta}),$$

(7)

commute with the operator of the field equation and satisfy the commutation rules $[X_a^{(\rho)}, X_b^{(\rho)}] = ic_{abc}X_c^{(\rho)}$ determined by the structure constants, c_{abc}, of the algebras $s(M) \sim i(M)$. In other words, the operators (7) are the basis-generators of a CR of the $s(M)$ algebra *induced* by the representation ρ of the $sl(2,\mathbb{C})$ algebra. These generators can be put in the covariant form (Cotăescu, 2000),

$$X_a^{(\rho)} = -ik_a^\mu D_\mu^{(\rho)} + \frac{1}{2}k_{a\,\mu;\nu}\,e_{\hat\alpha}^\mu\,e_{\hat\beta}^\nu\rho(S^{\hat\alpha\hat\beta}),$$

(8)

which represents the generalization to any representation ρ of the famous formula given by Carter and McLenaghan (Carter & McLenaghan, 1979) for the spinor representation $\rho_s = (\frac{1}{2},0)\oplus(0,\frac{1}{2})$.

A specific feature of the CRs is that their generators have, in general, point-dependent spin terms which do not commute with the orbital parts. However, there are tetrad-gauges in which at least the generators of a subgroup $G \subset I(M)$ may have point-independent spin terms commuting with the orbital parts. Then we say that the restriction to G of the CR $T^{(\rho)}$ is *manifest* covariant. Obviously, if $G = I(M)$ then the whole representation $T^{(\rho)}$ is manifest covariant (Cotăescu, 2000).

2.3 Isometries in natural frames

Whenever there are no spinors, the matter fields are vectors or tensors of different ranks and the whole theory is independent on the tetrad fields dealing with the natural frames only. In general, any tensor field, Θ, transforms under isometries as $\Theta \to \Theta' = T_\xi\Theta$, according to a tensor representation of the group $S(M)$ defined by the well-known rule in natural frames

$$\left[\frac{\partial\phi_\xi^\alpha(x)}{\partial x_\mu}\frac{\partial\phi_\xi^\beta(x)}{\partial x_\nu}\cdots\right](T_\xi\Theta)_{\alpha\beta\ldots}[\phi(x)] = \Theta_{\mu\nu\ldots}(x).$$

(9)

Hereby one derives the basis-generators of the tensor representation, $X_a = i\partial_{\xi^a}T_\xi|_{\xi=0}$, whose action reads

$$(X_a\Theta)_{\alpha\beta\ldots} = -i(k_a{}^\nu\Theta_{\alpha\beta\ldots;\nu} + k_a{}^\nu{}_{;\alpha}\Theta_{\nu\beta\ldots} + k_a{}^\nu{}_{;\beta}\Theta_{\alpha\nu\ldots}\cdots).$$

(10)

Our purpose is to show that the operators X_a can be written in a form which is equivalent to equation (8) of Carter and McLenaghan.

In order to accomplish this we start with the vector representation $\rho_v = (\frac{1}{2},\frac{1}{2})$ of the $SL(2,\mathbb{C})$ group, generated by the spin matrices $\rho_v(S^{\hat\alpha\hat\beta})$ which have the well-known matrix elements

$$[\rho_v(S^{\hat\alpha\hat\beta})]^{\hat\mu\,\cdot}_{\cdot\,\hat\nu} = i(\eta^{\hat\alpha\hat\mu}\delta_{\hat\nu}^{\hat\beta} - \eta^{\hat\beta\hat\mu}\delta_{\hat\nu}^{\hat\alpha}),$$

(11)

in local bases. Furthermore, we define the point-dependent spin matrices in natural frames whose matrix elements in the natural basis read

$$(\tilde{S}^{\mu\nu})^{\sigma\cdot}_{\cdot\tau} = e^{\mu}_{\hat{\alpha}} e^{\nu}_{\hat{\beta}} e^{\sigma}_{\hat{\gamma}} [\rho_v(S^{\hat{\alpha}\hat{\beta}})]^{\hat{\gamma}\cdot}_{\cdot\hat{\delta}} \hat{e}^{\hat{\delta}}_{\tau} = i(g^{\mu\sigma}\delta^{\nu}_{\tau} - g^{\nu\sigma}\delta^{\mu}_{\tau}).$$
(12)

These matrices represent the spin operators of the vector representation in natural frames. We observe that these are the basis-generators of the groups $G[g(x)] \sim G(\eta)$ which leave the metric tensor $g(x)$ invariant in each point x. Since the representations of these groups are point-wise equivalent with those of $G(\eta)$, one can show that in each point x the basis-generators $\tilde{S}^{\mu\nu}(x)$ satisfy the standard commutation rules of the vector representation ρ_v (but with $g(x)$ instead of η).

In general, the spin matrices of a tensor Θ of any rank, n, are the basis-generators of the representation $\rho_n = \rho_v^1 \otimes \rho_v^2 \otimes \rho_v^3 \otimes .... \otimes \rho_v^n$ which read $\rho_n(\tilde{S}) = \tilde{S}^1 \otimes I^2 \otimes I^3... + I^1 \otimes \tilde{S}^2 \otimes I^3... + ...$ Using these spin matrices a straightforward calculation shows that equation (8) can be rewritten in natural frames as (Cotăescu, 2009),

$$X^n_a = -ik^{\mu}_a \nabla_{\mu} + \frac{1}{2} k_{a\,\mu;\nu} \rho_n(\tilde{S}^{\mu\nu}),$$
(13)

where ∇_{μ} are the usual covariant derivatives. It is not difficult to verify that the action of these operators is just that given by equation (10) which means that X^n_a are the basis-generators of a tensor representation of rank n of the group $S(M)$.

Thus, it is clear that the tensor representations are *equivalent* with the CRs defined in local frames while the equation (13) represents the generalization to natural frames of the Carter and McLenaghan formula. We stress that this result is not trivial since it can not be seen as a simple basis transformation like in the usual tensor theory of the linear algebra. The principal conclusion here is that the Carter and McLenaghan formula is universal since it holds not only in local frames but in natural frames too (Cotăescu, 2009).

3. The de Sitter expanding universe

Our approach is helpful on the four-dimensional de Sitter space-time where all the usual free field equations can be analytically solved while the isometries give rise to the rich $so(1,4)$ algebra. Hereby we selected various sets of commuting operators determining the quantum modes of the scalar, vector and Dirac fields.

3.1 Natural and local frames

Let us consider (M,g) be the de Sitter space-time defined as the hyperboloid of radius $1/\omega$ (where ω denotes the Hubble de Sitter constant) embedded in the five-dimensional flat space-time (M^5, η^5) with Cartesian coordinates z^A (labeled by the indices $A, B,... = 0,1,2,3,4$) and the metric $\eta^5 = \text{diag}(1,-1,-1,-1,-1)$ (Birrel & Davies, 1982). The local charts (or natural frames) of coordinates $\{x\}$ can be easily introduced on (M,g) defining the sets of functions $z^A(x)$ able to solve the hyperboloid equation, $\eta^5_{AB} z^A(x) z^B(x) = -\omega^{-2}$. In a given chart the line element $ds^2 = \eta^5_{AB} dz^A dz^B = g_{\mu\nu}(x)dx^{\mu}dx^{\nu}$ defines the metric tensor of (M,g).

In what follows we restrict ourselves to consider only the *(co)moving* charts, $\{t, \vec{x}\}$ and $\{t_c, \vec{x}\}$ which have the same Cartesian space coordinates, x^i ($i,j,k... = 1,2,3$), but different time coordinates. The first chart is equipped with the *proper* time $t \in (-\infty, \infty)$ while

$t_c = -\omega^{-1} e^{-\omega t} \in (-\infty, 0]$ is the *conformal* time. These charts are defined by the functions (Birrel & Davies, 1982),

$$z^0 = \frac{e^{\omega t}}{2\omega}\left(1 + \omega^2 \vec{x}^2 - e^{-2\omega t}\right) = -\frac{1}{2\omega^2 t_c}\left[1 - \omega^2(t_c^2 - \vec{x}^2)\right] \tag{14}$$

$$z^4 = \frac{e^{\omega t}}{2\omega}\left(1 - \omega^2 \vec{x}^2 + e^{-2\omega t}\right) = -\frac{1}{2\omega^2 t_c}\left[1 + \omega^2(t_c^2 - \vec{x}^2)\right] \tag{15}$$

$$z^i = e^{\omega t} x^i = -\frac{1}{\omega t_c} x^i \tag{16}$$

and have the line elements

$$ds^2 = dt^2 - e^{2\omega t}\, d\vec{x} \cdot d\vec{x} = \frac{1}{\omega^2 t_c^2}\left(dt_c^2 - d\vec{x} \cdot d\vec{x}\right). \tag{17}$$

We remind the reader that these charts cover only the *expanding* portion of (M, g) known as the de Sitter expanding universe.

The theory of external symmetry in local frame depends on the choice of the tetrad-gauge. The simplest gauge in the chart $\{t, \vec{x}\}$ is the diagonal one in which the non-vanishing components of the tetrad fields are

$$\hat{e}^0_0 = e^0_0 = 1, \quad \hat{e}^i_j = e^{\omega t}\delta^i_j, \quad e^i_j = e^{-\omega t}\delta^i_j. \tag{18}$$

The corresponding gauge in the chart $\{t_c, \vec{x}\}$ is given by the non-vanishing tetrad components,

$$\hat{e}^0_0 = -\omega t_c, \quad \hat{e}^i_j = -\delta^i_j \omega t_c, \quad \hat{e}^0_0 = -\frac{1}{\omega t_c}, \quad \hat{e}^i_j = -\delta^i_j \frac{1}{\omega t_c}. \tag{19}$$

3.2 The Killing vectors of the $SO(1,4)$ isometries

The de Sitter manifold (M, g) is defined as a homogeneous space of the pseudo-orthogonal group $SO(1,4)$ which is in the same time the gauge group of the metric η^5 and the isometry group, $I(M)$, of the de Sitter space-time . The group of the external symmetry, $S(M) = \text{Spin}(\eta^5) = Sp(2,2)$, has the Lie algebra $s(M) = sp(2,2) \sim so(1,4)$ for which we use the covariant real parameters $\xi^{AB} = -\xi^{BA}$. In this parametrization, the Killing vectors corresponding to the $SO(1,4)$ isometries can be derived considering the natural representation carried by the space of the scalar functions over M^5. The basis-generators of this representation are the genuine orbital operators

$$L^5_{AB} = i\left[\eta^5_{AC} z^C \partial_B - \eta^5_{BC} z^C \partial_A\right] = -iK^C_{(AB)}\partial_C \tag{20}$$

which define the components of the Killing vectors $K_{(AB)}$ on (M^5, η^5). With their help we can derive the corresponding Killing vectors of (M, g), denoted by $k_{(AB)}$, using the obvious identities $k_{(AB)\mu}dx^\mu = K_{(AB)C}dz^C$.

In the chart $\{t, \vec{x}\}$ the Killing vectors of the de Sitter symmetry have the components,

$$k^0_{(0i)} = k^0_{(4i)} = x^i, \quad k^j_{(0i)} = k^j_{(4i)} - \frac{1}{\omega}\delta^j_i = \omega x^i x^j - \delta^j_i \vartheta, \tag{21}$$

$$k^0_{(ij)} = 0, \quad k^l_{(ij)} = \delta^l_j x^i - \delta^l_i x^j; \quad k^0_{(04)} = -\frac{1}{\omega}, \quad k^i_{(04)} = x^i. \tag{22}$$

while in the other moving chart, $\{t_c, \vec{x}\}$, these components become

$$k^0_{(0i)} = k^0_{(4i)} = \omega t_c x^i, \qquad k^j_{(0i)} = k^j_{(4i)} - \frac{1}{\omega}\delta^j_i = \omega x^i x^j - \delta^j_i \vartheta, \tag{23}$$

$$k^0_{(ij)} = 0, \quad k^l_{(ij)} = \delta^l_j x^i - \delta^l_i x^j; \qquad k^\mu_{(04)} = x^\mu, \tag{24}$$

where the function ϑ is defined as

$$\vartheta = \frac{1}{2\omega}(1 + \omega^2 \vec{x}^2 - e^{-2\omega t}) = \frac{1}{2\omega}[1 + \omega^2(\vec{x}^2 - t_c^2)], \tag{25}$$

3.3 The $so(1,4)$ generators of covariant representations

According to our general theory, the generators of the CRs $T^{(\rho)}$ of the group $S(M) = Sp(2,2)$, induced by the representations ρ of the $SL(2,\mathbb{C})$ group, constitute CRs of the $sp(2,2)$ algebra induced by the representations ρ of the $sl(2,\mathbb{C})$ algebra. Therefore, their commutation relations are determined by the structure constant of the group $Sp(2,2)$ and the principal invariants are the Casimir operators of the CRs which can be derived as those of the algebras $sp(2,2) \sim so(1,4)$.

In the covariant parametrization of the $sp(2,2)$ algebra adopted here, the generators $X^{(\rho)}_{(AB)}$ corresponding to the Killing vectors $k_{(AB)}$ result from equation (7) and the functions (5) with the new labels $a \to (AB)$. Using then the Killing vectors (21) and (22) and the tetrad-gauge (18) of the chart $\{t, \vec{x}\}$, after a little calculation, we find first the $sl(2,\mathbb{C})$ generators. These are the total angular momentum,

$$J^{(\rho)}_i \equiv \frac{1}{2}\varepsilon_{ijk}X^{(\rho)}_{(jk)} = -i\varepsilon_{ijk}x^j\partial_k + S^{(\rho)}_i, \tag{26}$$

and the generators of the Lorentz boosts

$$K^{(\rho)}_i \equiv X^{(\rho)}_{(0i)} = ix^i\partial_t + i\vartheta(x)\partial_i - i\omega x^i x^j\partial_j + e^{-\omega t}S^{(\rho)}_{0i} + \omega S^{(\rho)}_{ij}x^j, \tag{27}$$

where ϑ is defined by Eq. (25). In addition, there are three generators,

$$R^{(\rho)}_i \equiv X^{(\rho)}_{(i4)} = -K^{(\rho)}_i + \frac{1}{\omega}i\partial_i, \tag{28}$$

which play the role of a Runge-Lenz vector, in the sense that $\{J_i, R_i\}$ generate a $so(4)$ subalgebra. The energy (or Hamiltonian) operator,

$$H \equiv \omega X^{(\rho)}_{(04)} = i\partial_t - i\omega x^i\partial_i, \tag{29}$$

is given by the Killing vector $k_{(04)}$ which is time-like only for $\omega|\vec{x}|e^{\omega t} \leq 1$. Fortunately, this condition is accomplished everywhere inside the light-cone of an observer at rest in $\vec{x} = 0$. Therefore, the operator H is correctly defined.

The generators introduced above form the basis $\{H, J^{(\rho)}_i, K^{(\rho)}_i, R^{(\rho)}_i\}$ of the covariant representation of the $sp(2,2)$ algebra with the following commutation rules (Cotăescu, 2011a):

$$\left[J^{(\rho)}_i, J^{(\rho)}_j\right] = i\varepsilon_{ijk}J^{(\rho)}_k, \quad \left[J^{(\rho)}_i, R^{(\rho)}_j\right] = i\varepsilon_{ijk}R^{(\rho)}_k, \quad \left[K^{(\rho)}_i, K^{(\rho)}_j\right] = -i\varepsilon_{ijk}J^{(\rho)}_k \tag{30}$$

$$\left[J^{(\rho)}_i, K^{(\rho)}_j\right] = i\varepsilon_{ijk}K^{(\rho)}_k, \quad \left[R^{(\rho)}_i, R^{(\rho)}_j\right] = i\varepsilon_{ijk}J^{(\rho)}_k, \quad \left[R^{(\rho)}_i, K^{(\rho)}_j\right] = \frac{i}{\omega}\delta_{ij}H, \tag{31}$$

and

$$\left[H, J_i^{(\rho)}\right] = 0, \quad \left[H, K_i^{(\rho)}\right] = i\omega R_i^{(\rho)}, \quad \left[H, R_i^{(\rho)}\right] = i\omega K_i^{(\rho)}. \tag{32}$$

As mentioned in the previous section, it is useful to replace the operators $\vec{K}^{(\rho)}$ and $\vec{R}^{(\rho)}$ by the momentum operator $\vec{P}$ and its dual, $\vec{Q}^{(\rho)}$, whose components are defined as

$$P_i = \omega(R_i^{(\rho)} + K_i^{(\rho)}) = i\partial_i, \quad Q_i^{(\rho)} = \omega(R_i^{(\rho)} - K_i^{(\rho)}). \tag{33}$$

which have the remarkable properties (Cotăescu, 2011a)

$$[H, P_i] = i\omega P_i, \quad \left[H, Q_i^{(\rho)}\right] = -i\omega Q_i^{(\rho)}, \quad \left[Q_i^{(\rho)}, Q_j^{(\rho)}\right] = [P_i, P_j] = 0. \tag{34}$$

We obtain thus the basis $\{H, P_i, Q_i^{(\rho)}, J_i^{(\rho)}\}$ and the basis of the Poincaré type formed by $\{H, P_i, J_i^{(\rho)}, K_i^{(\rho)}\}$.

The last two bases bring together the conserved energy (29) and momentum (33a) which are the only genuine orbital operators, independent on ρ. What is specific for the de Sitter symmetry is that these operators can not be put simultaneously in diagonal form since they do not commute to each other, as it results from Eq. (34a). Therefore, there are no mass-shells.

4. The massive Klein-Gordon field

The quantum modes of the scalar field in moving frames of de Sitter manifolds are well-known from long time (Birrel & Davies, 1982) paying attention to the scalar propagators, known as two-point functions (Candelas & Raine, 1975; Chernikov & Tagiriv, 1968), we recover here using the canonical quantization (Cotăescu et al., 2008).

4.1 Scalar quantum mechanics

In what follows we study the scalar field minimally coupled to the de Sitter gravity using our recently proposed new quantum mechanics on spatially flat Robertson-Walker space-times in which we defined different time evolution pictures (Cotăescu, 2007).

4.1.1 Lagrangian theory

In an arbitrary chart $\{x\}$ of a curved manifold the action of a charged scalar field ϕ of mass m, minimally coupled to gravity, reads (Birrel & Davies, 1982),

$$S[\phi, \phi^*] = \int d^4x \sqrt{g}\, \mathcal{L} = \int d^4x \sqrt{g} \left(\partial^\mu \phi^* \partial_\mu \phi - m^2 \phi^* \phi\right), \tag{35}$$

where $g = |\det(g_{\mu\nu})|$. This action gives rise to the Klein-Gordon equation

$$\frac{1}{\sqrt{g}} \partial_\mu [\sqrt{g}\, g^{\mu\nu} \partial_\nu \phi] + m^2 \phi = 0. \tag{36}$$

The conserved quantities predicted by the Noether theorem can be calculated with the help of the stress-energy tensor

$$T_{\mu\nu} = \partial_\mu \phi^* \partial_\nu \phi + \partial_\nu \phi^* \partial_\mu \phi - g_{\mu\nu} \mathcal{L}. \tag{37}$$

Thus, for each isometry corresponding to a Killing vector $k_{(AB)}$ there exists the conserved current $\Theta^{\mu}[k_{(AB)}] = -T^{\mu}_{\cdot \nu} k^{\nu}_{(AB)}$ which satisfies $\Theta^{\mu}[k_{(AB)}]_{;\mu} = 0$ producing the conserved quantity

$$C[k_{(AB)}] = \int_{\Sigma} d\sigma_{\mu} \sqrt{g}\, \Theta^{\mu}[k_{(AB)}]\,, \tag{38}$$

on a given hypersurface $\Sigma \subset M$. Moreover, generalizing the form of the conserved electric charge due to the internal $U(1)$ symmetry one defines the relativistic scalar product of two scalar fields as

$$\langle \phi, \phi' \rangle = i \int_{\Sigma} d\sigma^{\mu} \sqrt{g}\, \phi^* \overset{\leftrightarrow}{\partial_{\mu}} \phi'\,, \tag{39}$$

using the notation $f \overset{\leftrightarrow}{\partial} h = f(\partial h) - h(\partial f)$. With this definition one obtains the following identities

$$C[k_{(AB)}] = \langle \phi, X_{(AB)} \phi \rangle \tag{40}$$

which can be proved for any Killing vector using the field equation (36) and the Green's theorem. These identities will be useful in quantization, giving directly the conserved one-particle operators of the quantum field theory (Cotăescu et al., 2008).

4.1.2 Time-evolution pictures on de Sitter space-time

Let us consider now the de Sitter expanding universe (M, g) and the chart $\{t, \vec{x}\}$ with FRW line element. We say that the *natural* time-evolution picture (NP) is the genuine quantum theory in this chart where the time evolution of the massive scalar field is governed by the Klein-Gordon equation

$$\left(\partial_t^2 - e^{-2\omega t} \Delta + 3\omega \partial_t + m^2 \right) \phi(x) = 0\,. \tag{41}$$

The solutions of this equation may be square integrable functions or tempered distributions with respect to the scalar product (39) that in NP and for $\Sigma = \mathbb{R}^3$ takes the form

$$\langle \phi, \phi' \rangle = i \int d^3x\, e^{3\omega t} \phi^*(x) \overset{\leftrightarrow}{\partial_t} \phi'(x)\,. \tag{42}$$

since in this chart $\sqrt{g} = e^{3\omega t}$.

The principal operators of NP are the isometry generators given by Eqs. (26)-(29) but calculated for the scalar representation $(0,0)$ whose generators vanish. Thus, these operators are just the genuine orbital generators of the natural representation which is equivalent to the scalar CR. In addition, we consider the coordinate operator, $\vec{X}$, defined as $(X^i \phi)(x) = x^i \phi(x)$, that obeys,

$$[P_i, X^j] = i\delta_{ij} I\,, \quad [H, X^i] = -i\omega X^i\,, \tag{43}$$

where I is the identity operator.

The NP can be changed using point-dependent operators which could be even non-unitary operators since the relativistic scalar product does not have a direct physical meaning as that of the non-relativistic quantum mechanics . We exploit this opportunity for introducing the new time-evolution picture, called the Schrödinger picture (SP), with the help of the transformation $\phi(x) \rightarrow \phi_S(x) = W(x)\phi(x)W^{-1}(x)$ produced by the operator of time dependent *dilatations* (Cotăescu, 2007),

$$W(x) = \exp\left[-\omega t(x^i \partial_i)\right], \quad W^+(x) = e^{3\omega t} W^{-1}(x), \tag{44}$$

which has the following convenient actions

$$W(x)F(x^i)W^{-1}(x) = F\left(e^{-\omega t}x^i\right), \quad W(x)G(\partial_i)W^{-1}(x) = G\left(e^{\omega t}\partial_i\right), \tag{45}$$

upon any analytical functions F and G. This transformation leads to the Klein-Gordon equation of the SP

$$\left[\left(\partial_t + \omega x^i \partial_i\right)^2 - \Delta + 3\omega(\partial_t + \omega x^i \partial_i) + m^2\right]\phi_S(x) = 0, \tag{46}$$

and allows us to define the scalar product of this picture,

$$\langle \phi_S, \phi_S' \rangle \equiv \langle \phi, \phi' \rangle = i \int d^3x \left[\phi_S^* \overset{\leftrightarrow}{\partial_t} \phi_S' + \omega x^i (\phi_S^* \overset{\leftrightarrow}{\partial_i} \phi_S')\right], \tag{47}$$

as it results from Eqs. (44).

The specific operators of the SP, denoted by H_S, P_S^i and X_S^i, are defined as

$$(H_S\phi_S)(x) = i\partial_t\phi_S(x), \quad (P_S^i\phi_S)(x) = -i\partial_i\phi_S(x), \quad (X_S^i\phi_S)(x) = x^i\phi_S(x), \tag{48}$$

The meaning of these operators can be understood in the NP. Indeed, performing the inverse transformation we recover the conserved energy operator $H = W^{-1}(x) H_S W(x)$ and we find the new interesting time-dependent operators of the NP,

$$X^i(t) = W^{-1}(x) X_S^i W(x) = e^{\omega t} X^i, \tag{49}$$

$$P^i(t) = W^{-1}(x) P_S^i W(x) = e^{-\omega t} P^i, \tag{50}$$

which satisfy the canonical commutation rules (Cotăescu, 2007),

$$[X^i(t), P^j(t)] = i\delta_{ij}I, \quad [H, X^i(t)] = [H, P^i(t)] = 0. \tag{51}$$

The angular momentum has the same expression in both these pictures since it commutes with $W(x)$. We note that even if $X^i(t)$ and $P^i(t)$ commute with H they can not be considered conserved operators since they do not commute with the Klein-Gordon operator.

In NP picture the eigenvalues problem $Hf_E(t,\vec{x}) = Ef_E(t,\vec{x})$ of the energy operator leads to energy eigenfunctions of the form

$$f_E(t,\vec{x}) = F[e^{\omega t}\vec{x}]e^{-iEt} \tag{52}$$

where F is an arbitrary function. This explains why in this picture one can not find energy eigenfunctions separating variables. However, in our SP these eigenfunctions become the new functions

$$f_E^S(t,\vec{x}) = W(x)f_E(t,\vec{x})W^{-1}(x) = F(\vec{x})e^{-iEt} \tag{53}$$

which have separated variables. This means that in SP new quantum modes could be derived using the method of separating variables in coordinates or even in momentum representation.

4.2 Scalar plane waves

As mentioned before, the specific feature of the quantum mechanics on M is that the conserved energy and momentum can not be measured simultaneously with desired accuracy. Consequently, there are no particular solutions of the Klein-Gordon equation with well-determined energy and momentum, being forced to consider different plane waves solutions depending either on momentum or on energy and momentum direction. Thus we shall work with two bases of fundamental solutions we call here the momentum and energy bases. The momentum basis is well-known (Birrel & Davies, 1982; Chernikov & Tagiriv, 1968) but the energy one is a new basis derived using our SP (Cotăescu et al., 2008).

4.2.1 The momentum basis

It is known that the Klein-Gordon equation (41) of NP can be analytically solved in terms of Bessel functions (Birrel & Davies, 1982). There are fundamental solutions determined as eigenfunctions of the set of commuting operators $\{P^i\}$ of NP whose eigenvalues p^i are the components of the momentum $\vec{p}$. Among different versions of solutions which are currently used we prefer the normalized solutions of positive frequencies that read

$$f_{\vec{p}}(x) = \frac{1}{2}\sqrt{\frac{\pi}{\omega}} \frac{1}{(2\pi)^{3/2}} e^{-3\omega t/2} Z_k \left(\frac{p}{\omega} e^{-\omega t}\right) e^{i\vec{p}\cdot\vec{x}}, \quad k = \sqrt{\mu^2 - \tfrac{9}{4}}, \tag{54}$$

where $p = |\vec{p}|$, the functions Z_k are defined in the Appendix A and we denote $\mu = \frac{m}{\omega}$. Obviously, the fundamental solutions of negative frequencies are $f_{\vec{p}}^*(x)$.

All these solutions satisfy the orthonormalization relations

$$\langle f_{\vec{p}}, f_{\vec{p}'}\rangle = -\langle f_{\vec{p}}^*, f_{\vec{p}'}^*\rangle = \delta^3(\vec{p} - \vec{p}'), \quad \langle f_{\vec{p}}, f_{\vec{p}'}^*\rangle = 0, \tag{55}$$

and the completeness condition

$$i \int d^3p\, f_{\vec{p}}^*(t, \vec{x}) \overset{\leftrightarrow}{\partial_t} f_{\vec{p}}(t, \vec{x}') = e^{-3\omega t}\delta^3(\vec{x} - \vec{x}'). \tag{56}$$

For this reason we say that the set $\{f_{\vec{p}} | \vec{p} \in \mathbb{R}_p^3\}$ forms the complete system of fundamental solutions of positive frequencies of the *momentum* basis of the Hilbert space $\mathbf{H}_{KG}^{(+)}$ of particle states. The solutions of negative frequencies, $\{f_{\vec{p}}^* | \vec{p} \in \mathbb{R}_p^3\}$, span an orthogonal Hilbert space, $\mathbf{H}_{KG}^{(-)}$, associated to the antiparticle states. We must stress that this separation of the positive and negative frequencies defines the Bunch-Davies vacuum which is known to be stable.

In this basis, the Klein-Gordon field can expanded in terms of plane waves of positive and negative frequencies in usual manner as

$$\begin{aligned}
\phi(x) &= \phi^{(+)}(x) + \phi^{(-)}(x) \\
&= \int d^3p \left[f_{\vec{p}}(x)a(\vec{p}) + f_{\vec{p}}^*(x)b^*(\vec{p}) \right]
\end{aligned} \tag{57}$$

where a and b are the particle and respectively antiparticle wave functions of the momentum representation. These can be calculated using the inversion formulas $a(\vec{p}) = \langle f_{\vec{p}}, \phi\rangle$ and $b(\vec{p}) = \langle f_{\vec{p}}, \phi^*\rangle$.

4.2.2 The energy basis

The plane waves of given energy have to be derived in the SP (Cotăescu, 2007) where the Klein-Gordon equation has the suitable form (46). We assume that in this picture the scalar field can be expanded as

$$\phi_S(x) = \phi_S^{(+)}(x) + \phi_S^{(-)}(x) = \int_0^\infty dE \int d^3q \left[\hat{\phi}_S^{(+)}(E,\vec{q})e^{-iEt+i\vec{q}\cdot\vec{x}} + \hat{\phi}_S^{(-)}(E,\vec{q})e^{iEt-i\vec{q}\cdot\vec{x}} \right] \quad (58)$$

where $\hat{\phi}_S^{(\pm)}$ behave as tempered distributions on the domain $\mathbb{R}_q^3$ such that the Green theorem may be used. Then we can replace the momentum operators P_S^i by q^i and the coordinate operators X_S^i by $i\partial_{q_i}$ obtaining the Klein-Gordon equation of the SP in momentum representation,

$$\left\{ \left[\pm iE + \omega \left(q^i\partial_{q_i} + 3 \right) \right]^2 - 3\omega \left[\pm iE + \omega \left(q^i\partial_{q_i} + 3 \right) \right] + \vec{q}^2 + m^2 \right\} \hat{\phi}_S^{(\pm)}(E,\vec{q}) = 0, \quad (59)$$

where E is the energy defined as the eigenvalue of H_S. We remind the reader that the operators P_S^i and X_S^i become in NP the time dependent operators (49) and respectively (50) while H_S is related to the conserved energy operator H. This means the energy E is a conserved quantity but the momentum $\vec{q}$ does not have this property. More specific, only the scalar momentum $q = |\vec{q}|$ is not conserved while the momentum direction is conserved since the operator (50) is parallel with the conserved momentum $\vec{P}$. For this reason we denote $\vec{q} = q\,\vec{n}$ observing that the differential operator of Eq. (59) is of radial type and reads $q^i\partial_{q_i} = q\,\partial_q$. Consequently, this operator acts only on the functions depending on q while the functions which depend on the momentum direction $\vec{n}$ behave as constants. Therefore, we have to look for solutions of the form

$$\hat{\phi}_S^{(+)}(E,\vec{q}) = h_S(E,q)\,a(E,\vec{n}), \quad (60)$$

$$\hat{\phi}_S^{(-)}(E,\vec{q}) = [h_S(E,q)]^*\,b^*(E,\vec{n}), \quad (61)$$

where the function h_S satisfies an equation derived from Eq. (59) that can be written simply using the new variable $s = \frac{q}{\omega}$ and the notation $\epsilon = \frac{E}{\omega}$. This equation,

$$\left[\frac{d^2}{ds^2} + \frac{2i\epsilon + 4}{s} \frac{d}{ds} + \frac{\mu^2 - \epsilon^2 + 3i\epsilon}{s^2} + 1 \right] h_S(\epsilon,s) = 0, \quad (62)$$

is of the Bessel type having solutions of the form $h_S(\epsilon,s) = \text{const } s^{-i\epsilon-3/2} Z_k(s)$. Collecting all the above results we derive the final expression of the Klein-Gordon field (58) as

$$\phi_S(x) = \int_0^\infty dE \int_{S^2} d\Omega_n \left\{ f_{E,\vec{n}}^S(x)a(E,\vec{n}) + [f_{E,\vec{n}}^S(x)]^*b^*(E,\vec{n}) \right\}, \quad (63)$$

where the integration covers the sphere $S^2 \subset \mathbb{R}_p^3$. The fundamental solutions $f_{E,\vec{n}}^S$ of positive frequencies, with energy E and momentum direction $\vec{n}$ result to have the integral representation

$$f_{E,\vec{n}}^S(x) = N_0 e^{-iEt} \int_0^\infty ds \sqrt{s}\, Z_k(s)\, e^{i\omega s\vec{n}\cdot\vec{x}-i\epsilon\ln s}, \quad (64)$$

where N_0 is a normalization constant.

For understanding the physical meaning of this result we must turn back to NP where the scalar field

$$\phi(x) = \int_0^\infty dE \int_{S^2} d\Omega_n \; \{f_{E,\vec{n}}(x)a(E,\vec{n}) + [f_{E,\vec{n}}(x)]^* b^*(E,\vec{n})\} \, , \tag{65}$$

is expressed in terms of the solutions of NP that can be put in the form

$$f_{E,\vec{n}}(t,\vec{x}) = W^{-1}(x) f^S_{E,\vec{n}}(t,\vec{x}) W(x) = f^S_{E,\vec{n}}(t, e^{\omega t}\vec{x})$$

$$= N_0 e^{-\frac{3}{2}\omega t} \int_0^\infty ds \, \sqrt{s} \, Z_k \left(se^{-\omega t}\right) e^{i\omega s \vec{n} \cdot \vec{x} - i\epsilon \ln s} \, , \tag{66}$$

changing the integration variable $e^{\omega t} s \to s$ in the integral (64). Using then the scalar product (108) and the method of the Appendix B we can show that the normalization constant

$$N_0 = \frac{1}{2} \sqrt{\frac{\omega}{2}} \frac{1}{(2\pi)^{3/2}} \, , \tag{67}$$

assures the desired orthonormalization relations

$$\langle f_{E,\vec{n}}, f_{E',\vec{n}'} \rangle = -\langle f^*_{E,\vec{n}}, f^*_{E',\vec{n}'} \rangle = \delta(E - E') \, \delta^2(\vec{n} - \vec{n}'), \quad \langle f_{E,\vec{n}}, f^*_{E',\vec{n}'} \rangle = 0, \tag{68}$$

and the completeness condition

$$i \int_0^\infty dE \int_{S^2} d\Omega_n \left\{ [f_{E,\vec{n}}(t,\vec{x})]^* \overset{\leftrightarrow}{\partial_t} f_{E,\vec{n}}(t,\vec{x}') \right\} = e^{-3\omega t} \delta^3(\vec{x} - \vec{x}') . \tag{69}$$

This means that the set of functions $\{f_{E,\vec{n}} | E \in \mathbb{R}^+, \vec{n} \in S^2\}$ constitutes the complete system of fundamental solutions of the *energy* basis of $\mathbf{H}^{(+)}_{KG}$. The set $\{f^*_{E,\vec{n}} | E \in \mathbb{R}^+, \vec{n} \in S^2\}$ forms the energy basis of $\mathbf{H}^{(-)}_{KG}$.

The last step is to calculate the transition coefficients between the momentum and energy bases of the NP that read (Cotăescu et al., 2008)

$$\langle f_{\vec{p}}, f_{E,\vec{n}} \rangle = \langle f_{E,\vec{n}}, f_{\vec{p}} \rangle^* = \frac{p^{-\frac{3}{2}}}{\sqrt{2\pi\omega}} \delta^2(\vec{n} - \vec{n}_p) e^{-i\frac{E}{\omega} \ln \frac{p}{\omega}} \, , \tag{70}$$

where $\vec{n}_p = \vec{p}/p$. With their help we deduce the transformations

$$a(\vec{p}) = \int_0^\infty dE \int_{S^2} d\Omega_n \langle f_{\vec{p}}, f_{E,\vec{n}} \rangle a(E,\vec{n}) = \frac{p^{-\frac{3}{2}}}{\sqrt{2\pi\omega}} \int_0^\infty dE \, e^{-i\frac{E}{\omega} \ln \frac{p}{\omega}} \, a(E,\vec{n}_p) \, , \tag{71}$$

$$a(E,\vec{n}) = \int d^3p \, \langle f_{E,\vec{n}}, f_{\vec{p}} \rangle a(\vec{p}) = \frac{1}{\sqrt{2\pi\omega}} \int_0^\infty dp \, \sqrt{p} \, e^{i\frac{E}{\omega} \ln \frac{p}{\omega}} \, a(p\vec{n}) \, , \tag{72}$$

and similarly for the wave functions b. These transformations do not mix the particle and antiparticle states such that we can conclude that the Bunch-Davies vacuum defined here is stable with respect to the basis transformations.

4.3 Quantization and one-particle operators

The quantization can be done in canonical manner considering that the wave functions a and b of the fields (57) and (65) become field operators (such that $b^* \to b^\dagger$). We assume that the particle (a, $a^\dagger$) and antiparticle (b, $b^\dagger$) operators fulfill the standard commutation relations in the momentum basis, from which the non-vanishing ones are

$$[a(\vec{p}), a^\dagger(\vec{p}')] = [b(\vec{p}), b^\dagger(\vec{p}')] = \delta^3(\vec{p} - \vec{p}'). \tag{73}$$

Then, from Eq. (71) it results that the field operators of the energy basis satisfy

$$[a(E, \vec{n}), a^\dagger(E', \vec{n}')] = [b(E, \vec{n}), b^\dagger(E', \vec{n}')] = \delta(E - E')\delta^2(\vec{n} - \vec{n}'), \tag{74}$$

and

$$[a(\vec{p}), a^\dagger(E, \vec{n})] = [b(\vec{p}), b^\dagger(E, \vec{n})] = \langle f_{\vec{p}}, f_{E, \vec{n}} \rangle, \tag{75}$$

while other commutators are vanishing. In this way the field ϕ is correctly quantized according to the *canonical* rule (Drell & Bjorken, 1965),

$$[\phi(t, \vec{x}), \pi(t, \vec{x}')] = e^{3\omega t} [\phi(t, \vec{x}), \partial_t \phi^\dagger(t, \vec{x}')] = i \delta^3(\vec{x} - \vec{x}'), \tag{76}$$

where $\pi = \sqrt{g}\, \partial_t \phi^\dagger$ is the momentum density derived from the action (35). All these operators act on the Fock space which has the unique Bunch-Davies vacuum state $|0\rangle$ accomplishing

$$a(\vec{p})|0\rangle = b(\vec{p})|0\rangle = 0, \quad \langle 0|a^\dagger(\vec{p}) = \langle 0|b^\dagger(\vec{p}) = 0, \tag{77}$$

and similarly for the energy basis. The sectors with a given number of particles have to be constructed using the standard methods, obtaining thus the generalized bases of momentum or energy.

The one-particle operators corresponding to the conserved operators can be calculated bearing in mind that for any self-adjoint generator X of the scalar representation of the group $I(M)$ there exists a *conserved* one-particle operator of the quantum field theory which can be calculated simply as

$$\mathcal{X} =: \langle \phi, X\phi \rangle : \tag{78}$$

respecting the normal ordering of the operator products. Hereby we recover the standard algebraic properties

$$[\mathcal{A}, \phi(x)] = -A\phi(x), \quad [\mathcal{A}, \mathcal{B}] =: \langle \phi, [A, B] \phi \rangle : \tag{79}$$

due to the canonical quantization adopted here. In other respects, the electric charge operator corresponding to the $U(1)$ internal symmetry (of Abelian gauge transformations $\phi \to e^{i\alpha I}\phi$) results from the Noether theorem to be $\mathcal{Q} =: \langle \phi, I\phi \rangle := : \langle \phi, \phi \rangle :$.

However, there are many other conserved operators which do not have corresponding differential operators at the level of quantum mechanics. The simplest examples are the operators of number of particles,

$$\mathcal{N}_{pa} = \int d^3 p\, a^\dagger(\vec{p})a(\vec{p}) = \int_0^\infty dE \int_{S^2} d\Omega_n a^\dagger(E, \vec{n})a(E, \vec{n}), \tag{80}$$

and that of antiparticles, $\mathcal{N}_{ap}$ (depending on b and $b^\dagger$), giving rise to the charge operator $\mathcal{Q} = \mathcal{N}_{pa} - \mathcal{N}_{ap}$ and that of the total number of particles, $\mathcal{N} = \mathcal{N}_{pa} + \mathcal{N}_{ap}$.

In what follows we focus on the conserved one-particle operators determining the momentum and energy bases. The diagonal operators of the momentum basis the are Q and the components of momentum operator,

$$\mathcal{P}^i =: \langle \phi, P^i \phi \rangle := \int d^3p \, p^i \left[a^\dagger(\vec{p}) a(\vec{p}) + b^\dagger(\vec{p}) b(\vec{p}) \right] . \tag{81}$$

In other words, the momentum basis is determined by the set of commuting operators $\{Q, \mathcal{P}^i\}$. The energy basis is formed by the common eigenvectors of the set of commuting operators $\{Q, \mathcal{H}, \tilde{\mathcal{P}}^i\}$, i.e. the charge, energy and momentum direction operators. The energy operator can be easily calculated since the solutions (66) are eigenfunctions of the operator H. In this way we find

$$\mathcal{H} =: \langle \phi, H\phi \rangle := \int_0^\infty dE \, E \int_{S^2} d\Omega_n \left[a^\dagger(E, \vec{n}) a(E, \vec{n}) + b^\dagger(E, \vec{n}) b(E, \vec{n}) \right] . \tag{82}$$

More interesting are the operators $\tilde{\mathcal{P}}^i$ of the momentum direction since they do not come from differential operators and, therefore, must be defined directly as

$$\tilde{\mathcal{P}}^i = \int_0^\infty dE \int_{S^2} d\Omega_n \, n^i \left[a^\dagger(E, \vec{n}) a(E, \vec{n}) + b^\dagger(E, \vec{n}) b(E, \vec{n}) \right] . \tag{83}$$

The above operators which satisfy simple commutation relations,

$$[\mathcal{H}, \mathcal{P}^i] = i\omega \mathcal{P}^i , \quad [\mathcal{H}, \tilde{\mathcal{P}}^i] = 0, \quad [Q, \mathcal{H}] = [Q, \mathcal{P}^i] = [Q, \tilde{\mathcal{P}}^i] = 0 , \tag{84}$$

are enough for defining the bases considered hare.

Our approach offers the opportunity to deduce mode expansions of conserved one-particle operators but in bases where these are not diagonal. For example, we can calculate the mode expansion of the energy operator in the momentum basis either starting with the identity

$$(Hf_{\vec{p}})(x) = -i\omega \left(p^i \partial_{p_i} + \frac{3}{2} \right) f_{\vec{p}}(x) \tag{85}$$

or using Eq. (72). The final result (Cotăescu et al., 2008),

$$\mathcal{H} = \frac{i\omega}{2} \int d^3p \, p^i \left[a^\dagger(\vec{p}) \overset{\leftrightarrow}{\partial}_{p_i} a(\vec{p}) + b^\dagger(\vec{p}) \overset{\leftrightarrow}{\partial}_{p_i} b(\vec{p}) \right] , \tag{86}$$

is similar to those obtained for other fields as we shall see later. This expansion has a remarkable property namely, the change of the phase factors,

$$f_{\vec{p}}(x) \to f_{\vec{p}}(x) e^{i\chi(\vec{p})}, \quad a(\vec{p}) \to e^{-i\chi(\vec{p})} a(\vec{p}), \quad b(\vec{p}) \to e^{-i\chi(\vec{p})} b(\vec{p}), \tag{87}$$

using a real phase function $\chi(\vec{p})$, preserve the form of the operators ϕ, Q and $\mathcal{P}^i$ but transforms the Hamiltonian operator as

$$\mathcal{H} \to \mathcal{H} + \omega \int d^3p \, [p^i \partial_{p^i} \chi(\vec{p})] \left[a^\dagger(\vec{p}) a(\vec{p}) + b^\dagger(\vec{p}) b(\vec{p}) \right] . \tag{88}$$

Our preliminary investigations indicate that this property may be helpful for avoid some mathematical difficulties related to the flat limit $\omega \sim 0$ (Cotăescu, 2011b).

We note that beside the above conserved operators we can introduce other one-particle operators extending the definition (78) to the non-conserved operators of our quantum mechanics. However, these operators will depend explicitly on time, their expressions being complicated and without an intuitive physical meaning.

4.4 Commutator and Green functions

In the quantum theory of fields the Green functions are related to the partial commutator functions (of positive or negative frequencies) defined as

$$D^{(\pm)}(x, x') = i[\phi^{(\pm)}(x), \phi^{(\pm)\dagger}(x')] \tag{89}$$

and the total one, $D = D^{(+)} + D^{(-)}$. These function are solutions of the Klein-Gordon equation in both the sets of variables and obey $[D^{(\pm)}(x, x')]^* = D^{(\mp)}(x, x')$ such that D results to be a real function. This property suggests us to restrict ourselves to study only the functions of positive frequencies,

$$D^{(+)}(x, x') = i \int d^3p \, f_{\vec{p}}(x) f_{\vec{p}}(x')^* = i \int_0^\infty dE \int_{S^2} d\Omega_n \, f_{E,\vec{n}}(x) f_{E,\vec{n}}(x')^*, \tag{90}$$

resulted from Eqs. (57) and (65). Both these versions lead to the final expression

$$D^{(+)}(x, x') = \frac{\pi}{4\omega} \frac{i}{(2\pi)^3} e^{-\frac{3}{2}\omega(t+t')} \int d^3p \, Z_k \left(\frac{p}{\omega} e^{-\omega t}\right) Z_k^* \left(\frac{p}{\omega} e^{-\omega t'}\right) e^{i\vec{p}\cdot(\vec{x}-\vec{x}')} \tag{91}$$

from which we understand that $D^{(+)}(x, x') = D^{(+)}(t, t', \vec{x} - \vec{x}')$ and may deduce what happens at equal time. First we observe that for $t' = t$ the values of the function $D^{(+)}(t, t, \vec{x} - \vec{x}')$ are c-numbers which means that $D(t, t, \vec{x} - \vec{x}') = 0$. Moreover, from Eqs. (56) or (69) we find

$$(\partial_t - \partial_{t'})D^{(+)}(t, t', \vec{x} - \vec{x}')\Big|_{t'=t} = e^{-3\omega t}\delta^3(\vec{x} - \vec{x}') \tag{92}$$

and similarly for $D^{(-)}$.

The commutator functions can be written in analytical forms since the mode integral (91) may be solved in terms of Gauss hypergeometric functions (Chernikov & Tagiriv, 1968). Indeed, in the chart $\{t_c, \vec{x}\}$ this integral becomes

$$D^{(+)}(t_c, t_c', \vec{x} - \vec{x}') = \frac{i\pi\omega^2}{4} \frac{e^{-\pi k}}{(2\pi)^3} (t_c t_c')^{\frac{3}{2}} \int d^3p \, e^{i(\vec{x}-\vec{x}')\cdot\vec{p}} H_{ik}^{(1)}(-pt_c) H_{ik}^{(1)}(-pt_c')^* \tag{93}$$

and can be solved as,

$$D^{(+)}(t_c, t_c', \vec{x} - \vec{x}') = \frac{im^2}{16\pi} e^{-\pi k} \text{sech}(\pi k) \, {}_2F_1\left(\frac{3}{2} + ik, \frac{3}{2} - ik; 2; 1 + \frac{y}{4}\right), \tag{94}$$

where the quantity

$$y(x, x') = \frac{(t_c - t_c' - i\epsilon)^2 - (\vec{x} - \vec{x}')^2}{t_c t_c'} \tag{95}$$

is related to the geodesic length between x and x' (Birrel & Davies, 1982).

In general, $G(x, x') = G(t, t', \vec{x} - \vec{x}')$ is a Green function of the Klein-Gordon equation if this obeys

$$\left[E_{KG}(x) - m^2\right] G(x, x') = -e^{-3\omega t}\delta^4(x - x'). \tag{96}$$

The properties of the commutator functions allow us to construct the Green function just as in the scalar theory on Minkowski space-time. We assume that the retarded, D_R, and advanced, D_A, Green functions read

$$D_R(t, t', \vec{x} - \vec{x}') = \theta(t - t')D(t, t', \vec{x} - \vec{x}'),$$ (97)

$$D_A(t, t', \vec{x} - \vec{x}') = -\theta(t' - t)D(t, t', \vec{x} - \vec{x}'),$$ (98)

while the Feynman propagator,

$$D_F(t, t', \vec{x} - \vec{x}') = i\langle 0|T[\phi(x)\phi^\dagger(x')]|0\rangle$$

$$= \theta(t - t')D^{(+)}(t, t', \vec{x} - \vec{x}') - \theta(t' - t)D^{(-)}(t, t', \vec{x} - \vec{x}'),$$ (99)

is defined as a causal Green function. It is not difficult to verify that all these functions satisfy Eq. (96) if one uses the identity $\partial_t^2[\theta(t)f(t)] = \delta(t)\partial_t f(t)$, the artifice $\partial_t f(t - t') = \frac{1}{2}(\partial_t - \partial_{t'})f(t - t')$ and Eq. (92).

5. The Dirac field

The first solutions of the free Dirac equation in the moving chart with proper time and spherical coordinates were derived in (Shishkin, 1991) and normalized in (Cotăescu et al., 2006). We derived other solutions of this equation but in moving charts with Cartesian coordinates where we considered the helicity basis in momentum representation (Cotăescu, 2002; Cotăescu & Crucean, 2008; Cotăescu, 2011b). These solutions are well-normalized, satisfy the usual completeness relations and correspond to a unique vacuum state.

5.1 Spinor quantum mechanics

Let ψ be a Dirac free field of mass m, defined on the space domain D, and $\overline{\psi} = \psi^+\gamma^0$ its Dirac adjoint. The tetrad gauge invariant action of the Dirac field minimally coupled with the gravitational field is

$$S[e, \psi] = \int d^4x \sqrt{g} \left\{ \frac{i}{2} [\overline{\psi}\gamma^{\hat{\alpha}} D_{\hat{\alpha}} \psi - (\overline{D_{\hat{\alpha}}\psi})\gamma^{\hat{\alpha}} \psi] - m\overline{\psi}\psi \right\}$$ (100)

where the Dirac matrices, $\gamma^{\hat{\alpha}}$, satisfy $\{\gamma^{\hat{\alpha}}, \gamma^{\hat{\beta}}\} = 2\eta^{\hat{\alpha}\hat{\beta}}$. The covariant derivatives in local frames, denoted simply by $D_{\hat{\alpha}}$, are given by Eq. (1) where we consider the spinor representation $\rho_s = (\frac{1}{2}, 0) \oplus (0, \frac{1}{2})$ of the $SL(2, \mathbb{C})$ group whose basis-generators in covariant parametrization are $S^{\hat{\alpha}\hat{\beta}} = \frac{i}{4}[\gamma^{\hat{\alpha}}, \gamma^{\hat{\beta}}]$. The operator of the Dirac equation $E_D\psi = m\psi$, derived from the action (100), reads $E_D = i\gamma^{\hat{\alpha}} D_{\hat{\alpha}}$. In other respects, from the conservation of the electric charge one deduces that when $e_i^0 = 0$ ($i, j, ... = 1, 2, 3$) the time-independent relativistic scalar product of two spinors,

$$\langle \psi, \psi' \rangle = \int_D d^3x \, \mu(x)\overline{\psi}(x)\gamma^0\psi'(x).$$ (101)

has the weight function $\mu = \sqrt{g}\, e_0^0$.

Our theory of external symmetry offers us the framework we need to calculate the conserved quantities predicted by the Noether theorem. Starting with the infinitesimal transformations of the one-parameter subgroup of $S(M)$ generated by X_a, we find that there exists the conserved current $\Theta^\mu[X_a]$ which satisfies $\Theta^\mu[X_a]_{;\mu} = 0$. For the action (100) this is

$$\Theta^\mu[X_a] = -\tilde{T}^\mu_{\cdot\nu} k_a^\nu + \frac{1}{4}\overline{\psi}\{\gamma^{\hat{a}}, S^{\hat{\beta}\hat{\gamma}}\}\psi\, e_{\hat{a}}^\mu\, \Omega_{a\,\hat{\beta}\hat{\gamma}} \qquad (102)$$

where

$$\tilde{T}^\mu_{\cdot\nu} = \frac{i}{2}\left[\overline{\psi}\gamma^{\hat{a}}e_{\hat{a}}^\mu\partial_\nu\psi - (\overline{\partial_\nu\psi})\gamma^{\hat{a}}e_{\hat{a}}^\mu\psi\right] \qquad (103)$$

is a notation for a part of the stress-energy tensor of the Dirac field. Finally, it is clear that the corresponding conserved quantity is the real number (Cotăescu, 2002),

$$\int_D d^3x \sqrt{g}\,\Theta^0[X_a] = \frac{1}{2}\left[\langle\psi, X_a\psi\rangle + \langle X_a\psi, \psi\rangle\right]. \qquad (104)$$

We note that it is premature to interpret this formula as an expectation value or to speak about Hermitian conjugation of the operators X_a with respect to the scalar product (101), before specifying the boundary conditions on D. What is important here is that this result is useful in quantization giving directly the one-particle operators of the quantum field theory.

On the de Sitter expanding universe we can chose the simple Cartesian gauge (18) of the chart $\{t, \vec{x}\}$ or the corresponding gauge (19) in the chart $\{t_c, \vec{x}\}$. Then the Dirac operator takes the forms

$$E_D = -i\omega t_c\left(\gamma^0\partial_{t_c} + \gamma^i\partial_i\right) + \frac{3i\omega}{2}\gamma^0 = i\gamma^0\partial_t + ie^{-\omega t}\gamma^i\partial_i + \frac{3i\omega}{2}\gamma^0 \qquad (105)$$

and the weight function of the scalar product (101) reads

$$\mu = (-\omega t_c)^{-3} = e^{3\omega t}. \qquad (106)$$

This operator commutes with the isometry generators (26)-(29) whose spin parts are given now by the matrices $S_{\hat{\alpha}\hat{\beta}}$. Thus we obtained the NP of the Dirac theory.

The SP can be introduced transforming the Dirac field using the same operator (44) as in the scalar case. In this picture the Dirac field becomes $\psi_S(x) = W(x)\psi(x)W^{-1}(x)$ while the genuine orbital operators (49), (50) and H remain the same as in the scalar case. Moreover, we obtain the free Dirac equation of the SP,

$$\left[i\gamma^0\partial_t + i\gamma^i\partial_i - m + i\gamma^0\omega\left(x^i\partial_i + \frac{3}{2}\right)\right]\psi_S(x) = 0, \qquad (107)$$

and the new form of the relativistic scalar product,

$$\langle\psi_S, \psi'_S\rangle = \langle\psi, \psi'\rangle = \int_D d^3x\,\bar{\psi}_S(x)\gamma^0\psi'_S(x), \qquad (108)$$

calculated according to Eqs. (101) and (44b). We observe that this is no longer dependent on $\sqrt{g}$, having thus the same form as in special relativity.

5.2 Polarized plane wave solutions

In what follows we present the principal polarized plane wave solutions of the free Dirac field minimally coupled to the de Sitter gravity. The polarization is described in the helicity basis such that we have to speak about the momentum-helicity basis and the energy-halicity one (Cotăescu, 2002; Cotăescu & Crucean, 2008). In addition, we derived the modes of the momentum-spin basis (Cotăescu, 2011b) but these exceed the space of this paper.

5.2.1 The momentum-helicity basis

The plane wave solutions of the Dirac equation with $m \neq 0$ may be eigenspinors of the momentum operators P^i corresponding to the eigenvalues p^i. Therefore, we assume that, in the standard representation of the Dirac matrices, with diagonal γ^0 (Thaler, 1992), these have the form

$$\psi_{\vec{p}}^{(+)} = \begin{pmatrix} f^+(t_c)\,\alpha(\vec{p}) \\ g^+(t_c)\,\dfrac{\vec{\sigma}\cdot\vec{p}}{p}\,\alpha(\vec{p}) \end{pmatrix} e^{i\vec{p}\cdot\vec{x}}, \quad \psi_{\vec{p}}^{(-)} = \begin{pmatrix} g^-(t_c)\,\dfrac{\vec{\sigma}\cdot\vec{p}}{p}\,\beta(\vec{p}) \\ f^-(t_c)\,\beta(\vec{p}) \end{pmatrix} e^{-i\vec{p}\cdot\vec{x}} \tag{109}$$

where σ_i denotes the Pauli matrices while α and β are arbitrary Pauli spinors depending on $\vec{p}$. Replacing these spinors in the Dirac equation given by (105) and denoting $\mu = \frac{m}{\omega}$ and $\nu_{\pm} = \frac{1}{2} \pm i\mu$, we find equations of the form (161) whose solutions can be written in terms of Hankel functions as

$$f^+ = (-f^-)^* = c\,t_c{}^2 e^{\frac{1}{2}\pi\mu} H_{\nu_-}^{(1)}(-pt_c) \tag{110}$$

$$g^+ = (-g^-)^* = c\,t_c{}^2 e^{-\frac{1}{2}\pi\mu} H_{\nu_+}^{(1)}(-pt_c). \tag{111}$$

The integration constant c will be calculated from the ortonormalization condition in the momentum scale.

The plane wave solutions are determined up to the significance of the Pauli spinors α and β. In general any pair of orthogonal spinors $\xi_\sigma(\vec{p})$ with polarizations $\sigma = \pm 1/2$ represents a good basis in the space of α-spinors. According to the standard interpretation of the negative frequency terms, the corresponding basis of the β-spinors is formed by the pair of orthogonal spinors defined as $\eta_\sigma(\vec{p}) = i\sigma_2[\xi_\sigma(\vec{p})]^*$. It remains to choose specific spinor bases, considering supplementary physical assumptions. Here we choose the *helicity* basis which is formed by the orthogonal Pauli spinors of helicity $\lambda = \pm\frac{1}{2}$ which satisfy the eigenvalues equations

$$\vec{\sigma}\cdot\vec{p}\,\xi_\lambda(\vec{p}) = 2p\lambda\,\xi_\lambda(\vec{p}), \quad \vec{\sigma}\cdot\vec{p}\,\eta_\lambda(\vec{p}) = -2p\lambda\,\eta_\lambda(\vec{p}), \tag{112}$$

and the orthonormalization condition $\xi_\lambda^+(\vec{p})\xi_{\lambda'}(\vec{p}) = \eta_\lambda^+(\vec{p})\eta_{\lambda'}(\vec{p}) = \delta_{\lambda\lambda'}$.

The desired particular solutions of the Dirac equation with $m \neq 0$ result from our starting formulas (109) where we insert the functions (110) and (111) and the spinors of the helicity basis (112). It remains to calculate the normalization constant c with respect to the scalar product (101) with the weight function (106). After a few manipulation, in the chart $\{t, \vec{x}\}$, it turns out that the Dirac field can be expanded as

$$\psi(t, \vec{x}) = \psi^{(+)}(t, \vec{x}) + \psi^{(-)}(t, \vec{x})$$

$$= \int d^3p \sum_\lambda \left[U_{\vec{p},\lambda}(x)a(\vec{p}, \lambda) + V_{\vec{p},\lambda}(x)b^*(\vec{p}, \lambda) \right], \tag{113}$$

in terms of the particle (a) and antiparticle (b) wave functions of the momentum representation. The fundamental spinors of positive and negative frequencies with momentum $\vec{p}$ and helicity λ read (Cotăescu, 2002)

$$U_{\vec{p},\lambda}(t, \vec{x}) = iN \begin{pmatrix} \frac{1}{2} e^{\frac{1}{2}\pi\mu} H_{\nu_-}^{(1)}(qe^{-\omega t})\,\xi_\lambda(\vec{p}) \\ \lambda\,e^{-\frac{1}{2}\pi\mu} H_{\nu_+}^{(1)}(qe^{-\omega t})\,\xi_\lambda(\vec{p}) \end{pmatrix} e^{i\vec{p}\cdot\vec{x} - 2\omega t} \tag{114}$$

$$V_{\vec{p},\lambda}(t, \vec{x}) = iN \begin{pmatrix} \lambda\,e^{-\frac{1}{2}\pi\mu} H_{\nu_-}^{(2)}(qe^{-\omega t})\,\eta_\lambda(\vec{p}) \\ -\frac{1}{2} e^{\frac{1}{2}\pi\mu} H_{\nu_+}^{(2)}(qe^{-\omega t})\,\eta_\lambda(\vec{p}) \end{pmatrix} e^{-i\vec{p}\cdot\vec{x} - 2\omega t}, \tag{115}$$

where we introduced the new parameter $q = \frac{p}{\omega}$ and

$$N = \frac{1}{(2\pi)^{3/2}} \sqrt{\pi q} \, . \tag{116}$$

These solutions are the common eigenspinors of the complete set of commuting operators $\{E_D, \vec{S}^2, P^i, W\}$ obeying

$$P^i U_{\vec{p},\lambda} = p^i U_{\vec{p},\lambda}, \qquad P^i V_{\vec{p},\lambda} = -p^i V_{\vec{p},\lambda}, \tag{117}$$

$$W U_{\vec{p},\lambda} = p\lambda U_{\vec{p},\lambda}, \qquad W V_{\vec{p},\lambda} = -p\lambda V_{\vec{p},\lambda}, \tag{118}$$

where $W = \vec{J} \cdot \vec{P} = \vec{S} \cdot \vec{P}$ is the helicity operator. For this reason we say that these spinors form the *momentum-helicity* basis.

In other respects, according to Eqs. (160) and (162), it is not hard to verify that these spinors are charge-conjugated to each other,

$$V_{\vec{p},\lambda} = (U_{\vec{p},\lambda})^c = C(\overline{U}_{\vec{p},\lambda})^T, \quad C = i\gamma^2 \gamma^0, \tag{119}$$

satisfy the ortonormalization relations,

$$\left\langle U_{\vec{p},\lambda}, U_{\vec{p}',\lambda'} \right\rangle = \left\langle V_{\vec{p},\lambda}, V_{\vec{p}',\lambda'} \right\rangle = \delta_{\lambda\lambda'} \delta^3(\vec{p} - \vec{p}'), \tag{120}$$

$$\left\langle U_{\vec{p},\lambda}, V_{\vec{p}',\lambda'} \right\rangle = \left\langle V_{\vec{p},\lambda}, U_{\vec{p}',\lambda'} \right\rangle = 0, \tag{121}$$

and represent a *complete* system of solutions in the sense that

$$\int d^3p \sum_\lambda \left[U_{\vec{p},\lambda}(t,\vec{x}) U_{\vec{p},\lambda}^+(t,\vec{x}') + V_{\vec{p},\lambda}(t,\vec{x}) V_{\vec{p},\lambda}^+(t,\vec{x}') \right] = e^{-3\omega t} \delta^3(\vec{x} - \vec{x}'). \tag{122}$$

Thus we can conclude that the separation of the positive and negative frequency modes performed here is point-independent and corresponds to a *stable* vacuum state which is of the Bunch-Davies type.

In the case of $m = 0$ (when $\mu = 0$) it is convenient to consider the chiral representation of the Dirac matrices (with diagonal γ^5) and the chart $\{t_c, \vec{x}\}$. We find that the fundamental solutions in momentum-helicity basis of the left-handed massless Dirac field (Cotăescu, 2002),

$$U_{\vec{p},\lambda}^0(t_c, \vec{x}) = \lim_{\mu \to 0} P_L U_{\vec{p},\lambda}(t_c, \vec{x}) = \left(\frac{-\omega t_c}{2\pi}\right)^{3/2} \left(\begin{array}{c} (\frac{1}{2} - \lambda) \xi_\lambda(\vec{p}) \\ 0 \end{array} \right) e^{-ipt_c + i\vec{p}\cdot\vec{x}}, \tag{123}$$

$$V_{\vec{p},\lambda}^0(t_c, \vec{x}) = \lim_{\mu \to 0} P_L V_{\vec{p},\lambda}(t_c, \vec{x}) = \left(\frac{-\omega t_c}{2\pi}\right)^{3/2} \left(\begin{array}{c} (\frac{1}{2} + \lambda) \eta_\lambda(\vec{p}) \\ 0 \end{array} \right) e^{ipt_c - i\vec{p}\cdot\vec{x}}, \tag{124}$$

where $P_L = \frac{1}{2}(1 - \gamma^5)$ is the left-handed projection matrix. These solutions are non-vanishing only for positive frequency and $\lambda = -\frac{1}{2}$ or negative frequency and $\lambda = \frac{1}{2}$, as in the flat case and, moreover, they have similar properties as in (119)-(118).

5.2.2 The energy-helicity basis

The energy basis formed by eigenspinors of the energy operator, H, must be studied in the SP where Eq. (107) can be solved in momentum representation (Cotăescu & Crucean, 2008). We start assuming that the spinors of the SP may be expanded in terms of plane waves of positive and negative frequencies as,

$$
\begin{aligned}
\psi_S(x) &= \psi_S^{(+)}(x) + \psi_S^{(-)}(x) \\
&= \int_0^\infty dE \int_{\hat{D}} d^3p \left[\hat{\psi}_S^{(+)}(E,\vec{p})\, e^{-i(Et - \vec{p}\cdot\vec{x})} + \hat{\psi}_S^{(-)}(E,\vec{p})\, e^{i(Et - \vec{p}\cdot\vec{x})} \right]
\end{aligned}
\tag{125}
$$

where $\hat{\psi}_S^{(\pm)}$ are spinors which behave as tempered distributions on the domain $\hat{D} = \mathbb{R}_p^3$ such that the Green theorem may be used. Then we can replace the momentum operators P_S^i by their eigenvalues p^i and the coordinate operators X_S^i by $i\partial_{p_i}$ obtaining the free Dirac equation of the SP in momentum representation,

$$
\left[\pm E\gamma^0 \mp \gamma^i p^i - m - i\gamma^0 \omega \left(p^i \partial_{p_i} + \frac{3}{2} \right) \right] \hat{\psi}_S^{(\pm)}(E,\vec{p}) = 0 ,
\tag{126}
$$

where E is the energy defined as the eigenvalue of H_S. Denoting $\vec{p} = p\,\vec{n}$ we observe that the differential operator of Eq. (126) is of radial type and reads $p^i \partial_{p_i} = p\,\partial_p$. Therefore, this operator acts on the functions which depend on p while the functions which depend only on the momentum direction $\vec{n}$ behave as constants.

Following the method of section 4.2.2 we derive the fundamental solutions of the *helicity* basis using the standard representation of the γ-matrices (with diagonal γ^0) (Thaler, 1992). The general solutions,

$$
\hat{\psi}_S^{(+)}(E,\vec{p}) = \sum_\lambda u^S(E,\vec{p},\lambda)\, a(E,\vec{n},\lambda) ,
\tag{127}
$$

$$
\hat{\psi}_S^{(-)}(E,\vec{p}) = \sum_\lambda v^S(E,\vec{p},\lambda)\, b^*(E,\vec{n},\lambda) ,
\tag{128}
$$

involve spinors of helicity $\lambda = \pm\frac{1}{2}$ and the particle and antiparticle wave functions, a and respectively b, which play here the role of constants since they do not depend on p. According to our previous results, the spinors of the momentum representation must have the form

$$
u^S(E,\vec{p},\lambda) = \begin{pmatrix} \frac{1}{2} f_E^{(+)}(p)\, \xi_\lambda(\vec{n}) \\ \lambda g_E^{(+)}(p)\, \xi_\lambda(\vec{n}) \end{pmatrix} , \quad
v^S(E,\vec{p},\lambda) = \begin{pmatrix} \lambda g_E^{(-)}(p)\, \eta_\lambda(\vec{n}) \\ -\frac{1}{2} f_E^{(-)}(p)\, \eta_\lambda(\vec{n}) \end{pmatrix} ,
\tag{129}
$$

where $\xi_\lambda(\vec{n})$ and $\eta_\lambda(\vec{n}) = i\sigma_2[\xi_\lambda(\vec{n})]^*$ denote now the Pauli spinors of the helicity basis introduced in section 5.2.1. (which depend only on the momentum direction $\vec{n}$). Furthermore, we derive the radial functions solving the system

$$
\left[i\omega \left(p\frac{d}{dp} + \frac{3}{2} \right) \mp (E - m) \right] f_E^{(\pm)}(p) = \mp p\, g_E^{(\pm)}(p) ,
\tag{130}
$$

$$
\left[i\omega \left(p\frac{d}{dp} + \frac{3}{2} \right) \mp (E + m) \right] g_E^{(\pm)}(p) = \mp p\, f_E^{(\pm)}(p) ,
\tag{131}
$$

resulted from Eq. (126). We find the solutions

$$f_E^{(+)}(p) = [-f_E^{(-)}(p)]^* = Cp^{-1-i\epsilon}e^{\frac{1}{2}\pi\mu}H_{\nu_-}^{(1)}(\tfrac{p}{\omega}),\qquad(132)$$

$$g_E^{(+)}(p) = [-g_E^{(-)}(p)]^* = Cp^{-1-i\epsilon}e^{-\frac{1}{2}\pi\mu}H_{\nu_+}^{(1)}(\tfrac{p}{\omega}).\qquad(133)$$

The normalization constant C has to assure the normalization in the energy scale.

Collecting all the above results we can write down the final expression of the Dirac field (125) in SP identifying the form of the fundamental spinors of given energy. Then we turn back to the NP where the Dirac field,

$$\psi(x) = \psi_S(t, e^{\omega t}\vec{x}) = \int_0^\infty dE \int_{S^2} d\Omega_n \sum_\lambda [U_{E,\vec{n},\lambda}(t,\vec{x})a(E,\vec{n},\lambda)\qquad(134)$$

$$+ V_{E,\vec{n},\lambda}(t,\vec{x})b^*(E,\vec{n},\lambda)],\qquad(135)$$

depends on the solutions written in NP,

$$U_{E,\vec{n},\lambda}(t,\vec{x}) = U_{E,\vec{n},\lambda}^S(t, e^{\omega t}\vec{x}),\quad V_{E,\vec{n},\lambda}(t,\vec{x}) = V_{E,\vec{n},\lambda}^S(t, e^{\omega t}\vec{x}).\qquad(136)$$

According to Eqs. (129), (132) and (133), we obtain the integral representations

$$U_{E,\vec{n},\lambda}(t,\vec{x}) = i\hat{N}e^{-2\omega t}\int_0^\infty s\,ds \begin{pmatrix} \frac{1}{2}e^{\frac{1}{2}\pi\mu}H_{\nu_-}^{(1)}(se^{-\omega t})\,\xi_\lambda(\vec{n}) \\ \lambda e^{-\frac{1}{2}\pi\mu}H_{\nu_+}^{(1)}(se^{-\omega t})\,\xi_\lambda(\vec{n}) \end{pmatrix} e^{i\omega s\vec{n}\cdot\vec{x}-i\epsilon\ln s},\qquad(137)$$

$$V_{E,\vec{n},\lambda}(t,\vec{x}) = i\hat{N}e^{-2\omega t}\int_0^\infty s\,ds \begin{pmatrix} \lambda e^{-\frac{1}{2}\pi\mu}H_{\nu_-}^{(2)}(se^{-\omega t})\,\eta_\lambda(\vec{n}) \\ -\frac{1}{2}e^{\frac{1}{2}\pi\mu}H_{\nu_+}^{(2)}(se^{-\omega t})\,\eta_\lambda(\vec{n}) \end{pmatrix} e^{-i\omega s\vec{n}\cdot\vec{x}+i\epsilon\ln s},\qquad(138)$$

where we denote the dimensionless integration variable by $s = \frac{p}{\omega}e^{\omega t}$ and take

$$\hat{N} = \frac{1}{(2\pi)^{3/2}}\frac{\omega}{\sqrt{2}}.\qquad(139)$$

We derived thus the *fundamental* spinor solutions of positive and, respectively, negative frequencies, with energy E, momentum direction $\vec{n}$ and helicity λ. These spinors are charge-conjugated to each other,

$$V_{E,\vec{n},\lambda} = (U_{E,\vec{n},\lambda})^c = \mathcal{C}(\overline{U}_{E,\vec{n},\lambda})^T,\quad \mathcal{C} = i\gamma^2\gamma^0,\qquad(140)$$

and satisfy the orthonormalization relations

$$\langle U_{E,\vec{n},\lambda}, U_{E,\vec{n}',\lambda'}\rangle = \langle V_{E,\vec{n},\lambda}, V_{E,\vec{n}',\lambda'}\rangle = \delta_{\lambda\lambda'}\delta(E-E')\,\delta^2(\vec{n}-\vec{n}'),\qquad(141)$$

$$\langle U_{E,\vec{n},\lambda}, V_{E,\vec{n}',\lambda'}\rangle = \langle V_{E,\vec{n},\lambda}, U_{E,\vec{n}',\lambda'}\rangle = 0.\qquad(142)$$

deduced as in the Appendix B. Moreover, the completeness relation

$$\int_0^\infty dE \int_{S^2} d\Omega_n \sum_\lambda \{U_{E,\vec{n},\lambda}(t,\vec{x})[U_{E,\vec{n},\lambda}(t,\vec{x}')]^+$$

$$+ V_{E,\vec{n},\lambda}(t,\vec{x})[V_{E,\vec{n},\lambda}(t,\vec{x}')]^+\} = e^{-3\omega t}\delta^3(\vec{x}-\vec{x}'),\qquad(143)$$

indicates that this system of solutions is complete. We say that this represents the *energy-helicity* basis.

Finally, we derive the transition coefficients transforming the momentum-helicity and the energy-helicity bases among themselves. After a few manipulation we find that these coefficients,

$$\left\langle U_{\vec{p},\lambda}, U_{E,\vec{n},\lambda'} \right\rangle = \left\langle V_{\vec{p},\lambda}, V_{E,\vec{n},\lambda'} \right\rangle^{*} = \delta_{\lambda\lambda'} \frac{p^{-\frac{3}{2}}}{\sqrt{2\pi\omega}} \delta^{2}(\vec{n} - \vec{n}_{p}) e^{-i\frac{E}{\omega}\ln\frac{p}{\omega}}, \tag{144}$$

$$\left\langle U_{\vec{p},\lambda}, V_{E,\vec{n},\lambda'} \right\rangle = \left\langle V_{\vec{p},\lambda}, U_{E,\vec{n},\lambda'} \right\rangle = 0, \tag{145}$$

are similar to those of the scalar modes (70). Therefore, the particle wave functions $a(\vec{p}, \lambda)$ and $a(E, \vec{n}, \lambda)$ are related among themselves by similar unitary transformations as (71) and (72) but conserving, in addition, the helicity. These transformations preserve the vacuum state since the antiparticle wave functions have the same properties and the particle and antiparticle Hilbert spaces remain orthogonal to each other, as it results from Eq. (145).

5.3 Quantization and propagators

The quantization of the Dirac field can be done easily in the helicity bases as well as in the spin one. We assume that the wave functions of the momentum-helicity basis, $a(\vec{p}, \lambda)$ and $b(\vec{p}, \lambda)$, become field operators (so that $b^{*} \to b^{\dagger}$) satisfying the standard anticommutation relations from which the non-vanishing ones are

$$\{a(\vec{p}, \lambda), a^{\dagger}(\vec{p}', \lambda')\} = \{b(\vec{p}, \lambda), b^{\dagger}(\vec{p}', \lambda')\} = \delta_{\lambda\lambda'}\delta^{3}(\vec{p} - \vec{p}'). \tag{146}$$

The corresponding anticommutation rules of the energy-helicity basis are

$$\{a(E, \vec{n}, \lambda), a^{\dagger}(E', \vec{n}', \lambda')\} = \{b(E, \vec{n}, \lambda), b^{\dagger}(E', \vec{n}', \lambda')\}$$
$$= \delta_{\lambda,\lambda'}\delta(E - E')\delta^{2}(\vec{n} - \vec{n}'). \tag{147}$$

We say that this quantization is *canonical* since the equal-time anticommutator takes the standard form (Drell & Bjorken, 1965)

$$\{\psi(t, \vec{x}), \overline{\psi}(t, \vec{x}')\} = e^{-3\omega t}\gamma^{0}\delta^{3}(\vec{x} - \vec{x}'), \tag{148}$$

as it results from Eq. (122). In addition, we know that the mode separation we use defines a stable vacuum state. Therefore, we have to construct the Fock space canonically, applying the creation operators upon the unique vacuum state $|0\rangle$.

The one-particle operators corresponding to the isometry generators can be calculated as in the scalar case using the definition $\mathcal{X} =: \langle \psi, X\psi \rangle :$. The operators which do not come from differential operators have to be defined directly giving their mode expansions (Cotăescu, 2002). It is remarkable that all these operators have similar properties to those of the scalar field presented here or of the vector fields studied in (Cotăescu, 2010; Cotăescu & Crucean, 2010). The most interesting result is the expansion of the energy operator in the momentum-helicity basis where we may use the identity

$$H U_{\vec{p},\lambda}(t, \vec{x}) = -i\omega \left(p^{i}\partial_{p^{i}} + \frac{3}{2} \right) U_{\vec{p},\lambda}(t, \vec{x}), \tag{149}$$

and the similar one for $V_{\vec{p},\lambda}$, leading to the expansion

$$\mathcal{H} = \frac{i\omega}{2} \int d^3 p\, p^i \sum_\lambda \left[a^\dagger(\vec{p},\lambda) \stackrel{\leftrightarrow}{\partial}_{p^i} a(\vec{p},\lambda) + b^\dagger(\vec{p},\lambda) \stackrel{\leftrightarrow}{\partial}_{p^i} b(\vec{p},\lambda) \right] \tag{150}$$

which depend on the phase factors of the field operators as in the scalar or vector cases.

The Green functions can be expressed in terms of anticommutator functions,

$$S^{(\pm)}(t,t',\vec{x}-\vec{x}') = i\{\psi^{(\pm)}(t,\vec{x}),\, \overline{\psi}^{(\pm)}(t',\vec{x}')\}, \tag{151}$$

and $S = S^{(+)} + S^{(-)}$ which can be written as mode integrals that can be analytically solved (Kosma & Prokopec, 2009). These functions are solutions of the Dirac equation in both their sets of coordinates and helped us to write down the Feynman propagator,

$$\begin{aligned} S_F(t,t',\vec{x}-\vec{x}') &= i\,\langle 0|\, T[\psi(x)\overline{\psi}(x')]\,|0\rangle \\ &= \theta(t-t')S^{(+)}(t,t',\vec{x}-\vec{x}') - \theta(t'-t)S^{(-)}(t,t',\vec{x}-\vec{x}'), \end{aligned} \tag{152}$$

and the retarded and advanced Green functions $S_R(t,t',\vec{x}-\vec{x}') = \theta(t-t')S(t,t',\vec{x}-\vec{x}')$ and respectively $S_A(t,t',\vec{x}-\vec{x}') = -\theta(t'-t)S(t,t',\vec{x}-\vec{x}')$, which satisfy the specific equation,

$$[E_D(x) - m]S_F(t,t',\vec{x}-\vec{x}') = -e^{-3\omega t}\delta^4(x-x'), \tag{153}$$

of the spinor Green functions on the de Sitter space-time (Cotăescu, 2002).

6. Concluding remarks

We presented here the complete quantum theory of the massive scalar and Dirac free fields minimally coupled to the gravity of the de Sitter expanding universe. Applying similar methods we succeeded to accomplish the theory of the Proca (Cotăescu, 2010) and Maxwell (Cotăescu & Crucean, 2010) fields on this background. The main points of our approach are the theory of external symmetry (Cotăescu, 2000; 2009) that provides us with the conserved operators of the fields with any spin and the Schrödinger time-evolution picture (Cotăescu, 2007) allowing us to derive new sets of fundamental solutions.

The wave functions defining quantum modes are solutions of the field equations and common eigenfunctions of suitable systems of commuting operators which represent conserved observables *globally* defined on the de Sitter manifold. All these observables form the global apparatus which *prepares* global quantum modes as it seems to be natural as long as the field equations are global too. In this manner, we obtain wave functions correctly normalized on the whole background such that the Hilbert spaces of the particle and respectively antiparticle states remain orthogonal to each other in *any* frame, assuring thus the stability of a vacuum state which is of the bunch-Davies type (Bunch & Davies, 1978).

The new energy bases introduced here completes the framework of the de Sitter quantum theory, being crucial for understanding how the energy and momentum can be measured simultaneously. We may convince that considering the simple example of a Klein-Gordon particle in the state

$$|\chi\rangle = \int d^3 p\, \chi(\vec{p}) a^\dagger(\mathbf{p})|0\rangle, \quad \chi(\vec{p}) = \rho(\vec{p})e^{-i\vartheta(\vec{p})}, \tag{154}$$

determined by the functions $\rho, \theta : \mathbb{R}_p^3 \to \mathbb{R}$. The normalization condition

$$1 = \langle \chi | \chi \rangle = \int d^3p \, |\chi(\vec{p})|^2 = \int d^3p \, |\rho(\vec{p})|^2 \tag{155}$$

shows that the function ρ must be square integrable on $\mathbb{R}_p^3$ while θ remains an arbitrary real function. Furthermore, according to Eqs. (81) and (86), we derive the expectation values of the non-commuting operators $\mathcal{P}^i$ and $\mathcal{H}$,

$$\langle \chi | \mathcal{P}^i | \chi \rangle = \int d^3p \, p^i \, |\rho(\vec{p})|^2 \,, \quad \langle \chi | \mathcal{H} | \chi \rangle = \omega \int d^3p \, [p^i \partial_{p_i} \vartheta(\vec{p})] \, |\rho(\vec{p})|^2 \,. \tag{156}$$

The expectation values of the momentum operators are independent on the phase θ while that the energy operator depends mainly on it. This means that we can prepare at anytime states with arbitrary desired expectation values of both these observables. Thus, in the particular case of $\theta(\vec{p}) = \epsilon \ln(p)$ we obtain $\langle \chi | \mathcal{H} | \chi \rangle = \omega \epsilon = E$ indifferent on the form of ρ if this obeys the condition (155). In other respects, we observe that the dispersion of the energy operator in this particular state,

$$\mathrm{disp}\mathcal{H} = \langle \chi | \mathcal{H}^2 | \chi \rangle - \langle \chi | \mathcal{H} | \chi \rangle^2 = \omega^2 \int d^3p \, |p^i \partial_{p_i} \rho(\vec{p})|^2 - \tfrac{9}{4}\omega^2 \,, \tag{157}$$

depends only on the momentum statistics given by the function ρ. Thus we can say that we meet a new quantum mechanics on the de Sitter expanding universe.

We conclude that our approach seems to be coherent at the level of the relativistic quantum mechanics where the conserved observables of different time-evolution pictures are correctly defined allowing us to derive complete sets of quantum modes. Consequently, the second quantization can be performed in canonical manner leading to quantum free fields, one-particle operators and Green functions with similar properties to those of special relativity. Under such circumstances, we believe that we constructed the appropriate framework for studying quantum effects of interacting fields on the de Sitter background. Assuming that the quantum transitions are measured by the *same* global apparatus which prepares the free quantum states we may use the perturbation theory for deriving transition amplitudes as in the flat case.

Finally we specify that our attempt to use quantum modes globally defined does not contradict the general concept of local measurements (Birrel & Davies, 1982) which is the only possible option when the isometries (or other symmetries) are absent and, consequently, the global apparatus does not work.

7. Appendix

Appendix A: Some properties of Hankel functions

Let us consider the functions Z_k depending on the Hankel functions $H_\nu^{(1,2)}(z)$ (Abramowitz & Stegun, 1964) as

$$Z_k(z) = e^{-\pi k/2} H_{ik}^{(1)}(z) \,, \quad Z_k^*(z) = e^{\pi k/2} H_{ik}^{(2)}(z) \,. \tag{158}$$

where $z, k \in \mathbb{R}$. Then, using the Wronskian W of the Bessel functions we find that

$$Z_k^*(z) \overset{\leftrightarrow}{\partial_z} Z_k(z) = W[H_{ik}^{(2)}, H_{ik}^{(1)}](z) = \frac{4i}{\pi z} \,. \tag{159}$$

A special case is of the Hankel functions $H_{\nu_\pm}^{(1,2)}(z)$ of indices $\nu_\pm = \frac{1}{2} \pm ik$ where $z, k \in \mathbb{R}$. These are related among themselves through

$$[H_{\nu_\pm}^{(1,2)}(z)]^* = H_{\nu_\mp}^{(2,1)}(z), \tag{160}$$

satisfy the equations

$$\left(\frac{d}{dz} + \frac{\nu_\pm}{z}\right) H_{\nu_\pm}^{(1)}(z) = ie^{\pm\pi k} H_{\nu_\mp}^{(1)}(z), \quad \left(\frac{d}{dz} + \frac{\nu_\pm}{z}\right) H_{\nu_\pm}^{(2)}(z) = -ie^{\mp\pi k} H_{\nu_\mp}^{(2)}(z) \tag{161}$$

and the identities

$$e^{\pm\pi k} H_{\nu_\mp}^{(1)}(z) H_{\nu_\pm}^{(2)}(z) + e^{\mp\pi k} H_{\nu_\pm}^{(1)}(z) H_{\nu_\mp}^{(2)}(z) = \frac{4}{\pi z}. \tag{162}$$

Appendix B: Normalization integrals

In spherical coordinates of the momentum space, $\vec{n} \sim (\theta_n, \phi_n)$, and the notation $\vec{q} = \omega s \vec{n}$, we have $d^3 q = q^2 dq\, d\Omega_n = \omega^3 s^2 ds\, d\Omega_n$ with $d\Omega_n = d(\cos\theta_n) d\phi_n$. Moreover, we can write

$$\delta^3(\vec{q} - \vec{q}') = \frac{1}{q^2} \delta(q - q')\delta^2(\vec{n} - \vec{n}') = \frac{1}{\omega^3 s^2} \delta(s - s')\delta^2(\vec{n} - \vec{n}'), \tag{163}$$

where we denoted $\delta^2(\vec{n} - \vec{n}') = \delta(\cos\theta_n - \cos\theta_n')\delta(\phi_n - \phi_n')$.

The normalization integrals can be calculated according to Eqs. (66) and (163), that yield

$$\langle f_{E,\vec{n}}, f_{E',\vec{n}'}\rangle = i \frac{N^2(2\pi)^3}{\omega^3} \delta^2(\vec{n} - \vec{n}') \int_0^\infty \frac{ds}{s} e^{i(\epsilon - \epsilon')\ln s} \left[Z_k^*(se^{-\omega t}) \overset{\leftrightarrow}{\partial_t} Z_k(se^{-\omega t})\right]. \tag{164}$$

The final result has to be obtained using Eq. (159) and the representation

$$\frac{1}{2\pi\omega} \int_0^\infty \frac{ds}{s} e^{\frac{i}{\omega}(E - E')\ln s} = \delta(E - E'). \tag{165}$$

8. References

Abramowitz, M. & Stegun, I. A. *Handbook of Mathematical Functions*, Dover

Allen, B. & Jacobson, T. (1986). Vector two-point functions in maximally symmetric spaces, *Commun. Math. Phys.*, 103, 669-692, ISSN 0010-3616

Birrel, N. D. & Davies, P. C. W. (1982). *Quantum Fields in Curved Space*, Cambridge University Press, Cambridge.

Bunch, T. S. & Davies, P. C. W. (1978). Quantum Field Theory in De Sitter Space: Renormalization by Point-Splitting, *Proc. R. Soc. Lond. A*, 360, 117-134, ISSN 1471-2946

Candelas, P. & D. J. Raine, D. J. (1975). General-relativistic quantum field theory: An exactly soluble model, *Phys. Rev. D*, 12, 965-974, ISSN 1550-7998

Carter, B. & McLenaghan, R. G. (1979). Generalized total angular momentum operator for the Dirac equation in curved space-time, *Phys. Rev. D*, 19, 1093-1097, ISSN 1550-7998

Chernikov, N. A. & Tagirov, E. A. (1968). Quantum theory of scalar field in de Sitter space-time, *Ann. Inst H. Poincaré* IX, 1147, ISSN 0246-0211

Cotăescu, I. I. (2000). External Symmetry in General Relativity, *J. Phys. A: Math. Gen.*, 33, 9177-9127, ISSN

Cotăescu, I. I. (2002). Polarized Dirac Fermions in de Sitter space-time, *Phys. Rev. D*, 65, 084008-084008-9, ISSN 1550-7998

Cotaescu, I. I.; Racoceanu, R.; Crucean, C. (2006). Remarks on the spherical waves of the Dirac field on de Sitter space-time, *Mod. Phys. Lett. A* 21, 1313-1318, ISSN 0217-7323

Cotăescu, I. I. (2007). The Schrödinger picture of the Dirac quantum mechanics on spatially flat Robertson-Walker backgrounds, *Mod. Phys. Lett. A*, 22, 2965-2969, ISSN 0217-7323

Cotăescu, I. I.; Crucean, C. ; Pop, A. (2007). The quantum theory of the scalar fields on the de Sitter expanding universe, *Int. J. Mod. Phys. A*, 23, 2563-2677, ISSN 0217-751X

Cotăescu, I. I. & Crucean, C.(2008). New Dirac quantum modes in moving frames of the de Sitter space-time *Int. J. Mod. Phys. A*, 23, 3707-3720, ISSN 0217-751X

Cotăescu, I. I. (2009). On the universality of the Carter and McLenaghan formula, *Europhys. Lett.*, 86, 20003-20003-p3, ISSN 0295-5075

Cotăescu, I. I. (2009). Polarized vector bosons on the de Sitter Expanding Universe, *Gen. Relativity Gravitation*, 42, 861-877 , ISSN 0001-7701

Cotăescu, I. I. & Crucean, C. (2010). The quantum theory of the free Maxwell field on the de Sitter expanding universe, *Prog. Theor. Phys*, 124, 1051-1066, ISSN 0033-068X

Cotăescu, I. I. (2011). The physical meaning of the de Sitter invariants, *Gen. Relativity Gravitation* 43, 1639-1656 , ISSN 0001-7701

Cotăescu, I. I. (2011). The free Dirac spinors of the spin basis on the de Sitter expanding universe, *Mod. Phys. Lett. A*, 26, 1613-1619 ,ISSN 0217-7323

Drell, S. & Bjorken, J. D. (1965). *Relativistic Quantum Fields* McGraw-Hill Book Co., New York

Koksma, J. F. & Prokopec, T. (2009). The fermion propagator in cosmological spaces with constant deceleration, *Class. Quantum Grav.*, 26, 125003, ISSN 0264-9381

Lawson, H. B. Jr. & Michaelson, M.-L.(1989). *Spin Geometry*, Princeton Univ. Press. Princeton

Shishkin, G. V. (1991). Some exact solutions of the Dirac equation in gravitational fields, *Class. Quantum Grav.*, 8, 175-185, ISSN 0264-9381

Thaller, B. (1992). *The Dirac Equation*, Springer Verlag, Berlin Heidelberg

Tung, W.-K. (1984). *Group Theory in Physics*, World Sci., Philadelphia

Unruh, W. G. (1976). Notes on black-hole evaporation *Phys. Rev. D*, 14, 879-892, ISSN 1550-7998

Wald, R. M. (1984). *General Relativity*, Univ. of Chicago Press, Chicago and London

8

Gravitational Quantisation and Dark Matter

Allan Ernest

Charles Sturt University
Australia

1. Introduction

This chapter looks at the concept of gravitational quantisation and the intriguing possibility that it may enable understanding one of the most mysterious problems in astrophysics: the nature and origin of dark matter. The concept of gravitational quantisation is relatively new, and from the traditional quantum mechanical viewpoint it raises questions about the applicability of quantum mechanics to gravitational fields, and also questions about the applicability of quantum mechanics on macroscopic scales, because the quantisation states of gravitational fields are sometimes large. Quantum physics was after all, originally developed to describe the behaviour of electrons in atoms and quickly became the recognised way to accurately model the physics of atomic-sized systems. Although it was an obvious extension that the structure of nuclei, which were even smaller, could also be well described using a wave-mechanical approach, for many years this was the limited domain in which quantum theory operated. Nevertheless the success of applying quantum physics to nuclear phenomena showed (1) that quantum theory was appropriate to potentials other than electrical (in this case the strong nuclear force) and (2) that it provided a correct description of nature over a range of scales (< ~2 fm for nuclear structure compared to > ~50 pm for atoms). The success in modelling nuclear phenomena did not necessarily mean however that quantum theory was suitable for describing nature over all scales and that it applied to all other types of potentials. For example, the existence of a region where quantum physics breaks down and classical physics takes over still remains a debated issue, but if such a region exists, at what scale does it happen? And secondly, for what other potentials or pseudo-potentials might quantum theory be valid and how would this validity be demonstrated?

With respect to the first question, we argue that there are at present no experiments that invalidate quantum theory at any scale, and that quantum-based predictions of classical or macroscopic measurements are expected to equally agree with those of classical calculations, provided careful attention is paid to effects such as decoherence (Zurek, 1981, 1982, Chou et al., 2011, Lamine et el., 2011, Jaekel et al., 2006, Schlosshauer, 2007). Indeed isolating a consistent value for the scale of the so-called mesoscopic region between the quantum and classical domain, above which classical theory would dominate, has proved elusive and experiments have continued to demonstrate the applicability of quantum physics at macroscopic scales. However, as the many newly-observed macroscopic quantum phenomena demonstrate, this does not preclude the possibility that quantum theory may lead to additional novel macroscopic phenomena that have no classical analogue. Examples include experiments involving superconductivity and superconducting interference devices, and the experiments with entangled photons, atoms and molecules (Nakamura et al., 1999, Ketterle, 2002, Van der Wal et al., 2000, Zbinden et al., 2001, Schmitt-Manderbach et al., 2007).

With respect to the second point, recent experiments have also observed quantum effects in potentials and pseudo potentials other than electromagnetic or nuclear. In seminal work by Nesvizhevsky et al., 2002, quantised states of the gravitational field were observed in the laboratory. These remarkable experiments demonstrate for the first time that particles form gravitational eigenstates in gravitational potential wells, and hence that particles in gravitational potentials conform to the laws of quantum physics in the same way that electrons do in the electrical potentials around nuclei (despite any apparent but ongoing inconsistency between quantum mechanics and general relativity). If relatively pure gravitational quantum eigenstates can form in a laboratory situation then the question arises as to whether such eigenstates might exist naturally elsewhere in the universe, and if so, what their theoretical properties might be.

Research in the area of quantum gravity has been active for some time (for general reviews see DeWitt and Esposito, 2007, Rovelli, 2008 and Woodard, 2009). It should be noted that the aim in this chapter is not to develop a theory of quantum gravity. Quantum gravity seeks to produce a unified theory of quantum physics and general relativity under all conditions, particularly in regions of strong gravity where classical Newtonian approaches break down. Such a theory does not yet exist and is not needed in the current context, where a Newtonian formulation of gravitational quantum theory is used. The purpose here is to examine properties of the predicted quantum based eigenstates that exist predominantly in the weak gravity regions of (possibly deep) gravitational wells and to study the behaviour of particles in these regions using the traditional quantum eigenspectral decomposition of the particle wavefunctions in terms of their energy eigenstate basis vectors. That is, we do not include eigenstates that might have significant amplitude fractions in regions of strong gravity such as near black holes (and such states should not be needed as they form a small fraction of the eigenspectral decomposition for particles in weak gravity regions), and we assume that those states that are included in the decomposition may be approximated by ignoring any small fraction of their eigenstate function that does encroach on such regions. It will turn out that the use of a quantum gravitational approach introduces novel properties to particles that enable dark matter to arise as a natural consequence of cosmic evolution without the need for new particles or physics beyond traditional quantum theory. Coincidentally we will see that not only can gravitational quantisation potentially solve the dark matter problem, but also that it compels the introduction of a new paradigm for the macroscopic description of particles and their interaction properties.

The first evidence that dark matter might exist appeared over 70 years ago with Zwicky's observations of high rotation velocities of galaxies in the Virgo cluster which pointed to excess unseen mass (Zwicky, 1937). About 30 years later (Rubin, 1970) showed that the orbital speeds of stars and gas within galaxies did not fall off with radial position in a Keplarian manner as expected, but maintained a constant velocity as far out as could be measured. These galactic rotation curves seemed to clearly show that galaxies also contained mass beyond that that would be expected from their visible component. Significantly this "missing" mass is not a small fraction of the visible component. Instead it dominates, making over 80% of the expected 27% matter content of the universe, and much more than this in some galaxies. Because these controversial observations seemed inexplicable and had such radical implications for the understanding of the universe they were initially treated with scepticism. Evidence has continued however to point to a universe whose dominant matter content remains a mystery. Strong evidence supporting this hypothesis also comes from observations of gravitational lensing and also those from the Wilkinson Microwave Anisotropy Probe (WMAP). These 'dark' particles are essentially invisible: as far as observations reveal, they do not radiate energy, are transparent to electromagnetic radiation and weakly interacting with ordinary (baryonic) particles.

Cosmologists' currently favoured solution is the cold dark matter theory (CDM) or the more recent modification that includes dark energy, lambda cold dark matter (LCDM) (Primack, 2001). The theory is based then on the hypothetical existence of an as yet undiscovered weakly interacting elementary particle. Although no such particle is predicted from the standard model of particle physics, there are extensions to the standard model such as supersymmetry (Feng, 2010) that potentially have theoretically predicted particles that might function as long-lived weakly interacting particles. Given that such a particle exists, numerical simulations of LCDM have been developed, and these have been very successful in predicting the very large scale structure of the universe (see Croft and Estathiou, 1993, Colberg et al., 1997). In these simulations the structure of the universe grows by gravitational coalescence of dark matter particles to initially form small agglomerations that then combine in a process of hierarchical merging (Knebe, 1998, Diemand and Moore, 2009) to form the clusters and large galaxies seen today. Numerical simulations initially using particle masses of $\sim 10^{41}$ kg (because of computational constraints) and later smaller masses both give similar and excellent prediction of large scale universal structure.

Despite the successes of LCDM cosmology on large scales, it faces some very serious challenges on the galactic scale and below (Kroupa et al., 2010). The problems have been so difficult that some cosmologists have questioned the validity of the entire LCDM theory. The numerical simulations of CDM produce steep cusp-like density profiles at low galactic halo radii, an overabundance of satellite galaxies and problems with the predicted angular momentum (ibid.). Additionally, the observation of fully formed galaxies like the Milky Way at very early times in cosmic history presents a formation mechanism problem since it is difficult to understand how this could have happened via hierarchical merging, a cornerstone of LCDM cosmology for understanding galaxy formation, in the limited time available (ibid.).

There have been several proposed alternate theories to the idea that the dark matter observations arise as the result of hidden mass. One group of these alternative solutions is based on the notion that the equations describing gravity on large scales may need to be modified. The idea, originally due to Milgrom (1983) - Modified Newtonian Dynamics or MOND has now many variations that include relativity and generalisations of quantum gravity. These theories all have the effect of modifying gravity in such a way as to explain the galactic rotation curves and thus also remove the need for the explanation of the dark matter observations in terms of unseen excess mass. However evidence that the "dark matter" observations are truly the result of the existence of extra unseen matter continues to grow and it has become increasingly difficult to explain dark matter in terms of modified gravity. One group of relatively recent observations compelling for the bona-fide existence of dark matter are those of galactic cluster collisions such as the famous Bullet cluster, 1E 0657-56. In the Bullet cluster evidence is seen of the separation of dark matter from the visible components as a result of the collision. The dominant mass component, the dark matter, along with the individual galaxies within the cluster have passed through one another largely unaffected, but the gas components have interacted and are seen between the two now separated dark matter/galactic cluster components (Clowe et al., 2006).

The suggestion that gravitational quantisation might play a direct role in the solution to the dark matter problem was first made by Ernest (2001) and later using a fundamentally different approach by Chavda and Chavda (2002). In the former case (the approach in this chapter) Ernest considered the effects that the eigenspectral array describing the wavefunctions of traditional particles has on their interaction properties, while Chavda and Chavda consider the bound quantum eigenstate of two micro black holes and suggest that due to their binding, the constituent black holes of such a system will not lose mass in the expected manner (i.e from Hawking radiation). In this way micro black holes formed in the early universe could be still present today rather than having evaporated at earlier times and hence function as an alternative dark matter candidate like the neutralino.

2. Naturally occurring gravitational eigenstates: Quantum wavefunctions with limited eigenspectral range and the connection with dark matter

2.1 Conditions for naturally occurring gravitational eigenstates

Predicting the behaviour of particles in gravitational fields using quantum mechanics is in principle simple. One applies the non-relativistic Schrodinger equation to a Newtonian potential to yield its eigenstates and energies. Once the eigenstates are known, the temporal evolution of any particle in that potential is determined from the temporal evolution of the eigenspectral array that describes the initial wavefunction, essentially what is classically the initial conditions of the particle. Several authors have developed theoretical and numerical solutions to these types of equations under various conditions (Bernstein et al., 1998, Doran et al., 2005, Vachaspati, 2005, Gossel et al., 2010) and used them to make predictions about various physical phenomena ranging from astrophysical processes such as rates of accretion onto black holes, to understanding the behaviour of a gravitational Bose-Einstein condensate.

Very little work has been done however to investigate the potential existence or theoretical properties of gravitational eigenstates and their consequences for astrophysics (Ernest 2009a, 2009b). The simplest gravitational eigenstate system would feasibly consist of two neutral, spin-zero elementary-type particles in a bound state. The problem for such a system is that the binding energy is unrealistically small. For example, if the masses were $\sim 10^{-27}$ kg, the most highly bound state, with principle quantum number $n = 1$ has a binding energy of $\sim -10^{-69}$ eV which is minuscule compared to typical universally pervading energies, for example, cosmic microwave background photons ($\sim 2 \times 10^{-4}$ eV) or the observed dark energy component of ~ 1 GeV m-3 (Tegmark et al., 2004).

The easiest way to increase the binding energy is to increase one or more of the two masses of this two component system. By pushing the component masses to 'grains' of $\sim 10^{-13}$ kg each, the energy of the $n = 1$ state becomes a healthy -10 eV. But there are things to consider than just the binding energy. Given typical densities of $10^3 -$

10^4 kg m^{-3} for the masses, the physical size of each is $\sim 10^{-6}$ m. The initial and simplest mathematical approach involves two single point potentials and clearly requires that the 'physical extent' of the masses should not encroach into any significant fraction of the space occupied by the eigenfunction, otherwise the effective potential between the two changes, and the description in terms of a simple Schrödinger equation breaks down. For $n = 1$, the scale of the -10 eV energy eigenstate above is $\sim 10^{-19}$ m so that this condition is clearly not satisfied. For higher n states, the position of the wavefunction is shifted to larger radii, which alleviates this difficulty, but the binding energy of the eigenstate approaches zero. One way to circumvent these difficulties is to consider a small elementary type particle (say of mass 1.7×10^{-27} kg) bound in the central potential well of a much larger mass and consider only large values of n, and particularly the high angular momentum states where $l \sim n$ (see Table 1)

Central mass M (kg)	n value at eigenstate energy ~ 1 eV	Eigenstate 'size' for $l \sim n$ (m)	Physical radius of central mass (m) assuming density $\sim 5 \times 10^3$ kg m^{-3}
10^{10}	7.6×10^2	3.5×10^{-9}	7.6×10^1
10^{20}	7.6×10^{12}	3.5×10^1	1.6×10^5
6×10^{24} (M_{Earth})	4.6×10^{17}	2.1×10^6	6.4×10^6
10^{30} (M_{Sun})	7.6×10^{22}	3.5×10^{11}	3.5×10^8
10^{40}	7.6×10^{32}	3.5×10^{21}	7.6×10^{11}

Table 1. Eigenstate 'size' versus central mass size for a fixed binding energy (1 eV $\sim$ 5000 K) and 'orbit mass' $= 1.7 \times 10^{-27}$ kg.

With a central point mass potential well, for a given fixed binding energy E_n, the quantum number n is proportional to the central mass M. (see solutions to equation (1) below). Additionally, for the high angular momentum states ($l \sim n$), the average effective eigenstate radius and thickness are proportional to M and $M^{1/2}$ respectively, while the physical size of the central mass grows only as $M^{1/3}$. Thus although the value of n required to form an eigenstate which 'clears' the physical extent of the central mass increases as the central mass is increased, the critical value of n required to do this corresponds to a more highly bound structure the larger M is. Thus by increasing the central mass sufficiently it is possible to obtain a well-bound structure and maintain an eigenstate which does not encroach on the physical extent of the central mass. This behaviour is illustrated in Table 1. What constitutes 'well-bound' is arguable, but for a binding energy of 1eV and an 'orbit mass' of $\sim 10^{-27}$kg, Table 1 shows that the required central mass needs to be of the order or greater than the mass of the Earth. Certainly for the solar mass ($\sim 10^{30}$ kg) and above the condition is easily achieved and the analysis can be carried out using a single particle simple Newtonian-type Schrodinger equation.

2.2 Connection with dark matter

What connection does dark matter have with gravitational eigenstates? Quantum mechanics is simply an alternative way to model nature, most useful on small scales. But we do expect that quantum physics should equally well model all classical macroscopic phenomena, predicting essentially identical results, a belief echoed in the correspondence principle. Yet quantum theory has already yielded many novel macroscopic non-classical phenomena, so could a quantum mechanical description of the motion of elementary particles in a galactic potential also yield new insights? Remarkably, a quantum description of gravitationally bound particles does indeed predict new and intriguing effects. The reason comes from the predictions that quantum theory makes about variations in the interaction cross sections of particles based on their eigenspectra, which can exhibit unique properties in the case of macroscopic gravity.

2.2.1 Cross sections and eigenspectra

The concept of describing interactions in terms of a cross section has been one of the most fundamental and useful in physics. Simply put, cross sections measure the effective area that one particle (the 'target' particle) presents to another (the 'bullet') when interacting. Additionally, it is a measure of the rate at which a reaction occurs. In the laboratory, measurements of cross sections are made using a localised 'beam' of bullets, measuring the rate at which they 'hit' the target, and performing an analysis generally based on the assumption of a uniform incoming 'plane wave' of particles. But whilst it may seem reasonable, there is an implicit assumption in this about the nature of wavefunctions representing the particles. There is also often an assumption that a measured cross section is somehow independent of all but the most evident characteristics of the wavefunctions representing the particles involved in the measurement.

Some aspects of the wavefunction that affect the cross section are obvious. For example classically the 'chance' (and hence rate) at which two particles placed in a box will interact depends on the size of the box. This is trivially allowed for in the experiment by including the 'beam intensity' in the analysis. Likewise no one finds it surprising that cross sections depend on temperature because it is clear we are changing the fundamental nature of the interaction. Importantly though, this represents a specific example of how the wavefunction characteristics, in this case the eigenspectral wavelengths, affect the relevant overlap integrals that determine the rate of reaction and the cross section. But there are also subtleties of the wavefunctional form that can lead to dramatic changes in the resulting cross section of a particular interaction. In short, the measured cross section for any interaction is intimately linked to the eigenspectral array of the wavefunctions representing the particles involved. If the eigenspectral array from which the particle wavefunction is composed contains a significant fraction of states that are weakly interacting (aka 'dark eigenstates'), then the measured cross section for that interaction will be much reduced.

2.2.2 Dark eigenstates of the eigenspectral ensemble of a large gravity well

The remarkable thing about the eigenspectral ensemble of a large gravitational structure such as a galactic halo is that, in addition to the vast array of states which would normally make up the quantum description of any localised 'visible' particle, it also contains vast numbers of gravitational eigenstates that are weakly interacting or 'dark'. These dark states are the highly excited, high angular momentum (high (n,l)) states of Figure 1 (those closest to the left hand curve of the (n,l) diagram, that is the dotted curve $p = 1$ where we have

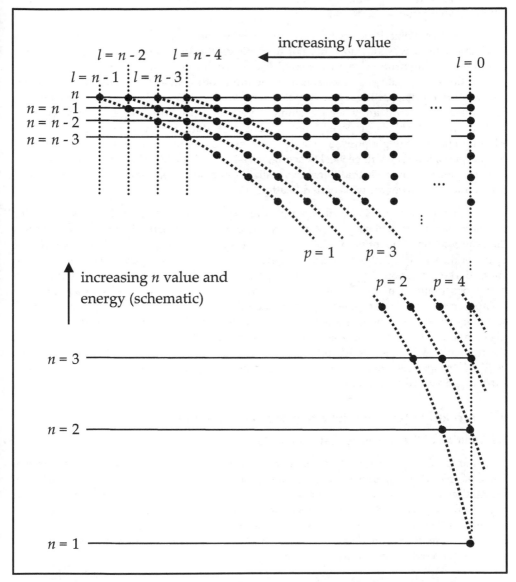

Fig. 1. Schematic (n,l) diagram showing the high n, l, m -valued stationary states, drawn to emphasize the values of the parameter $p = n - l$ (dotted curves). Each solid circle on the diagram represents $2l + 1$, z-projection (m-valued) substates.

introduced the notation $p \equiv n - l$) and are somewhat analogous to the Rydberg states in atoms, but have much long lifetimes, extreme stability and are much more robust because of the physical scale of the potential. A wavefunction whose eigenspectral array on the (n,l) diagram determines it as existing as a relatively localised particle in classical phase space predisposes it to having similar interaction cross sections to those determined in standard laboratory measurements.

We stress this point in another way: The localisation of a particle in phase space is determined by a wavefunction that not only yields the localisation, but also an average position and momentum in that space. There are however many other eigenspectral distributions that could yield the same average momentum and position in phase space but that would describe particles with very different cross sections. It is possible in principle to form an eigenspectral array with a similar average phase space location as a localised particle, but which is composed predominantly of dark gravitational quantum states. Although particles might possess a similar average momentum and position as their visible counterparts, but they will be weakly acting, invisible and unable to reach equilibrium with visible localised matter via traditional thermalizing interaction processes. Particles with such distributions can form the basis of dark matter.

Predominantly dark gravitational eigenspectral arrays represent an attractive solution for the dark matter problem because (1) they can potentially arise naturally in the universe as a direct prediction of quantum mechanics and (2) they can be used to explain the nature and origin of dark matter without the need for any new physics or new particles. As will be seen later, 'dark-gravitational-eigenstate' dark matter appears to possess properties that fit with many observations of dark matter behaviour, including an inability to gravitationally collapse or coalesce, a transparency to light and other parts of the electromagnetic spectrum, and does not suffer from the same problems associated with other CDM WIMP candidates. Eigenstructure halos formed from vast arrays of particles with predominantly high-l, dark eigenspectra enable the success of LCDM cosmology on large scales to be retained while yielding properties that remove the difficulties faced by LCDM on the galactic cluster/galaxy scale and below.

3. Schrödinger-type solutions to gravitational potentials

3.1 Solution of the simple central point potential

More complex arrangements of gravitational binding than that discussed in 2.1 are possible, for example agglomerations of elementary neutral particles bound in a 'collective' gravitational potential well in a similar manner to the structure of nuclear material. We begin here however with the Schrödinger approach to the central point potential and make refinements on this. The analysis for this has been done elsewhere and we give a brief summary here. (For more detail see Ernest, 2009a.) The gravitational form of Schrodinger's equation for a two particle system is written as

$$-\frac{\hbar^2}{2\mu}\nabla^2\psi - \frac{Gm_0 M_0}{r}\psi = i\hbar\frac{\partial\psi}{\partial t} \tag{1}$$

where M_0 and m_0 are the masses of the large central and small 'orbit' point particles respectively, μ is the reduced mass and the other symbols have their normal meanings. Separation of variables leads to the eigenenergies E_n and normalised eigenfunctions $u_{n,l,m}(\mathbf{r},t)$, which are given by $E_n = -\mu G^2 m_0{}^2 M_0{}^2/2\hbar^2 n^2$ and $u_{n,l,m}(\mathbf{r},t) = R_{n,l}(r)Y_{l,m}(\theta,\phi)$, where n, l and m are the standard quantum numbers. $Y_{l,m}(\theta,\phi)$ are the spherical harmonics. The limits of the angular eigenfunction components for large m and l are not significant for the present discussion and have been dealt with previously (Ernest, 2009a). The radial component $R_{n,l}(r) = N_{nl}(2r/nb_0)^l \exp(-r/nb_0)L_{n-l-1}^{2l+1}(2r/nb_0)$ is written in terms of Laguerre polynomials

$$L_{n-l-1}^{2l+1}(2r/nb_0) = (n+l)! \sum_{k=0}^{n-l-1} (-1)^{k+2l} (2r/nb_0)^k (n-l-1-k)!(2l+1+k)!k! \qquad \text{where}$$

$b_0 = \hbar^2/G\mu m_0 M_0$ is a scale factor analogous to the atomic case, and $N_{nl} = (2/nb_0)^{3/2}((n-l-1)!/2n(n+l)!)^{1/2}$ is a normalising constant. The radial eigenfunctions are very significant in developing the present theory and we concentrate on these from now on. The states are shown in Figure 1 where, for a central point-mass potential as in this diagram, the set of all l and m values is degenerate for any given value of n.

In Figure 1 the l-m-degenerate n levels are shown in proportion schematically to their state energy $\propto 1/n^2$ but it should be noted that their average radial positions behave in the opposite way: approximately $\propto n^2$. For the high-l, high-n states the average radial position is accurately written as $\langle r_n \rangle = n^2 b_0$. The radial eigenfunction extent $r_{max} - r_{min}$ spreads out as a square root dependence on the value of p, approximately centred on $n^2 b_0$. The position of the eigenfunctions is important because a sufficient density of occupied states can in principle modify the potential so it is no longer that of a central point-mass. The deviation from a point potential can be approximately allowed for by incorporating b_0 as a variable dependent on the enclosed variable mass $M(r)$. Each $M(r)$ gives an eigenvalue series of energies $E_{n_1}^{(M_1)}$, $E_{n_2}^{(M_2)}$, $E_{n_3}^{(M_3)}$ ... and the appropriate n chosen from each series $\{n_1\}$, $\{n_2\}$, $\{n_3\}$, ... using the value of $M(r)$ determined from its functional form. This method works but it is not the optimal approach.

3.2 Solutions for halo mass distributions and logarithmic potentials

Solutions to the gravitational central-point-potential Schrodinger equation given above can be shown to contain some states that possess the fundamental properties required for dark matter. The galaxy and clusters however are not point potentials but have extended mass distributions. As mentioned earlier, the radial velocity profiles of galaxies for example show constant velocities with radius as far out as can measured rather than the expected Keplarian decline. This implies a radial density profile that varies as $1/r^2$. The force is then proportional to the enclosed mass $M_{enclosed}$ which is therefore given by

$$F = \frac{Gm_0 M_{enclosed}}{r^2} = \frac{Gm_0}{r^2}\int_0^r \rho(r')4\pi r'^2\, dr' = \frac{Gm_0}{r^2}\int_0^r \frac{k}{r'^2}4\pi r'^2\, dr' = \frac{Gm_0 M_0}{R_0\, r}$$

where $k = M_0/R_0$ is the density proportion constant and R_0 is the hypothetical radius of the halo whose *total* mass out to R_0 is now M_0. Since $F = -\nabla V$ we have that

$$V(r) = -\int_\infty^{R_0} -GM_0/r^2 dr - \int_{R_0}^r -GM_0/R_0 r\, dr \quad \text{and}$$

$$V(r) = -\frac{Gm_0M_0}{r} \qquad (r \geq R_0)$$

$$= -\frac{Gm_0M_0}{R_0}\ln\left(\frac{R_0e}{r}\right) (r < R_0) \qquad (2)$$

where again m_0 is the 'orbit mass and e the natural logarithm base. This hybrid $1/r$-logarithmic potential differs from the equivalent point-mass, $1/r$ point potential substantially at low r although in principle the characteristics of the high-l, n quantum states of $1/r$ potentials that make them suitable as fundamental components of dark matter eigenspectral arrays will have equivalent states in the eigenstate ensembles of the real logarithmic halo potentials. In the central point-mass potential case it was possible (Ernest 2009a, 2009b) to obtain approximations for the energies and wavefunction forms of the eigenstates which can be used to obtain quantitative values for interaction rates etc. It is clearly important to be able to develop approximations for similar states in the real logarithmic potentials of the galactic halo case which we now do below.

For wavefunctions inside the halo, the logarithmic potential Schrödinger equation becomes

$$-\frac{\hbar^2}{2\mu}\nabla^2\psi - \frac{Gm_0M_0}{R_0}\ln\left(\frac{R_0}{r}\right)\psi = i\hbar\frac{\partial\psi}{\partial t} \qquad (3)$$

which, on separation of variables, gives the equation for the radial component $u_{n,l}(r)$ of the wavefunction as

$$-\frac{\hbar^2}{2\mu}\frac{\partial^2 u_{n,l}(r)}{\partial r^2} - \left(\frac{Gm_0M_0}{R_0}\ln\left(\frac{R_0}{r}\right) - E_{n,l}\right)u_{n,l}(r) = 0 \qquad (4)$$

The Schrödinger equation is known to be analytically solvable for $1/r$ and $-r^2$ potentials but equations with logarithmic potentials require approximation techniques. The equation can be recast into a standard form as

$$-\frac{\hbar^2}{2m_0}\frac{d^2 u_{n,l}(r)}{dr^2} + \left(-\frac{Gm_0M_0}{R_0}\ln\left(\frac{R_0e}{r}\right) + \frac{\hbar^2 l(l+1)}{2m_0r^2} - E_{n,l}\right)u_{n,l}(r) = 0 \qquad (5)$$

where l is the angular momentum quantum number. A few attempts have been made to solve this equation. Ciftci et al. (2003) developed approximate solutions for various power-law and logarithmic potentials based on extensions of the Laguerre solutions applicable to $1/r$ potentials. Their procedure however, when applied to the present case, involves summations (over n) with prohibitively large numbers of terms of Laguerre polynomial coefficients.

The technique used here involves first finding the eigenenergies of the states that correspond to the highest angular momentum for any given n value, that is, all the states that have $p = 1$ (i.e. $l = n-1$) for each n. These eigenstates have a thin single peaked radial form. For $m = m_{max} = \pm l$ the states are ring-like or circular. The energy may be deduced semi-classically noting that all the probability density lies at essentially the same radius r_n, giving the state 'potential' energy as $E_U(r_n) \approx -GM_0m_0 / R_0 \ln(R_0e / r_n)$. The number of polar oscillations is $l_{max} \approx n$, giving $\lambda \approx 2\pi r_n/n$, and momentum $m_0 v = \hbar n/r_n$ using the de Broglie relation. The equivalent "kinetic" energy is then $E_K \approx Gm_0 M_0/2R_0$ and the net energy of the state is therefore

$$E_{n,p=1}(r_n) = E_{n,l=n-1}(r_n) \approx -\frac{Gm_0 M_0}{R_0}\ln\left(\frac{R_0\sqrt{e}}{r_n}\right) \tag{6}$$

As the quantum number m decreases, the state form spreads gradually in the $\pm$ polar direction yielding thin annular shells, ultimately creating a closed spherical shell for $m = 0$. These states simply represent variations in the polar-azimuthal orientation and so (6) corresponds to the energy of all m sublevels of the l value. The energy of the lower m-states is the same as for the maximum value of m, both by symmetry and because classically lower m-values behave simply as mixtures of tilted circular states of equivalent energy.

For the circular states the velocity v can also be written as a (halo-constant) $v = \sqrt{(GM_0/R_0)}$, by equating the centripetal and gravitational force based on the enclosed mass. Combined with $m_0 v = \hbar n/r_n$, this allows r_n to be expressed in terms of n and leads to an expression for the energy level levels $E_{n,p=1}$ for the circular ($p=1$) states (compared with $E_{n,\,all\,p} = -\mu G^2 m_0^2 M^2/2\hbar^2 n^2$ in the centralised point-mass case). r_n and $E_{n,p=1}$ are given by

$$r_n = n\hbar\sqrt{\frac{R_0}{Gm_0^2 M_0}}$$

$$E_{n,p=1}(r_n) = E_{n,l=n-1}(r_n) \approx -\frac{Gm_0 M_0}{2R_0}\ln\left(\frac{Gm_0^2 M_0 R_0 e}{n^2\hbar^2}\right) \tag{7}$$

where the $E_{n,l}$ are the energy eigenvalues which will in general depend on both n and l.

Each of these states is itself the lowest energy eigenvalue in a set of eigenvalues all of which have the same value of l, say $l = l'$ and increasing values of n and p, that is $p = 1, 2, 3 \dots$. The energies are $E_1 = E_{n=l'+1,l=l',p=1}$, $E_2 = E_{n=l'+2,l=l',p=2}$, $E_3 = E_{n=l'+3,l=l',p=3}$ $\cdots$ with E_1 given by equation (7).

The quantity $\hbar^2 l(l+1)/2mr^2$ in equation (5) may be thought of as the 'centrifugal potential' and for each value of l we define the effective potential energy as

$$V_{effective}(r,l) = -\frac{Gm_0 M_0}{R_0}\ln\left(\frac{R_0 e}{r}\right) + \frac{\hbar^2 l(l+1)}{2m_0 r^2}, \quad r \le R_0$$

$$V_{effective}(r,l) = -\frac{Gm_0 M_0}{r} + \frac{\hbar^2 l(l+1)}{2m_0 r^2}, \quad r > R_0 \tag{8}$$

plots of which are shown in Figure 2.

The maximum energy depths E_d of the well minima and their corresponding radial positions are given by

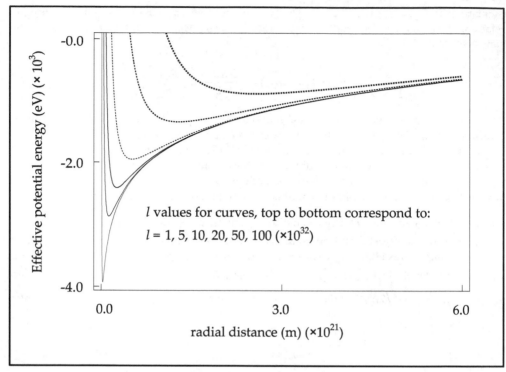

Fig. 2. Effective halo potentials for various l. $M_0 = 6.0 \times 10^{42}$ kg, $m_0 = 1.6 \times 10^{-27}$ kg, $R_0 = 3.0 \times 10^{21}$ m; curves, top to bottom, correspond to $l = 1, 5, 10, 20, 50, 100$ $\left(\times 10^{32} \right)$

$$E_d = -\frac{Gm_0 M_0}{2R_0} \ln\left(\frac{Gm_0{}^2 M_0 R_0 e}{l(l+1)\hbar^2} \right) \tag{9}$$

and

$$r_d = \sqrt{\frac{l(l+1)R_0\hbar^2}{Gm_0{}^2 M_0}} \tag{10}$$

Within each well (characterised by its l-value) the lowest the energy state (labelled say by $v = 0$) is a $p = 1$ eigenstate and corresponds to the n-value $n = l + 1$. For n states not too much greater than l we use a simple harmonic Taylor series approximation around the minimum, which is analytically solvable to obtain a set of energy eigenvalues and eigenfunctions for any $n > l + 1$ provided n is not too far above l. (For typical galactic halo parameters the harmonic approximation gives better than 1% accuracy to the potential over the region of the well encompassed by eigenstates with p from 1 to 10^{20} - see Figure 3).

The harmonic form of the Schrödinger equation is

$$-\frac{\hbar^2}{2m_0}\frac{d^2 u_v(r)}{dr^2} + \frac{1}{2}m_0\omega^2 r^2 u_v(r) - E_v u_v(r) = 0 \tag{11}$$

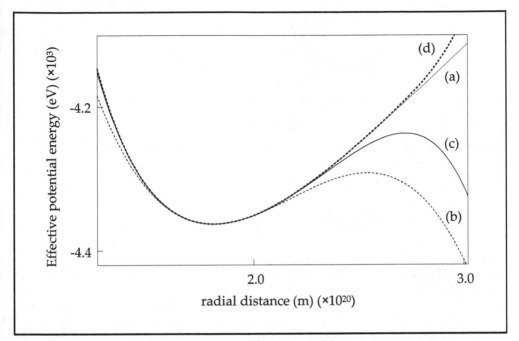

Fig. 3. Taylor approximations to the $l = 3.0 \times 10^{33}$ effective halo potential. $M_0 = 6.0 \times 10^{42}$ kg, $m_0 = 1.6 \times 10^{-27}$ kg, $R_0 = 3.0 \times 10^{21}$ m, Curve: (a) exact potential, (b) 3 term approximation (harmonic potential), (c) 5 term approximation, (d) 10 term approximation.

and putting $r' = r - r_d$ leads to an equation of the form of equation (11) equivalent to (5) as

$$-\frac{\hbar^2}{2m_0}\frac{d^2 y_v(r')}{dr'^2} + \left(-\frac{Gm_0 M_0}{2R_0 r_d^2} + \frac{3l(l+1)\hbar^2}{2m_0 r_d^4}\right)r'^2 y_v(r')$$
$$-\left(\frac{Gm_0 M_0}{R_0}\ln\left(\frac{eR_0}{r_d}\right) - \frac{l(l+1)\hbar^2}{2m_0 r_d^2} + E_n\right)y_v(r') = 0$$

(12)

provided that

$$\omega = \sqrt{-\frac{GM_0}{R_0 r_d^2} + \frac{3l(l+1)\hbar^2}{m_0^2 r_d^4}} = \frac{Gm_0 M_0}{R_0 \hbar}\sqrt{\frac{2}{l(l+1)}}$$

(13)

and

$$E_n = E_v + \frac{l(l+1)\hbar^2}{2m_0 r_d^2} - \frac{Gm_0 M_0}{R_0}\ln\left(\frac{eR_0}{r_d}\right) = E_v - \frac{Gm_0 M_0}{2R_0}\ln\left(\frac{Gm_0^2 M_0 R_0 e}{l(l+1)\hbar^2}\right)$$

(14)

The eigenvalue and eigenfunction solutions to (11) are those for the standard harmonic oscillator, given by

$$E_v = \left(v + \frac{1}{2}\right)\hbar\omega, \quad v = 0,1,2,...$$

$$u_v(r) = \left(\frac{m_0\omega}{\pi\hbar}\right)^{1/4} \frac{1}{\sqrt{2^v v!}} H_v(\xi)\exp\left(-\xi^2/2\right)$$

(15)

where $H_v(\xi)$ are the Hermite polynomials and $\xi = (r - r_d)\sqrt{m_0\omega/\hbar}$. Combining equations (14) and (15), noting that v is related directly to the p-value for the state by $v = p - 1 = n - l - 1$ and taking care to distinguish between quantum parameters that refer to the well compared to the states within that well, we can write a general formula for the eigenvalues of any high angular momentum (low-p) state of a logarithmic centrifugal well in terms of the quantum numbers n and l as

$$E_{n,l} = \left(\frac{Gm_0 M_0}{2R_0}\right)\left(\frac{\sqrt{2}(2n - 2l - 1)}{\sqrt{l(l+1)}} - \ln\left(\frac{Gm_0^2 M_0 R_0 e}{l(l+1)\hbar^2}\right)\right)$$

(16)

A schematic state diagram similar to that for the central point-mass state ensemble is shown in Figure 4, with a superimposed schematic for the harmonic approximation to the centrifugal well for an angular momentum quantum number l. The levels ($v = 0, 1, 2 ...$) of the centrifugal well schematic correspond to the increasing values of p and n and constant l, beginning with the state $n = l + 1$, $p = 1$. The lines of constant-n are no longer horizontal because of the additional dependence of the energy on l. It is of interest to compare the energy level differences between two adjacent l-values with two adjacent n-values since this gives a measure of the degree to which the presence of a logarithmic potential affects the otherwise degenerate l-states of the classic central point-mass potential. Equation (16) shows that for any given l, the state energy increases linearly with n as anticipated from the discussion of the circular states discussed earlier in this section. Indeed as l approaches n, equation (16) approaches equation (7), as expected. Putting $l \approx \sqrt{l(l+1)}$ and noting that l is close to n, partial differentiation of equation (16) with respect to n and l gives the energy spacing per unit change in n and l as: $\partial E_{n,l}/\partial n \approx \sqrt{2}Gm_0 M_0/lR_0$ and $\partial E_{n,l}/\partial l \approx -Gm_0 M_0(\sqrt{2}n - l)/l^2 R_0$ respectively. For $l \sim n$, $(\sqrt{2}n - l) \sim (\sqrt{2} - 1)l$, hence $\partial E_{n,l}/\partial l \approx \partial E_{n,l}/\partial n \times (\sqrt{2} - 1)/\sqrt{2} \approx 0.3\partial E_{n,l}/\partial n$ so that the energy spacing for adjacent l-states in the extended $1/r^2$ density distribution is a substantial fraction of the energy spacing of adjacent n-states, as illustrated in Figure 4, and in contrast to the angular momentum degeneracy observed for the eigenstates in the central point-potential. Also since $\partial E_{n,l}/\partial l$ is always negative, state energy increases as l decreases (at least for the $\sim 10^{20}$ p-values that are within the range of validity of the harmonic approximation) so that the low angular momentum states are less well bound than the high angular momentum ones for any given n. For the typical halo parameters $M_0/R_0 = 2.0 \times 10^{21}$ kg/m, $m_0 = 1.67 \times 10^{-27}$ kg , and taking $r = 1.0 \times 10^{21}$ m ($l \sim 6 \times 10^{33}$) as a typical outer halo position, the energy spacing for adjacent n states is $\sim 4.8 \times 10^{-32}$ eV, and for l states is $\sim 1.6 \times 10^{-32}$ eV . If the same procedure used here is applied to a standard central point-mass potential then the spacing between the harmonic

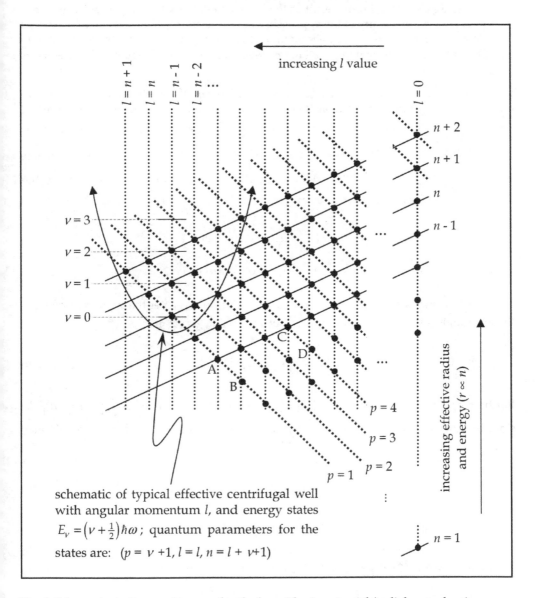

Fig. 4. Schematic (n, l) state diagram for the logarithmic potential (radial mass density $\propto 1 / r^2$) showing the high n, l-valued stationary states. A typical effective centrifugal l-well of momentum $\sqrt{l(l+1)}\hbar$ is shown superimposed on the state diagram to illustrate how the higher energy states $E_\nu = (\nu + 1/2)\hbar\omega$, $\nu = 0,1,2...$ corresponding to $(p = \nu + 1, l = l, n = l + \nu + 1)$ are calculated for each l-well. Each solid circle on the diagram represents $2l + 1$, z-projection (m) substates.

levels gives essentially the same result as that obtained directly from equation (1) which is again a check on the validity of the approach. That is the energy of a particular vibrational level in one l-harmonic well becomes equal to the energy of one level above or below in the adjacent, $l \pm 1$ well, making the energy levels for different l, same n degenerate as expected.

It is also possible to write down the eigenfunctions of the high angular momentum eigenstates in a logarithmic potential. It is necessary to generalise the normalisation constant $(m_0\omega/\pi\hbar)^{1/4}/\sqrt{2^\nu \nu!}$ in equation (15). Writing the second of equations (15) as

$u_\nu(r) = N H_\nu(\xi)\exp(-\xi^2/2)$ where $N = (m_0\omega/\pi\hbar)^{1/4}/\sqrt{2^\nu \nu!}$, and $\xi = \alpha r - \beta$, where $\alpha = (2l(l+1))^{1/4}/r_d$ and $\beta = (2l(l+1))^{1/4}$, we modify the procedure of Schiff (1968) to find N. The integral for calculating the normalising constant is given by

$$N^2 \int_0^\infty |u_\nu(\alpha r - \beta)|^2 r^2 dr = 1$$

$$\frac{N^2}{\alpha^3} \int_0^\infty |u_\nu(\alpha r - \beta)|^2 (\alpha r)^2 d(\alpha r) = 1 \tag{17}$$

Now $\alpha r \gg 1$ and is approximately constant and equal to αr_d over the significant range of the integrand and it can therefore be taken outside the integral to give

$$\frac{N^2 r_d^2}{\alpha} \alpha \int_0^\infty |u_\nu(\alpha r - \beta)|^2 dr' = 1 \tag{18}$$

Furthermore the integral $\int_{-\infty}^\infty |u_\nu(\alpha r)|^2 d(\alpha r)$ is essentially the same as $\int_0^\infty |u_\nu(\alpha r - \beta)|^2 dr$ since it is merely a displacement of the harmonic function from $r = 0$ to $r = r_d$ so (17) may be written as

$$\frac{N^2 r_d^2}{\alpha} \int_{-\infty}^\infty H_\nu(\alpha r)^2 \exp(-(\alpha r)^2) = 1 \tag{19}$$

Using the Hermite generating function relation $e^{-s^2 + 2s\alpha r} = \sum_{i=0}^\infty H_\nu(\alpha r)s^i/i!$, expanding the exponential as a power series and equating series coefficients, enables the integral in equation (18) as $\pi^{1/2} 2^\nu \nu!$ (see Schiff, 1968 for details) and gives the normalising constant N as

$$N = \frac{1}{r_d}\left(\frac{\alpha}{\sqrt{\pi} 2^\nu \nu!}\right)^{1/2} = \left(\frac{(2l(l+1))^{1/4}}{r_d^3 \sqrt{\pi} 2^\nu \nu!}\right)^{1/2} \tag{20}$$

Thus the eigenfunctions $u_{l,p}(r)$ (or $u_{n,l}(r)$) may be written in terms of the well parameters r_d and l, as

$$u_{l,p}(r) = \left(\frac{(2l(l+1))^{1/4}}{r_d^3 \sqrt{\pi}\, 2^{p-1}(p-1)!} \right)^{1/2} \times$$

$$H_{p-1}\left((2l(l+1))^{1/4} \left(\frac{r}{r_d} - 1 \right) \right) \exp\left(-\frac{(l(l+1))^{1/2}}{\sqrt{2}} \left(\frac{r}{r_d} - 1 \right)^2 \right)$$

$$u_{n,l}(r) = \left(\frac{(2l(l+1))^{1/4}}{r_d^3 \sqrt{\pi}\, 2^{n-l-1}(n-l-1)!} \right)^{1/2} \times$$

$$H_{n-l-1}\left((2l(l+1))^{1/4} \left(\frac{r}{r_d} - 1 \right) \right) \exp\left(-\frac{(l(l+1))^{1/2}}{\sqrt{2}} \left(\frac{r}{r_d} - 1 \right)^2 \right)$$

(21)

where r_d is defined as in equation (10).

4. High angular momentum states: Longevity, darkness, transparency, stability and weak interaction with low l-states

High angular momentum, high-n gravitational eigenstates make excellent dark matter candidates. Any particle, even traditional baryons or electrons, placed into these states or into wavefunctions whose eigenspectral composition is rich in high-n,l states will be dark, weakly interacting and unable to gravitationally collapse in the traditional classical sense. Why is this?

Quantum theory provides standard ways to calculate the interaction properties of eigenstates such as spontaneous and stimulated emission and absorption rates, particle interaction cross-sections etc. These rates depend, among other things, on the overlap integral (matrix element) for the interaction, which itself depends on the initial and final states and on the interaction Hamiltonian, generally expressed as a combination of spatial and momentum (differential) operators. In the case of radiative dipole decay for example the rate $A_{i,f}$ is given by (Ernest, 2009b):

$$A_{i,f} = \frac{\omega_{if}^3 \left| \langle f | e\mathbf{r} | i \rangle \right|^2}{3\varepsilon_0 \pi \hbar c^3} = \frac{\omega_{if}^3 \Pi_{if}^2}{3\varepsilon_0 \pi \hbar c^3}$$

(22)

where e is the electronic charge, the ε_0 electrical permittivity, $\left| \langle f | e\mathbf{r} | i \rangle \right| = \Pi_{if}$ the absolute value of the dipole matrix element for spontaneous decay for the transition i to f, ω_{if} the transition angular frequency, and the other symbols have their usual meanings. $\left| \langle f | e\mathbf{r} | i \rangle \right| = \Pi_{if}$ is given (Ernest, 2009b) as normal by

$$\Pi_{if} = \left| \int_0^\infty \int_0^\pi \int_0^{2\pi} R_{nf,lf}^* Y_{lf,mf}^* \, e\mathbf{r}\, R_{ni,li} Y_{li,mi}\, r^2 \sin^2(\theta)\cos(\phi)\, d\phi\, d\theta\, dr \right|$$

$$= \sqrt{\left(\Pi_{ifx}^2 + \Pi_{ify}^2 + \Pi_{ifz}^2 \right)}$$

(23)

where Π_{ifx}, Π_{ify} and Π_{ifz} are standard Cartesian components, expressible in terms of angular and radial overlap integrals $I_{\theta,\phi,x}, I_{\theta,\phi,y}, I_{\theta,\phi,z}, I_r$ over the initial, i, and final, f, spherical harmonic $Y_{m,l}(\theta,\phi)$ and radial component $u_{n,l}(r)$ eigenfunctions:

$$I_{\theta\phi x} = \int_0^\pi \int_0^{2\pi} Y^*_{lf,mf} \, Y_{li,mi} \sin^2(\theta)\cos(\phi)d\phi d\theta$$

$$I_{\theta\phi y} = \int_0^\pi \int_0^{2\pi} Y^*_{lf,mf} \, Y_{li,mi} \sin^2(\theta)\sin(\phi)d\phi d\theta$$

$$I_{\theta\phi z} = \int_0^\pi \int_0^{2\pi} Y^*_{lf,mf} \, Y_{li,mi} \cos(\theta)\sin(\theta)d\phi d\theta \qquad (24)$$

$$I_r = \int_0^\infty u^*_{nf,lf}(r) r^3 u_{ni,li}(r)dr$$

Ernest (2009b) calculated state to state dipole decay times $1/A_{i,f}$ based on a central point-mass potential. Using the logarithmic potential developed here, decay times remain very similar. Times are affected by the choice of the density parameter M_0/R_0 but if the same enclosed mass is used the energy spacing between levels is only marginally different. For example the differences in energy spacing (occurring as a result of differences in the shape of the well for the two different potentials) between two adjacent $p = 1$ states with $\Delta p = 0$ (such as $A \rightarrow B$ on Figure 4) for logarithmic (real halo) versus point-mass potentials produce differences in decay times of less than 5% for the same enclosed mass at typical galactic halo radii. As a result particles occupying these states (or predominant mixtures of them) in the logarithmic potentials corresponding to actual galactic halos do not emit radiation, are stable over cosmic lifetimes and do not undergo gravitational collapse. We restate this important result:

- The lifetimes of the high angular momentum states are far greater than the age of the universe and the states are stable and do not emit radiation.

Similar arguments apply to $\Delta p = 0$ transitions originating on the 'deeper' $p = 2,3,4...$ states (such as $C \rightarrow D$ on Figure 4), provided that p is a negligible fraction of n. These lifetimes remain long even when the number of available decay channels is taken into account (selection rules require that $\Delta l = \pm 1$ and $\Delta m = 0, \pm 1$). This was discussed extensively by Ernest (2009b) and is related to the effects that eigenfunction orthogonality, large radial position and differences in spatial oscillation frequency (SOF) have on the overlap integral. (Where transitions are involved that 'cross' lines of constant p on the state diagram, it can be shown (Ernest, 2009b) that the overlap integrals become exceedingly small.) We do not repeat the arguments in detail here but note that the results can be applied in the same way with essentially the same conclusions to the states of the logarithmic potential wells.

The extreme longevity of the low-p states has implications for the rates of stimulated emission and absorption. The probability per unit time P_{if} of stimulated emission or absorption of radiation from a state i to a state f is given by

$$P_{if} = \frac{\pi e^2}{3\varepsilon_0 \hbar^2}\left|\langle f|\mathbf{r}|i\rangle\right|^2 \rho\left(\omega_{if}\right) = \frac{\pi \Pi_{if}^2}{3\varepsilon_0 \hbar^2}\rho\left(\omega_{if}\right) = \frac{4\pi^2 \Pi_{if}^2}{3\varepsilon_0 \hbar^2 c}I\left(\omega_{if}\right) \qquad (25)$$

where $\rho(\omega_{if})$ is the radiation energy density per unit angular frequency and $I(\omega_{if})$ is the corresponding beam intensity per unit angular frequency. Assuming the upper and lower degeneracies are similar one can calculate the absorption and scattering times for photons as they pass through a typical halo. The details of this calculation are derived explicitly in Ernest (2009b). The calculation derived there shows that photons in virtually all known regions of the electromagnetic spectrum will pass through the halo without scattering via low-p eigenstates, so that halos composed of such states will be completely transparent and not subject to the usual processes of Compton or Raleigh scattering as are traditional localised 'free' particles represented by visible eigenspectral ensembles. Again these calculations extend immediately to the eigenstates of the logarithmic potentials with essentially the same results. We therefore also explicitly note that:

- A halo consisting of ordinary particles composed predominantly of low-p eigenspectral components will be transparent to virtually all regions of the electromagnetic spectrum.

Interactions of low-p eigenstates with other particles involve different Hamiltonian operators and are not subject to the same stringent selection rules as with photon interactions, particularly for three body 'collisions'. We therefore expect that low-p eigenstates will not be as 'dark' for particles as they are for photons. It turns out however that it is still difficult for state transfer across lines of constant p, particularly when p is small. The reasons behind this relate to several properties of the low-p states. These include their limited radial range, the limited number of, and wavelength of, their spatial oscillations, and the effect of differences in the SOF between the initial and final quantum states. These aspects were discussed by Ernest (2009b). It was seen in figure 9 of that paper that there is a relationship between the relative spatial oscillation frequency (RSOF) of the initial and final states and the size of the overlap integral. If one state has $p = 1$ then the value of $\log(-\log(I_r))$ is linearly related to $\log(\text{RSOF})$. Recent work has shown that a similar relationship exists for $p - 2$ states and should hold in general provided p is a negligible fraction of n. This relationship means that if either of the states has a low-p value, the size of the overlap integral diminishes rapidly with any difference in p-values, that is the degree of 'p-crossing'. Furthermore because the interaction Hamiltonian does not have a significant effect on the spatial oscillation frequencies of the states, it means that similar relationships should exist for other types of interactions such as those involving particles. This is a direct consequence of the orthogonality and limited radial extent of the low-p eigenstates.

Figure 5 shows a general summary of the different ways in which 'visible' matter (that is photons and traditional particles in thermalized, broad-range mixtures of halo eigenstates) with low-p states is limited by effects such as spatial oscillation frequency and spatial overlap. In case (a) we have small changes in both p and n, and transitions are possible, but the size of the change is so small that changes in energy or momentum of the perturbing entity, and state dispersion effects over cosmic history, are negligible. In case (b) again the change in p is small, i.e. no significant 'p-crossing' and the initial and final spatial oscillation frequencies are similar (RSOF~1). Transitions are also theoretically possible in this case but here Δn is made large enough to enable observable changes in the momentum and/or energy of the perturbing entity. However exclusively in the low-p regime, the Δn required for such measurable angular momentum or energy changes requires eigenstate functions that are spatially distinct making $I_r \approx 0$ irrespective of the Hamiltonian involved. In case (c) the initial and final states are no longer spatially distinct (i.e. forced to overlap by

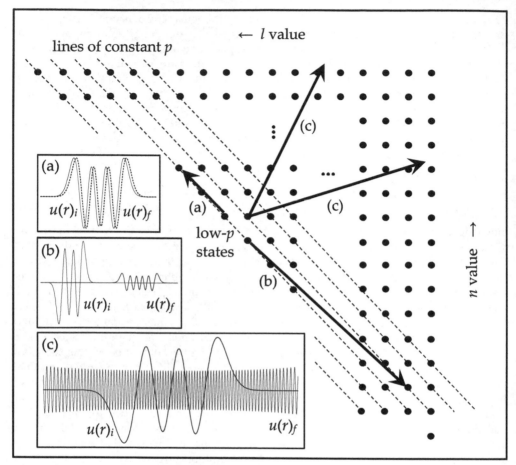

Fig. 5. Schematic (n,l) state diagram showing how characteristics of low-p initial $(u(r)_i)$ states, and final $(u(r)_f)$ states lead to weak interaction:

case (a) small change in p $\Rightarrow$ similar SOF; transitions possible, $I_R \neq 0$ but ΔE, $\Delta(mv) \sim 0$

case (b) small change in p $\Rightarrow$ similar SOF; Δn, $\Delta l \gg 1$ $\Rightarrow$ ΔE, $\Delta(mv) \gg 0$, but initial final states spatially distinct $\Rightarrow$ $I_r \sim 0$

case (c) large change in p $\Rightarrow$ large differences in SOF; Δn, $\Delta l \gg 1$ $\Rightarrow$ ΔE, $\Delta(mv) \gg 0$, but large relative SOF $\Rightarrow$ $I_r \sim 0$

the choice of the final state) but still maintain large differences in Δl and Δn. This however requires significant 'p-crossing' and, in a similar way to figure 9 of Ernest (2009b), we expect the resulting differences in spatial oscillation frequency to result in $I_R \approx 0$, irrespective of the interaction Hamiltonian involved. Again we note that:

- A halo consisting of ordinary particles composed predominantly of low-p eigenspectral components will be weakly interacting with particles and therefore have difficulty in thermalising and hence redistributing its eigenspectral distribution to that of traditional localised, Maxwellian eigenspectral compositions.

For interactions between identical particles there is the possibility that exchange degeneracy enables particles in very different p states to still interact because of 'SOF swapping'. This can be seen by writing down the symmetric/antisymmetric form of the two-particle (eigenstate+perturbing particle) overlap integral. The SOF of the initial and final eigenstates could theoretically work together with the SOF of the eigenspectral components of the initial and final perturbing particle states to allow a p-crossing transition even when the SOF of the initial eigenstate is very different from the final. Optical thickness follows from such a particle interaction because low-p eigenspectral components may then be transferred to more strongly interacting components. This possibility enables the prospect of limited interchange of baryons in dark states with traditional 'Maxwellian' matter during cosmic history. Potentially it offers a solution to the significant astrophysical problems of a continued source of star forming material in galaxies, the disk-halo conspiracy (mass distribution follows radial luminosity), the production of ionised gas from precessional galactic jets as observed in M42 at temperatures consistent with those expected from outer halo interactions, and the hot interacting gas in the Bullet cluster collision.

5. A formation scenario

A short account of the possible qualitative model of the formation of dark halos was given by Ernest (2006) and we summarise that approach here. There is still considerable work to be done in developing and testing the validity of the approach but the qualitative scenario does provide a general basis for how formation would need to proceed. Given the properties of particles in dark gravitational eigenstates, it is reasonable to expect that once particles have eigenspectral distributions that are dark, that, aside from for the types of particle interactions discussed above, they could largely remain in these distributions over cosmic times. Clearly the processes involved are dynamic and on-going, and the proportion of particles with dark eigenspectral distributions depends on the rates at which distributions are transferred back and forward between dark and visible states during cosmic history. Detailed calculations of these processes are potentially very difficult because not only do the transition rates depend on their closeness to the $p = 1$ diagonal, but the dynamic redistribution of matter concurrently changes the shape of the potential well and hence the overlap integrals involved in the rate calculations.

In the formation scenario proposed, each massive dark 'eigenstructure' halo is occupied by baryons and electrons in a self-assembled massive gravitational potential, each of which is initially seeded by the potential well of a supermassive black hole formed at the last, e^+/e^- phase transition at $t \sim 0.75$ s in the early universe. Such black holes were themselves originally one of the first candidates proposed as a solution to the dark matter problem, but it was shown that their abundance could only provide up to 10^{-7} of the closure density (Hall and Hsu, 1990). However in the present scenario we note that if a PBH is sufficiently large it can continue to accrete baryonic matter. Under normal circumstances this is countered by photon pressure via baryon-photon oscillation, but this process is critically dependent on the cross section for Compton scattering which is significantly reduced if the eigenspectral distribution of the captured baryons is biased toward dark states. In the present scenario therefore, the black holes formed at the last phase transition act as the seed potentials to capture baryons and electrons that then transfer to dark eigenspectral distributions thereby insulating them from the baryon-photon oscillation process and enabling them to add to the captured baryonic dark halo, concurrently increasing the density contrast and well potential, and enhancing further matter capture in a continuing cycle.

It is well known that primordial black holes (PBH) can form from over-density regions at the various phase transitions in the early universe (Carr, 1975, Jedamzik and Niemeyer, 1999). It is expected that such black holes will form with masses up to the horizon mass M_H at the time of the phase transition. The PBHs required for eigenstructure seeding need to be massive and we are interested in the black holes formed at the e^+/e^- phase transition at $t \sim 0.75$ s. Numerical calculations by Hawke and Stewart (2002) suggest a lower limit to M_{PBH} as $\Box \, 10^{-4} M_H$ while Carr gives an approximate upper limit on PBH mass as

$$M_{PBH} \sim 1 M_S \left(\frac{T}{100 \, \text{MeV}} \right)^{-2} \left(\frac{g_{eff}}{10.75} \right)^{-\frac{1}{2}} \tag{26}$$

where T is the temperature at the phase transition (~ 1 MeV at $t \sim 0.75$ s), g_{eff} is the number of degrees of freedom ($43/4$ at the last phase transition) and M_S is the solar mass. This suggests a maximum black hole mass $\Box \, 10^4 M_S$ while Afshordi et al (2003) consider up to $10^6 M_S$.

The capture of baryonic matter is in a sense similar to the capture of electrons by ions to form atoms and a simple version of the Saha equation

$$\frac{n_e n_i}{n_a (n_a - n_i)} = \frac{2 \pi m_e k T}{\hbar^3} \exp \left(-\frac{E_i}{kT} \right) \tag{27}$$

suggests that particles might undergo a form of 'gravitational recombination' provided that the energy level concerned is deeper than the corresponding energy related to the temperature at the time involved. This fixes a particular radius within the seed well at which the condition $|E_i| > f k T$ (where $f \sim 2 - 10$) is satisfied. Ernest (2006) considered seed masses of between 10^4 and $10^6 M_S$, and derived an equation for the temporal development of a so-called thermal radius r_{th} at which this condition was satisfied:

$$\frac{\partial r_{th}(t)}{\partial t} - k_1 \frac{r_{th}(t)^3}{t^2} - \frac{r_{th}(t)}{2t} = 0 \tag{28}$$

where k_1 is a constant relating to relevant parameters such as the effective enclosed mass, temperature etc., and the solution a function of $r_{th}(t_0)$, the initial size of the thermal radius (Ernest 2006). If the matter is effectively captured inside r_{th} then, as the surrounding universe expands and the temperature drops, the contrast density increases and r_{th} increases with time. Depending on the chosen value of f, results show that provided $M_S \sim 5 \times 10^4 M_S$ or greater, r_{th} can increase rapidly (until equation 27 breaks down due to the presence of adjacent halos) and can accommodate a halo mass of 10^{42} kg, generally before the completion of nucleosynthesis. Whether or not particles can effectively transfer to dark states within this time is still an open question because the model is still to be developed in detail and requires calculation of deep-state stimulated transition rates (the prevailing transition process in this strongly radiation-dominated era). It does suggest however that it might be possible to maintain consistency with measured nucleosynthesis ratios and still maintain baryonic densities applicable to the total matter content of the universe: i.e. dark plus visible matter, either (1) because the baryons have transferred to weakly interacting states before nucleosynthesis completion or (2) because of the extreme inhomogeneity introduced by the eigenstructure halos. (Mass captured in the halos is sequentially removed from the more rapidly expanding, necessarily under-density, mid-halo regions as expansion proceeds.) Interestingly however, from Carr's black hole number density-mass spectrum equation, it can be calculated that the value r_{th} takes for the separation of such adjacent halos at the formation time translates, after universal expansion, into the order of magnitude for the number density of present day galaxies (Ernest, 2006).

In a traditional p^+/e^- recombining plasma once the temperature has dropped well below the ionisation level, virtually all electrons and protons have formed into atoms, so we might initially expect that today all the baryons and electrons of the universe would have gravitationally combined into halo eigenstructures and possibly also collapsed into dark eigenstates. There would then be no visible matter in the universe, particularly in the present scenario where halos grow until they overlap. During the formation process however, as discussed above there are weak, relatively 'field-free' corner regions in between adjacent halos and some baryons and electrons would have remained as thermalised distributions in these regions. The probability of any individual particle transferring to a dark state is necessarily small and so we expect that some as yet unknown fraction of matter in the halo itself as well as the matter in these field-free corner regions will form the basis of the visible matter we observe today. An interesting consequence from this is the prediction of the visible to total matter ratio. If we assume that halos fill space like oranges in a box and imagine that growth stops when the halo edges meet with the remaining matter left over in the corners then the ratio of this left over matter to the total matter should give us a ratio of the visible to total matter. Bertschinger (1985) has looked at the capture of baryonic matter by a central potential and has shown that such capture can result in self-similar density profiles which are reasonably consistent with the logarithmic potential, $1/r^2$ profiles observed today. Assuming such a density profile, we get a ratio for visible to total matter of 0.23 for loose random packing and 0.186 for close random packing. This can be compared with the WMAP result which measures the visible to total matter ratio of the universe via baryon-photon oscillation as 0.18 (Tegmark etal., 2004).

It was estimated that initially the eigenstructure radius would have be formed from the horizon size $\square\, 3\times10^8\,\text{m}$ to $\square\, 10^{13}\,\text{m}$. It is expected that various processes would act on the eigenstates after baryonic capture. These would be required to expand the halos to their presently observed sizes. We imagine the matter at radius $r\,\square\, 10^{13}\,\text{m}$ in a Maxwellian distribution which we approximate by a single momentum $p=\sqrt{3mk_BT}$ (where k_B is Boltzmann's constant). Given an isotropic distribution of particles then particles enter the n,l diagram at a position that corresponds to their angular momentum. For the temperature conditions at initial formation this suggests the majority of particles will have values of $l\sim r\sqrt{3mk_BT}/\hbar$ up to $l\,\square\, 8\times10^{22}$ for $r\,\square\, 3\times10^8\,\text{m}$, $(T\sim10^{10}\,\text{K})$ and $l\,\square\, 5\times10^{26}$ for $r\,\square\, 10^{13}\,\text{m}$ $(T\sim4\times10^8\,\text{K})$.

We want to estimate the rate at which such particles might be promoted up or down via stimulated radiation processes. The stimulated transition rate is given by equation 24 which depends on the overlap integral Π_{if} and on the energy density per unit angular frequency $\rho(\omega_{if})$ at the transition frequency which relates to the temperature at the formation time. For $T \sim 10^{10}$ K this peaks at $\omega \sim 4 \times 10^{21}$ Hz, but this frequency corresponds to large changes in p values ($\Delta l = \pm 1$) that do not have favourable Π_{if} values when the initial value of $p(= p_i)$ is low. Hence there is an optimum value of Δn that will yield the maximum transfer rate. For smaller changes in p the value of Π_{if} is higher (for $p_i = 1$ where $\Delta p = 1$ necessarily, there is virtually 100% in-phase overlap in the dipole matrix element and $\Pi_{if} \approx er$) but $\rho(\omega_{if})$ is much smaller. Nevertheless at these high temperatures there is a peak rate $\sim 10^{34}$ s^{-1} that occurs for $\Delta p = 1$ transitions originating on $n_i = 1, p_i = 1$. Likewise similar $\Delta p = 1$ transitions originating on the deeper states where $p_i \neq 1$ will show similar rates. Additionally however if the states are very deep, the rapid decrease in Π_{if} with increase in Δp (Ernest, 2009b) does not occur, and promotion demotion rates may be even greater. Thus although it was seen that the state dispersion due to radiation such as that from the CMB does not appreciably alter the relative eigenstate position on the state diagram for present day halos, in the early times of formation the rates are such that these halos could have been easily expanded. It is relatively easy to track the rates during cosmic history and indications are that radiation expansion of halos may have occurred up to as late in cosmic history as the beginning of the period of reionization. Although accurate calculations require detailed knowledge of the energy level spacings which have only recently been determined for logarithmic potentials. Further work is being carried out in this area.

6. Observations and predictions

It is difficult for any model of dark matter to predict observable phenomena because by the nature of dark matter it is not very observable! Perhaps the biggest departure of the present approach from traditional cosmic concordance is the early formation of massive eigenstructures. Whilst within the present theory traditional LCDM is left intact on the largest scales (indeed we know that eigenstructures themselves will function effectively in describing large scale structure as the original numerical modelling was carried out with similar mass 'particles'). One would anticipate that such a change would leave some imprint on for example the CMB. The halo separation at decoupling corresponds to fluctuations in the anisotropy spectrum at $l \Box 10^4$. Unfortunately at such high l values the finite time for atomic recombination has most likely smoothed out peaks due to the individual halos. One might see evidence in the Lyman-α forest, but since it is anticipated that every eigenstructure forms into a galaxy, this is really just noting the observed effects that galactic halos already have on the Lyman-α forest.

One possible observation might reveal the eigenstate nature of dark matter. We know that in the $1/r^2$ density profiles the total energy of an eigenstate minus its potential energy is effectively a measure of its kinetic energy. This kinetic energy is constant with radius and determined only by eigenstate particle mass and the ratio M_0/R_0. Furthermore we know that although these high-l eigenstates are very dark with respect to photon interactions, their interactions with other particles may not be so ineffective, because of relaxed selection rules. It is possible that dark baryons slowly 'leak' into the visible regime over time via collisions and interactions with particles and other eigenstates. If particle/dark-eigenstate collisions do occur then the eigenspectral distribution will shift to a lower-l, visible composition. It is suggested this is one of the primary origins of the hot x-ray gas seen in halos, and the radio and other emission from black hole jets as they precess in the halo and the hot x-ray emission seen in cluster collisions such as the famous "Bullet cluster".

Firstly we note that the halo x-ray emission is at an approximately constant temperature across the radius of the halo, consistent with the radial dependence of the effective kinetic energy of eigenstates. The equivalent kinetic energy of these eigenstates averaged over the electron and proton component is most simply calculated from $\bar{E}_K = GM_0(m_p + m_e)/4R_0$ which for the Milky Way corresponds to a temperature $= 2\bar{E}_K / 3k_B = 1.3 \times 10^6 \, K$. The Milky Way exhibits a range of x-ray energies but the best estimates of the diffuse halo gas temperature are $1.3 - 1.5 \times 10^6 \, K$ (Kappes et al., 2003). If the hypothesis of energy transfer from dark to visible states is correct, one might expect to see, over a range of different halos, a linear correlation between two quantities M_0/R_0 and the x-ray temperature, both of which should be measurable, the first by lensing and the second by x-ray satellites. In some halos a variation in velocity profiles with radius is observed and there might be a correlation between the halo temperature and the local the equivalent eigenstate kinetic energy as a function of radius within a single halo. It is also significant that some observed x-ray intensities mimic halo dark matter density profiles (ibid.) the generality of which could be tested.

7. Conclusion

Quantum theory predicts the existence of well-bound, dark gravitational eigenstates in potential wells like those associated with galactic or cluster halos. By allowing the possibility that these states could be incorporated into the eigenspectra of what would normally be visible elementary particles, it enables them to function as dark matter candidates. This then enables the nature and origin of dark matter to be understood without the need for new particles or new physics. Gravitational eigenstates have already been experimentally observed in the laboratory and there is no reason to deny their existence in large potential wells. Many of the properties of dark matter then arise as a natural consequence of (i) the functional properties of the wavefunctions corresponding to the dark eigenstates or (ii) the envisaged formation scenario.

Perhaps the most exciting aspect of gravitational eigenstates though will be realised if these states do turn out to be responsible for dark matter, for in doing this they will provide us not only with a solution to a long standing problem in astrophysics but also with a more generalised way to describe and understand the nature of matter on macroscopic scales.

8. References

Afshordi, N., McDonald, P. and Spergel, D. N., (2003). Primordial Black Holes as Dark Matter: The Power Spectrum and Evaporation of Early Structures. *ApJ*, 594, pp. L71-L74

Bernstein, D. H., Giladi, E. and Jones, K. R.W., (1998). Eigenstates of the gravitational Schrodinger equation. *MPLA*, 13, 29, pp. 2327-2336

Bertschinger E., (1985). Self-similar secondary infall and accretion in an Einstein-de Sitter universe. *ApJS*. 58, pp. 39-66

Carr B. J., (1975). The primordial black hole mass spectrum. *ApJ.*, 201, pp. 1-19. doi: 10.1086/153853

Chavda, L. K. and Chavda, A.L., (2002). Dark matter and stable bound states of primordial black holes. *Class. Quantum Grav.* 19 2927 doi: 10.1088/0264-9381/19/11/311

Chou C. H., Hu B. L. and Subaşi Y., (2011). Macroscopic quantum phenomena from the large N perspective, *J. Phys.: Conf. Ser.* 306 012002, doi: 10.1088/1742-6596/306/1/012002

Ciftci, H., Ateser, E. and Koru, H., (2003). The power law and the logarithmic potentials. *J. Phys. A*, 36, 3821, (13 pp.), doi:10.1088/0305-4470/36/13/313

Clowe, D., Bradač, M., Gonzalez, A. H., Markevitch M., Randall, S. W., Jones C. and Zaritsky, D., (2006). A Direct Empirical Proof of the Existence of Dark Matter, *ApJ* 648, L109, (5 pp.), doi: 10.1086/508162

Colberg, J. M., White, S. D. M., Jenkins, A and Pearce, F. R., (1997). Linking cluster formation to large scale structure. arXiv:astro-ph/9711040v1. (6 pp.), Available from http://arxiv.org/PS_cache/astro-ph/pdf/9711/9711040v1.pdf

Croft, R. A. C., and Efstathiou, G., (1993). Large-scale structure and motions from simulated galaxy clusters. astro-ph/9310016 11. (5 pp.), Available from http://arxiv.org/PS_cache/astro-ph/pdf/9310/9310016v1.pdf

DeWitt, B. S. and Esposito, G., (2007). An introduction to quantum gravity. arXiv:hep-th/0711.2445v1., (68 pp.), Available from http://arxiv.org/PS_cache/arxiv/pdf/0711/0711.2445v1.pdf

Diemand, J. and Moore, B., (2009). The structure and evolution of cold dark matter halos. arXiv:astr-ph/0906.4340v1, (17 pp.), Available from http://arxiv.org/PS_cache/arxiv/pdf/0906/0906.4340v1.pdf

Doran, C., Lazenby, A., Dolan, S. and Hinder, I., (2005). Fermionic absorption cross section of a Schwarzschild black hole. Phys. Rev. D, 41, 124020, (6 pp.)

Dvali, G., Gomez, C. and Mukhanov, S., (2011). Black Hole Masses are Quantized. arXiv:hep-ph/1106.5894v1, pp. 1-23, Available from http://arxiv.org/PS_cache/arxiv/pdf/1106/1106.5894v1.pdf

Ernest, A. D., (2001). Dark matter and galactic halos - a quantum approach. arXiv:astro-ph/0108319, pp. 1-53, Available from http://arxiv.org/abs/astro-ph/0108319

Ernest, A. D., (2006). A Quantum approach to dark matter, In: *Dark Matter: New Research*, Ed J. Val Blain, pp. 91-147, NOVA Science Publishers, ISBN: 1-59454-549-9, New York

Ernest, A. D., (2009a). Gravitational eigenstates in weak gravity I: Dipole decay rates of charged particles. *J. Phys. A: Math. Theor.* 42, 115207, (16 pp.)

Ernest, A. D., (2009b). Gravitational eigenstates in weak gravity II: Futher approximate methods for decay rates. *J. Phys. A: Math. Theor.* 42, 115208, (22 pp.)

Feng, J. L., (2010). Dark Matter Candidates from Particle Physics and Methods of Detection. arXiv:astro-ph/1003.0904. (50 pp.), Available from http://arxiv.org/PS_cache/arxiv/pdf/1003/1003.0904v2.pdf

Gossel, G. H., Berengut, J.C. and Flambaum, V.V., (2010). Energy levels of a scalar particle in a static gravitational field close to the black hole limit. arXiv:gr-qc/1006.5541, (4 pp.), Available from http://arxiv.org/abs/1006.5541

Hall, L. J. and Hsu, S. D., (1990). Cosmological production of black holes. *Phys. Rev. Lett.* 64, pp. 2848-2851, doi:10.1103/PhysRevLett. 64.2848

Hawke, I. and Stewart, J. M.,(2002). The dynamics of primordial black-hole formation. *Class. Quantum Grav*, 19, pp. 3687-3707, doi:10.1088/0264-9381/19/14/310

Jaekel, M. T., Lamine, B., Lambrecht, A., Reynaud, S. and Maia Neto, P., (2006). Quantum decoherence and gravitational waves, In: *Beyond the Quantum*, Theo M Nieuwenhuizen et. al. (Eds.), pp. 125-134, Eproceedings, Lorentz Center Leiden, The Netherlands, May 29 - June 2, 2006, Eds.,World Scientific, 10.1142/9789812771186_0010

Jedamzik, K. and Niemeyer, J. C., (1999). Primordial black hole formation during first-order phase transitions. *Phys.Rev.D*, 59, 124014, (7 pp.) doi:10.1103/PhysRevD.59.124014

Kappes, M., Kerp, J. and Richter, P., (2003). The Composition of the Interstellar Medium towards the Lockman Hole. *A&A*, 405, pp. 607-616, doi: 10.1051/0004-6361:20030610

Ketterle, W., (2002). Nobel lecture: When atoms behave as waves: Bose-Einstein condensation and the atom laser. *Rev. Mod. Phys.* 74, pp. 1131-1152

Knebe, A., (1998). Virialisation of galaxy clusters in numerical simulations. arxiv.org/9811159v1 (12pp.) Available from
http://arxiv.org/PS_cache/astro-ph/pdf/9811/9811159v1.pdf

Kroupa, P., Famaey, B., de Boer1, K. S., Dabringhausen, J., Pawlowski1, M. S., Boily, C.M., Jerjen, H., Forbes D., Hensler, G. and Metz1 M. (2010). Local-Group tests of dark-matter concordance cosmology. *A&A*, 523, A32, (26 pp.) DOI 10.1051/0004-6361/201014892

Lamine, B., Herve, R., Jaekel, M.T., Lambrecht, A. and Reynaud, S., (2011). Large-scale EPR correlations and cosmic gravitational waves. *EPL*, 95, 20004, (9 pp.)

Milgrom, M., (1983). A modification of the Newtonian dynamics as a possible alternative to the hidden mass hypothesis. *ApJ* 270, pp. 365-370

Nakamura, Y., Pashkin, Yu. A., and Tsai, J. S., (1999). Coherent control of macroscopic quantum states in a single-Cooper-pair box. *Nature* 398, pp. 786-788

Nesvizhevsky V. V., Borner H. G., Petukhov A. K. et al., (2002). Quantum states of neutrons in the Earth's gravitational field. *Nature* 415, pp 297-9

Primack, J. R., (2001). The nature of dark matter. arXiv:astro-ph/0112255 v1. (22 pp.) Available from
http://arxiv.org/PS_cache/astro-ph/pdf/0112/0112255v2.pdf

Reynaud, S., Maia Neto, P. A., Lambrecht, A. and Jaekel, M.-T., (2001). Gravitational decoherence of planetary motions. *Eyrophys. Lett*, 54, pp. 135-141. doi:10.1209/epl/i2001-00286-8

Rovelli, C., (2008). Loop Quantum Gravity. Living Rev. Relativity, 11, 5, ISSN 1433-8351. Available from
http://www.livingreviews.org/lrr-2008-5

Rubin, V. and Ford, W. K. Jr., (1970). Rotation of the Andromeda Nebula from a Spectroscopic Survey of Emission Regions. ApJ. 159, pp. 379-403. doi:10.1086/150317

Schiff, L. I., (1968). *Quantum Mechanics* (3rd Ed.), McGraw-Hill, ISBN-13: 978-0070856431, New York

Schlosshauer, Maximilian A. (2007). *Decoherence and the Quantum-To-Classical Transition*, Springer, ISBN 978-3-540-35773-5

Schmitt-Manderbach, T., Weier, H., Fürst, M., Ursin, R., Tiefenbacher, F., Scheidl, Th., Perdigues, J., Sodnik, Z., Rarity, J. G., Zeilinger, A. and Weinfurter, H., (2007).

Experimental Demonstration of Free-Space Decoy-State Quantum Key Distribution over 144 km. *Phys. Rev. Lett.* 98, 010504, (4 pp.)

Snowden, S.L., Freyberg, M.J., Kuntz, K.D., and Sanders, W.T., (2000). A Catalog of Soft X-Ray Shadows, and More Contemplation of the ¼ keV Background. *ApJS*, 128, 171, (64 pp.), doi:10.1086/313378

Tegmark, M. *et al.* [SDSS Collaboration], (2004). Cosmological parameters from SDSS and WMAP. *Phys. Rev.* D 69, 103501, (26 pp.)

Vachaspati, T., (2009). Schrodinger Picture of Quantum Gravitational Collapse. Class. Quantum Grav. 26, 215007 (14 pp.), doi:10.1088/0264-9381/26/21/215007

Van der Wal, Caspar H., ter Haar1, C. J., Wilhelm, F. K., Schouten, R. N., Harmans, C. J. P. M., Orlando, T. P., Lloyd, Seth and Mooij, J. E., (2000). Quantum Superposition of Macroscopic Persistent-Current States. *Science*, 290, no. 5492, pp. 773-777, DOI: 10.1126/science.290.5492.773

Woodard, R. P., (2009). How far are we from the quantum theory of gravity? *Rep. Prog. Phys.* 72, 126002, (106 pp.), doi: 10.1088/0034-4885/72/12/126002

Zbinden, H., Brendel, J., Tittel, W. and Gisin, N., (2001). Experimental test of relativistic quantum state collapse with moving reference frames. *J. Phys. A: Math. Gen.* 34, pp. 7103-7110, doi: 10.1088/0305-4470/34/35/334

Zurek A.J., (1981). Pointer basis of quantum apparatus: Into what mixture does the wave packet collapse?, *Phys. Rev.* D 24, pp. 1516-1525.

Zurek A.J., (1982). Environment-induced superselection rules. *Phys. Rev.* D 26, pp. 1862-1880.

Zwicky, F., (1937). On the masses of nebulae and clusters of nebulae. ApJ. 86, pp. 217-246

9

The Equivalence Theorem in the Generalized Gravity of $f(R)$-Type and Canonical Quantization

Y. Ezawa[1] and Y. Ohkuwa[2]
[1]Dept. of Physics, Ehime University, Matsuyama,
[2]Section of Mathematical Science, Dept. of Social Medicine,
Faculty of Medicine, University of Miyazaki,
Japan

1. Introduction

Since the discovery of the accelerated expansion of the universe(Fadely et al., 2009; Fu et al., 2008; Hicken et al., 2009; Kessler et al., 2009; Massey et al., 2007; Mantz et al., 2009; Percival et al., 2009; Reid et al., 2009; Riess et al., 2009; Schrabback et al., 2009; Suyu et al., 2009; Vikhlinin et al., 2009),(Bennet et al., 2010; Jarosik et al., 2010; Komatsu et al., 2010; Larson et al., 2010), much attention has been attracted to the generalized gravity theories of the $f(R)$-type(Caroll et al., 2004; Sotiriou & Faraoni, 2008; Nojiri & Odintsov, 2006). Before the discovery, such theories have been interested in because of its theoretical advantages: The theory of the graviton is renormalizable(Utiyama & DeWitt, 1962; Stelle, 1977). It seems to be possible to avoid the initial singularity of the universe which is the prediction of the theorem by Hawking(Hawking & Ellis, 1973) (Nariai, 1971; Nariai & Tomita, 1971). And inflationary model without inflaton field is possible(Starobinsky, 1980).

There is a well-known equivalence theorem between this type of theories and Einstein gravity with a scalar field(Barrow & Cotsakis, 1988; Maeda, 1989; Teyssandier & Tourrence, 1983; Wands, 1994; Witt, 1984). The theorem states that two types of theories related by a suitable conformal transformation are equivalent in the sense that the field equations of both theories lead to the same paths. Many investigations have been devoted to this issue(Magnano & Sokolowski, 1994; Sotiriou & Faraoni, 2008). In this work, we first review classical aspects of the theorem by deriving it in a self-contained and pedagogical way. Then we describe the problems of to what extent the equivalence holds. Main problems are: (i) Is the surface term given by Gibbons and Hawking (Gibbons & Hawking, 1977) which is necessary in Einstein gravity also necessary in the $f(R)$-type gravity? (ii) Does the equivalence hold also in quantum theory? (iii) Which metric is physical, i.e., which metric should be identified with the observed one? Next we solve the problem of the surface terms or the variational conditions. The surface term is not necessary since we can impose the variational conditions at the time boundaries that the metric and its "time derivative" can be put to be vanishing. This simplicity could be added to the advantages of $f(R)$-type gravity. Quantum aspects of the theorem are then summarized when we quantize the theory canonically in the framework of the generalized Ostrogradski formalism (Ezawa et al., 2006) which is a natural generalization to the system in a curved spacetime. The main result is that if the $f(R)$-type theory is quantized canonically, Einstein gravity with a scalar field has to be quantized non-canonically. Brief comments are given on the problem (iii).

In section 2, the Lagrangian density and field equations for the $f(R)$-type gravity are summarized. In section 3, the equivalence theorem is derived in a pedagogical way. In section 4, the problems concerning the equivalence theorem are pointed out, especially to what extent the equivalence holds. In section 5, the issue of surface term is clarified. Section 6 is devoted to a description of the canonical formalism of the $f(R)$-type gravity and the classicaly equivalent Einstein gravity with a scalar field. Summary and discussions are given in section 7. Conformal transformations of geometrical quantities are summarized in the appendix.

2. Generalized gravity of $f(R)$-type

Generalized gravity of $f(R)$-type is one of the higher curvature gravity(HCG) theories in which the action is given by

$$S = \int d^D x \mathcal{L} = \int d^D x \sqrt{-g} f(R). \tag{2.1}$$

The spacetime is taken to be D-dimensional. Here $g \equiv \det g_{\mu\nu}$ and R is the D-dimensional scalar curvature. Taking the variational conditions at the hypersurfaces Σ_{t_1} and Σ_{t_2} (Σ_t is the hypersurface $t = constant$) as

$$\delta g_{\mu\nu} = 0 \quad \text{and} \quad \delta \dot{g}_{\mu\nu} = 0, \tag{2.2}$$

field equations are derived by the variational principle as follows:

$$-\frac{\delta \mathcal{L}}{\delta g_{\mu\nu}(x)} = \sqrt{-g} \left[f'(R) R^{\mu\nu} - \frac{1}{2} f(R) g^{\mu\nu} - \nabla^\mu \nabla^\nu f'(R) + g^{\mu\nu} \Box f'(R) \right] = 0, \tag{2.3a}$$

or

$$G_{\mu\nu} = \frac{1}{f'(R)} \left[\frac{1}{2} \left(f(R) - R f'(R) \right) g_{\mu\nu} - (g_{\mu\nu} \Box - \nabla_\mu \nabla_\nu) f'(R) \right], \tag{2.3b}$$

where a prime represents the differentiation with respect to R, ∇_μ the covariant derivative with respect to the metric $g_{\mu\nu}$ and $G_{\mu\nu}$ is the D-dimensional Einstein tensor. Equations (2.3a,b) are the 4-th order partial differential equations, so the above variational conditions are allowed. Further discussions on this issue will be given in Section 5.

Here we comment on the dimensionality of $f(R)$. Comparing the action S with the Einstein-Hilbert one

$$S_{E-H} = \frac{1}{2\kappa_D^2} \int d^D x \sqrt{-g} R, \tag{2.4}$$

where $\kappa_D \equiv \sqrt{8\pi G_D}$ with G_D the D-dimensional gravitational constant, we obtain the dimension of $f'(R)$ to be equal to that of κ_D^{-2}, so that

$$[f'(R)] = [\kappa_D^{-2}] = [L^{2-D}]. \tag{2.5}$$

It is well known that this type of theory is transformed to Einstein gravity with a scalar field by a conformal transformation, which is usually referred to as equivalence theorem. We will review and clarify the content of the theorem.

3. Equivalence theorem

The theorem concerns with the conformal transformation

$$\tilde{g}_{\mu\nu} \equiv \Omega^2 g_{\mu\nu}. \tag{3.1}$$

In terms of the transformed Einstein tensor, field equations (2.3b) are written as

$$\tilde{G}_{\mu\nu} = \frac{1}{f'(R)}\nabla_\mu\nabla_\nu f'(R) - (d-1)\nabla_\mu\nabla_\nu(\ln\Omega) - g_{\mu\nu}\left[\frac{1}{f'(R)}\Box f'(R) - (d-1)\Box(\ln\Omega)\right]$$

$$+(d-1)\partial_\mu(\ln\Omega)\partial_\nu(\ln\Omega) + g_{\mu\nu}\left[\frac{f(R) - Rf'(R)}{2f'(R)} + \frac{(d-1)(d-2)}{2}\partial_\lambda(\ln\Omega)\partial^\lambda(\ln\Omega)\right], \tag{3.2}$$

where we put $D \equiv 1 + d$ (i.e. d is the dimension of the space). Eqs.(3.2) are the field equations after the conformal transformation. If they are the equations for Einstein gravity with a scalar field, 2nd order derivatives on the right hand side should vanish. From this requirement, Ω is determined to be

$$\Omega^2 = \left[2\kappa_D^2 f'(R)\right]^{2/(d-1)}. \tag{3.3}$$

The coefficient of $f'(R)$ in the square bracket, which can be any constant, was chosen to be $2\kappa_D^2$ in order to make Ω to be dimensionless and equal to unity for Einstein gravity. So, (3.1) takes the following form

$$\tilde{g}_{\mu\nu} = \left[2\kappa_D^2 f'(R)\right]^{2/(d-1)} g_{\mu\nu}. \tag{3.4}$$

Scalar field is defined as

$$\kappa_D\,\tilde{\phi} \equiv \sqrt{d(d-1)}\ln\Omega = \sqrt{d/(d-1)}\ln[2\kappa_D^2 f'(R)], \tag{3.5a}$$

or

$$f'(R) = \frac{1}{2\kappa_D^2}\exp\left(\sqrt{(d-1)/d}\,\kappa_D\,\tilde{\phi}\right),$$

$$\ln\Omega = \frac{1}{\sqrt{d(d-1)}}\kappa_D\,\tilde{\phi}. \tag{3.5b}$$

The coefficient of $\ln\Omega$, or equivalently $\ln[2\kappa_D^2 f'(R)]$, in (3.5a) was chosen for the right-hand side of (3.2) to take the usual form of scalar field source. Solving (3.5) for R, we denote the solution as

$$R = r(\tilde{\phi}). \tag{3.6}$$

In terms of $\tilde{\phi}$, (3.2) takes the following form

$$\tilde{G}_{\mu\nu} = \kappa_D^2\left[\partial_\mu\tilde{\phi}\partial_\nu\tilde{\phi} + \tilde{g}_{\mu\nu}\left(-\frac{1}{2}\partial_\lambda\tilde{\phi}\tilde{\partial}^\lambda\tilde{\phi} - V(\tilde{\phi})\right)\right], \tag{3.7}$$

where $\tilde{\partial}\tilde{\phi} \equiv \tilde{g}^{\lambda\rho}\partial_\rho\tilde{\phi}$ and

$$V(\tilde{\phi}) \equiv -f\left(r(\tilde{\phi})\right)\exp\left(-\frac{d+1}{\sqrt{d(d-1)}}\kappa_D\,\tilde{\phi}\right) + \frac{1}{2\kappa_D^2}\,r(\tilde{\phi})\exp\left(-\frac{2}{\sqrt{d(d-1)}}\kappa_D\,\tilde{\phi}\right). \tag{3.8}$$

Field equation for the scalar field is obtained by taking the trace of (3.2) as

$$\tilde{\Box}\tilde{\phi} = -\frac{\kappa_D}{\sqrt{d(d-1)}}\exp\left(-\frac{d+1}{\sqrt{d(d-1)}}\kappa_D\tilde{\phi}\right)\left[(d+1)f(r(\tilde{\phi})) - \kappa_D^{-2}\,r(\tilde{\phi})\exp\left(\sqrt{(d-1)/d}\,\kappa_D\tilde{\phi}\right)\right].$$
$$\tag{3.9}$$

Equations (3.7) and (3.9) are obtained also by the variational principle with the following Lagrangian density:

$$\tilde{\mathcal{L}} = \tilde{\mathcal{L}}_G + \tilde{\mathcal{L}}_{\tilde{\phi}}, \tag{3.10}$$

where

$$\tilde{\mathcal{L}}_G = \frac{1}{16\pi G_D}\sqrt{-\tilde{g}}\tilde{R}, \quad \tilde{\mathcal{L}}_{\tilde{\phi}} = \sqrt{-\tilde{g}}\left[-\frac{1}{2}\partial_\lambda\tilde{\phi}\tilde{\partial}^\lambda\tilde{\phi} - V(\tilde{\phi})\right]. \tag{3.11}$$

Here

$$\sqrt{-\tilde{g}} = \left[2\kappa_D^2 f'(R)\right]^{(d+1)/(d-1)}\sqrt{-g}, \tag{3.12}$$

and

$$\tilde{R} = \left[2\kappa_D^2 f'(R)\right]^{-2/(d-1)}\left[R - \frac{2d}{d-1}\left(\frac{1}{f'(R)}\Box f'(R) - \frac{1}{2}\frac{1}{f'(R)^2}\partial_\lambda f'(R)\partial^\lambda f'(R)\right)\right]. \tag{3.13}$$

$\tilde{\mathcal{L}}_{\tilde{\phi}}$ is given by terms in the parenthesis multiplying $\tilde{g}_{\mu\nu}$ in (3.7) and $V(\tilde{\phi})$ is given by (3.8). It is noted that this Lagrangian density $\tilde{\mathcal{L}}$ is not equal to the Lagrangian density $\mathcal{L}$ in (2.1) which, in terms of the transformed variables $\tilde{g}_{\mu\nu}$ and $\tilde{\phi}$, is expressed as

$$\mathcal{L} = \sqrt{-\tilde{g}}f\left(r(\tilde{\phi})\right)\exp\left(-\frac{d+1}{\sqrt{d(d-1)}}\kappa_D\tilde{\phi}\right).$$

Thus from the field equations (2.3b) for the $f(R)$-type gravity, field equations for $\tilde{g}_{\mu\nu}$ with the source of the scalar field and the field equation for the scalar field are derived. So the equivalence seems to be shown. However, eqs.(2.3b) are 10 4-th order differential equations for 10 component $g_{\mu\nu}$, so that, to obtain a unique set of solutions, 40 initial conditions seem to be required. On the other hand eqs.(3.7) are 10 2nd order differential equations for 10 component $\tilde{g}_{\mu\nu}$, only 20 initial conditions are required to have a set of unique solution. Similarly, eq.(3.9) requires only 2 initial conditions. Therefore equivalence does not appear to hold if the initial conditions are taken into account. This apparent breakdown comes from the fact that the 40 initial conditions are not independent, which is easily seen in canonical formalism (see section 5).

The above result that the variational equations of both theories coincide is usually stated as "HCG described by the Lagrangian density $\mathcal{L}$ is equivalent to Einstein gravity with a scalar field described by the Lagrangian density $\tilde{\mathcal{L}}$" and is referred to as the equivalence theorem. Note, however, that the variational equations hold on the paths that make the action stationary. Ref.(Magnano & Sokolowski, 1994)is recommended as a good review on the equivalence theorem. For recent investigations, see Ref.(Faraoni & Nadeau, 2007)and references cited in these references. We use the following usual terminology on this issue:

$$\begin{cases} \text{descriptions with } \mathcal{L} : \text{descriptions in the Jordan frame} \\ \text{descriptions with } \tilde{\mathcal{L}} : \text{descriptions in the Einstein frame} \end{cases}$$

4. Problems

We have seen that the equivalence of the two theories hold at least on the classical paths which can be determined by the variational principle. However, there would be problems on the other kinds of equivalence. In order to examine these problems, we note the following:

$$\begin{bmatrix} \text{1. The theories are not conformally invariant.} \\ \text{2. The physical metric is identified with the one determined from observations.} \end{bmatrix}$$

Unsettled problems include the following:

(I) To what extent the equivalence would hold?

(I-1) In the Einstein frame, it is well known that the surface term given by Gibbons and Hawking (GH term) is necessary. It is often argued that, from the equivalence point of view, surface term is necessary also in the Jordan frame(Dyer & Hinterbichler, 2009). However, this equivalence is not taken for granted, but should be examined carefully. The examination is given in the next section.

(I-2) Would the equivalence hold also in quantum theory? If the equivalence holds in the canonical quantum theories, fundamental Poisson brackets should be equivalent. That is, the fundamental Poisson brackets in one frame should be derived from those of the other frame.

(II) Which metric is physical in the sense that should be identified with the observed one? This problem has been investigated from various aspects(Magnano & Sokolowski, 1994). If the metric in the Einstein frame is physical(Chiba, 2003), HCG has no essential meaning and it appears by the choice of unphysical frame. If the metric in the Jordan frame is physical, the equivalence theorem states that the metric in this frame has one more scalar degrees of freedom which could be observed as non-transverse-traceless polarization of gravitational waves(Alves et al. 2009, 2010)in future observations. Furthermore, equivalence theorem states that, instead of treating the complicated Jordan frame, we can use the simpler and familiar Einstein frame for calculation. However, for comparison with observations, the results should be expressed in the words of Jordan frame. It should be noted only one of the metrics is physical. In the following, assuming that the metric in the Jordan frame is physical, we restrict ourselves to the description of problem (I).

5. Surface terms

5.1 General considerations

We first consider discrete systems whose Lagrangians contain the time derivatives of the generalized coordinates q^i up to the n-th order $q^{i(n)}$. If the n-th order derivatives are contained non-linearly the equations of motion are $2n$-th order differential equations. Then $2n$ conditions are necessary for each q^i to determine the solution uniquely. These conditions can be given by

$2n$ initial conditions or n boundary conditions at two times, t_1 and t_2. The latter conditions can be taken to be the values of the generalized coordinates themselves and their time derivatives up to the $(n-1)$-th order. Then we can take the variational conditions (boundary conditions) as

$$\delta q^{i(k)}(t_1) = \delta q^{i(k)}(t_2) = 0, \quad (k = 0, 1, \cdots, n-1). \tag{5.1}$$

Therefore no boundary terms are necessary.

On the other hand, if the n-th order derivatives are contained linearly, equations of motion are at most $(2n-1)$-th order differential equations. Then at least one condition in (5.1) does not hold generally. Therefore special solutions are required to satisfy all the conditions in (5.1) and to eliminate generally the corresponding variations at the boundaries, boundary terms are necessary. In other words, in order that the equations of motion and the variational conditions are compatible, boundary terms are required.

For continuous systems, or fields, we can proceed similarly, i.e. if the Lagrangian contains the highest order derivatives linearly, surface terms are required to eliminate some of the variations of derivatives at the boundaries.

5.2 $f(R)$-type gravity

In this theory, the Lagrangian density contains the components of the metric, the generalized coordinates, and their derivatives up to the second order in a non-linear way. So from the general considerations above, no surface terms are necessary. Concrete situations are as follows.

The variational principle leads to the field equations which are 4-th order differential equations, (3.2), so that 40 conditions are formally required to decide the solution for the metric uniquely, although they are not independent. These conditions can be taken to be the initial functions of the components of the metric $g_{\mu\nu}$ itself and their derivatives up to the 3rd order, or $g_{\mu\nu}$ and their first order derivatives at 2 times $t = t_1$ and $t = t_2$. The latter conditions correspond to the variational conditions at the time boundaries. That is, at 2 time boundaries $t = t_1$ and $t = t_2$, variational conditions are taken as $\delta g_{\mu\nu} = 0$ and $\delta \dot{g}_{\mu\nu} = 0$ given by (2.2). In fact the Lagrangian density contains up to the 2nd order derivatives non-linearly, no surface term is necessary.

5.3 Einstein gravity with a scalar field

In this theory, the gravity theory is the Einstein one and if we start from the Lagrangian density $\tilde{\mathcal{L}}$, (3.10), whose gravitational part $\tilde{\mathcal{L}}_G$ contains the second order derivatives of the metric linearly, surface term e.g. the GH term, is necessary from the above considerations. Some arguments exist that if we require the equivalence also in the boundary terms, surface term is necessary also in the $f(R)$-type gravity(Dyer & Hinterbichler, 2009).This is not the case. This equivalence should be examined carefully. The situation can be seen by examining the variation. If the theory is obtained from the $f(R)$-type theory by the conformal transformation, $\tilde{g}_{\mu\nu} = \left[2\kappa_D^2 f'(R)\right]^{2/(d-1)} g_{\mu\nu}$ and if we express the variation of this quantity and $\tilde{\phi}$ in terms of the variations in the Jordan frame, we have the following relations:

$$
\begin{cases}
\delta \tilde{g}_{\mu\nu} = \left[2\kappa_D^2 f'(R) \right]^{2/(d-1)} \delta g_{\mu\nu} + \dfrac{4\kappa_D^2}{d-1} \left[2\kappa_D^2 f'(R) \right]^{-(d-3)/(d-1)} g_{\mu\nu}\, \delta f'(R), \\[2ex]
\delta \tilde{\phi} = \kappa_D^{-1} \sqrt{d/(d-1)}\, \dfrac{1}{f'(R)}\, \delta f'(R),
\end{cases}
\tag{5.2}
$$

where

$$
\delta f'(R) = \frac{\partial f'}{\partial g_{\alpha\beta}} \delta g_{\alpha\beta} + \frac{\partial f'}{\partial(\partial_\lambda g_{\alpha\beta})} \delta(\partial_\lambda g_{\alpha\beta}) + \frac{\partial f'}{\partial(\partial_\lambda \partial_\rho g_{\alpha\beta})} \delta(\partial_\lambda \partial_\rho g_{\alpha\beta}).
\tag{5.3}
$$

Therefore, if both sets of the variational conditions

$$
\delta \tilde{g}_{\mu\nu} = \delta \tilde{\phi} = 0,
\tag{5.4}
$$

which are usually taken for $\tilde{\mathcal{L}}$ and (2.2), $\delta g_{\mu\nu} = \delta \dot{g}_{\mu\nu} = 0$, are imposed, we have

$$
\delta \ddot{g}_{\mu\nu} = 0,
\tag{5.5}
$$

at the boundary. However, this is not generally possible, but would require specific solutions as noted above. That is, the variational conditions, which require the GH term in the Einstein gravity with a scalar field, are different from those in the $f(R)$-type theory. To compare the surface terms, the variational conditions have to be carefully treated.

The above situation is related to the fact that the conformal transformation is not the transformation of the generalized coordinates, $g_{\mu\nu}$, but the transformation depending on the 2nd order derivatives of them. Comparison of the surface terms is made as follows. When $\tilde{\mathcal{L}}$ is expressed in terms of the metric in the Jordan frame, $g_{\mu\nu}$, it is written as follows:

$$
\tilde{\mathcal{L}} = \mathcal{L} - \partial_\lambda \left(\frac{2d}{d-1} \sqrt{-g}\, \partial^\lambda f'(R) \right).
\tag{5.6}
$$

Since $\mathcal{L}$ requires no surface term when the variational condition (2.2) are taken, the second term on the right-hand side is the surface term which is different from the GH term. This is an example that surface terms depend on the boundary conditions.

6. Canonical formalism

The canonical formalism belongs to classical physics. However, most quantum theory is obtained by canonical quantization which requires that commutation relations among the fundamental quantities are proportional to the corresponding Poisson brackets, e.g. for one dimensional system

$$
[\hat{q}, \hat{p}] = i\hbar \{q, p\}_{PB},
$$

where a hat represents an operator. It is noted that one of the proportional factor i assures that the observables are Hermitian operators and the other $\hbar$ adjusts the dimensionality, a very natural proportional factors.

Canonical quantum theories are very successful and only well-known failure is the theory of gravitons in general relativity. On the other hand, the canonical quantum theory of gravitons in $f(R)$-type gravity is known to be renormalizable(Utiyama & DeWitt, 1962; Stelle, 1977). This suggests a possibility that the equivalence theorem would be violated in quantum theory. The violation might come from the fact that classical equivalence means the equivalence along the classical paths. While, the Poisson brackets require derivatives in all directions in the phase space. The laws of usual canonical quantum theory describe the dynamics of matter and radiation which have duality of waves and particles assured by experiments. On the other hand, gravity describes the dynamics of spacetime. However, no nature of spacetime similar to the duality has been observed. Investigation of quantum gravity arises from various motivations. For example, since the gravity mediates interactions of elementary particles, it would be natural that the gravity is also described quantum mechanically. A preferable possibility that fundamental laws of nature would take forms of quantum theory is also one of them. The canonical quantum theory would be the first candidate for quantum gravity. Therefore a canonical formalism of gravity is very important. In this section results on a canonical formalism, a generalization of the Ostrogradski formalism, are reviewed. In the following, we use a unit for which $2\kappa_D^2 = 1$.

6.1 Canonical formalism in the Einstein frame

We adopt the ADM method for the gravitational field(Arnowitt et al., 2006), so the procedure is well known.

(1) Gravitational field

The spacetime is supposed to be constructed from the hypersurfaces Σ_t with $t = constant$ (foliation of spacetime). The dynamics of the spacetime determines the evolution of the hypersurface. So the generalized coordinates are the metric of the d-dimensional hypersurface $\tilde{h}_{ij}(\mathbf{x}, t)$.

Since $\tilde{R}$ contains 2nd order time derivatives linearly, we first make a partial integration to transform the Lagrangian density of the gravitational part in (3.11) to the following GH form:

$$\tilde{\mathcal{L}}_h = \sqrt{\tilde{h}}\,\tilde{N}\left[\tilde{K}_{ij}\tilde{K}^{ij} - \tilde{K}^2 + \tilde{\mathbf{R}}\right], \tag{6.1}$$

where $\tilde{K}$ is the trace of the extrinsic curvature $\tilde{K}_{ij}(\tilde{K} \equiv \tilde{h}^{ij}\tilde{K}_{ij})$ and $\tilde{\mathbf{R}}$ is the scalar curvature constructed from $\tilde{h}_{ij}$.

Canonical formalism is obtained by the Legendre transformation as usual. The momenta $\tilde{\pi}^{ij}$ canonically conjugate to $\tilde{h}_{ij}$ are defined as

$$\tilde{\pi}^{ij} \equiv \frac{\partial \tilde{\mathcal{L}}_h}{\partial(\partial_0 \tilde{h}_{ij})} = \sqrt{\tilde{h}}\left[\tilde{K}^{ij} - \tilde{h}^{ij}\tilde{K}\right], \tag{6.2}$$

where $\tilde{h} \equiv \det \tilde{h}_{ij}$ and $\tilde{N}$ is the lapse function. The extrinsic curvature $\tilde{K}_{ij}$ with respect to $\tilde{h}_{ij}$ is defined as

$$\tilde{K}_{ij} \equiv \frac{1}{2}\tilde{N}^{-1}\left(\partial_0 \tilde{h}_{ij} - \tilde{N}_{i;j} - \tilde{N}_{j;i}\right), \tag{6.3}$$

$\tilde{N}_i$ is the shift vector. A semicolon ; represents a covariant derivative with respect to $\tilde{h}_{ij}$. Solving (6.3) for $\tilde{K}_{ij}$, we have

$$\tilde{K}^{ij} = \frac{1}{\sqrt{\tilde{h}}}\left[\tilde{\pi}^{ij} - \frac{1}{d-1}\tilde{h}^{ij}\tilde{\pi}\right] \quad \text{and} \quad \tilde{K} = -\frac{\tilde{\pi}}{(d-1)\sqrt{\tilde{h}}}. \tag{6.4}$$

Hamiltonian density is given as

$$\begin{aligned}
\tilde{\mathcal{H}}_h &= \tilde{\pi}^{ij}\dot{\tilde{h}}_{ij} - \tilde{\mathcal{L}}_h \\
&= \tilde{N}\left[G_{ijkl}\tilde{\pi}^{ij}\tilde{\pi}^{kl} - \sqrt{\tilde{h}}\tilde{\mathbf{R}}\right] + 2(\tilde{\pi}^{ij}\tilde{N}_i)_{;j} - 2\tilde{\pi}^{ij}_{\ ;j}\tilde{N}_i.
\end{aligned} \tag{6.5}$$

where

$$G_{ijkl} \equiv \frac{1}{2\sqrt{\tilde{h}}}\left(\tilde{h}^{ik}\tilde{h}^{jl} + \tilde{h}^{il}\tilde{h}^{jk} - \frac{2}{d-1}\tilde{h}^{ij}\tilde{h}^{kl}\right), \tag{6.6}$$

is sometimes referred to as supermetric. In deriving (6.5), we used the expression for $\tilde{\mathcal{L}}_h$, expressed in terms of canonical variables, as follows

$$\tilde{\mathcal{L}}_h = \frac{\tilde{N}}{\sqrt{\tilde{h}}}\left[\tilde{\pi}^{ij}\tilde{\pi}_{ij} - \frac{1}{d-1}\tilde{\pi}^2 + \tilde{h}\tilde{\mathbf{R}}\right]. \tag{6.7}$$

(2) Scalar field

The generalized coordinate is $\tilde{\phi}(\mathbf{x}, t)$. Momenta canonically conjugate to $\tilde{\phi}$ is defined as usual by

$$\tilde{\pi}(\mathbf{x}, t) \equiv \frac{\partial\tilde{\mathcal{L}}_\phi}{\partial(\partial_0\tilde{\phi}(\mathbf{x}, t))} = -\sqrt{\tilde{g}}\,\tilde{g}^{0\mu}\partial_\mu\tilde{\phi} = \tilde{N}^{-1}\sqrt{\tilde{h}}\left[\partial_0\tilde{\phi} - \tilde{N}^i\partial_i\tilde{\phi}\right], \tag{6.8a}$$

so

$$\partial_0\tilde{\phi} = \frac{\tilde{N}}{\sqrt{\tilde{h}}}\left[\tilde{\pi} + \tilde{N}^{-1}\sqrt{\tilde{h}}\tilde{N}^i\partial_i\tilde{\phi}\right] = \frac{\tilde{N}}{\sqrt{\tilde{h}}}\tilde{\pi} + \tilde{N}^i\partial_i\tilde{\phi}. \tag{6.8b}$$

In terms of canonical variables, $\tilde{\mathcal{L}}_\phi$ is expressed as follows

$$\tilde{\mathcal{L}}_\phi = \tilde{N}\left[\frac{1}{2\sqrt{\tilde{h}}}\tilde{\pi}^2 - \frac{1}{2}\sqrt{\tilde{h}}\tilde{h}^{ij}\partial_i\tilde{\phi}\partial_j\tilde{\phi} - V(\tilde{\phi})\right].$$

Using this, we have the following expression for the Hamiltonian density

$$\tilde{\mathcal{H}}_\phi = \tilde{\pi}\dot{\tilde{\phi}} - \tilde{\mathcal{L}}_\phi = \frac{\tilde{N}}{2\sqrt{\tilde{h}}}\tilde{\pi}^2 + \tilde{N}^i\partial_i\tilde{\phi}\,\tilde{\pi} + \frac{1}{2}\tilde{N}\sqrt{\tilde{h}}\tilde{h}^{ij}\partial_i\tilde{\phi}\partial_j\tilde{\phi} + V(\tilde{\phi}). \tag{6.9}$$

(3) Fundamental Poisson brackets

Nonvanishing fundamental Poisson brackets in the Einstein frame are given as

$$\{\tilde{h}_{ij}(\mathbf{x}, t), \tilde{\pi}^{kl}(\mathbf{y}, t)\}_{PB} = \delta^{kl}_{(ij)}\delta(\mathbf{x} - \mathbf{y}) \quad \text{and} \quad \{\tilde{\phi}(\mathbf{x}, t), \tilde{\pi}(\mathbf{y}, t)\}_{PB} = \delta(\mathbf{x} - \mathbf{y}), \tag{6.10}$$

where (ij) expresses the symmetrization and not the symmetric part.

6.2 Canonical formalism in the Jordan frame

There are several canonical formalisms for generalized gravity theories in the Jordan frame. Among them formalism given by Buchbinder and Lyakhovich(Buchbinder & Lyakhovich, 1987) is logically very simple. However, concrete calculation is somewhat cumbersome partly due to arbitrariness although it allows a wide application. In addition, the Hamiltonian is generally transformed under the transformation of generalized coordinates that does not depend on time explicitly. Here we use the formalism which is a generalization of the well-known one given by Ostrogradski(Ezawa et al., 2010). For comparison of typical formalisms, see (Deruelle et al., 2009).

(1) Generalized coordinates

In this frame, we also use the foliation of the spacetime. Since the $f(R)$-type gravity is a higher-derivative theory, we follow the modified Ostrogradski formalism in which the time derivatives in the Ostrogradski formalism is replaced by Lie derivatives along the timelike normal n to the hypersurface Σ_t in the ADM formalism (Ezawa et al., 2006,2010). So the generalized coordinates are

$$h_{ij}(\mathbf{x}, t) \quad \text{and} \quad K_{ij}(\mathbf{x}, t) = \frac{1}{2}\mathcal{L}_n h_{ij}(\mathbf{x}, t) \equiv Q_{ij}(\mathbf{x}, t). \tag{6.11}$$

Here contravariant and covariant components of n are given as follows:

$$n^\mu = N^{-1}(1, -N^i) \quad \text{and} \quad n_\mu = N(-1, 0, 0, 0). \tag{6.12}$$

(2) Conjugate momenta

Denoting the momenta canonically conjugate to these generalized coordinates as π^{ij} and Π^{ij} respectively, we have from the modified Ostrogradski transformation

$$\begin{cases} \pi^{ij} = -\sqrt{h}\left[f'(R)Q^{ij} + h^{ij}f''(R)\mathcal{L}_n R\right], \\ \Pi^{ij} = 2\sqrt{h}f'(R)h^{ij}. \end{cases} \tag{6.13}$$

From (6.13), it is seen that Π^{ij} has only the trace part, so it is expressed as

$$\Pi^{ij} = \frac{1}{d}\Pi h^{ij} \quad \text{and} \quad \Pi = 2d\sqrt{h}f'(R). \tag{6.14}$$

From the second equation, we have

$$f'(R) = \frac{\Pi}{2d\sqrt{h}} \quad \text{or} \quad R = f'^{-1}(\Pi/2d\sqrt{h}) \equiv \psi(\Pi/2d\sqrt{h}). \tag{6.15}$$

Correspondingly, it is also seen from (6.13) that the traceless part of Q_{ij} is related to that of π^{ij}, and we have

$$Q^{ij} = -\frac{2}{P}\pi^{tij} + \frac{1}{d}h^{ij}Q, \tag{6.16}$$

where

$$P \equiv \frac{\Pi}{d}, \tag{6.17}$$

and

$$\pi^{\dagger ij} \equiv \pi^{ij} - \frac{1}{d}h^{ij}\pi \tag{6.18}$$

is the traceless part. A dagger is used to represent the traceless part. (Q, P) is one of the canonical pairs. In terms of these variables, the scalar curvature is expressed as follows

$$R = 2h^{ij}\mathcal{L}_n Q_{ij} + Q^2 - 3Q_{ij}Q^{ij} + \mathbf{R} - 2\Delta(\ln N). \tag{6.19}$$

(3) Hamiltonian density

In the modified Ostrogradski formalism, Hamiltonian density is defined as

$$\mathcal{H} \equiv \pi^{ij}\dot{h}_{ij} + \Pi^{ij}\dot{Q}_{ij} - \mathcal{L}. \tag{6.20}$$

Using

$$\mathcal{L}_n Q_{ij} = N^{-1}(\partial_0 Q_{ij} - N^k Q_{ij;k} - N^k_{;i}Q_{kj} - N^k_{;j}Q_{ik} - N^{-1}\partial_i N \partial_j N) \tag{6.21}$$

and eqs. (6.14)−(6.19), we have an explicit expression for $\mathcal{H}$ as follows:

$$\mathcal{H} = N\left[\frac{2}{P}\pi^{\dagger ij}\pi^{\dagger}_{ij} + \frac{2}{d}Q\pi + \frac{1}{2}P\psi(P/2\sqrt{h}) - \frac{d-3}{2d}Q^2P - \frac{1}{2}RP + \Delta P - \sqrt{h}f\left(\psi(P/2\sqrt{h})\right)\right]$$

$$+ N^k\left[2\pi^{\dagger\ ij}_{kj} - \frac{2}{d}\pi_{;k} + P\partial_k Q - \frac{2}{d}(QP)_{;k}\right]$$

$$+ \left[-2N_j\pi^{\dagger ij} + \frac{2}{d}N^i(\pi + QP) + \partial^i NP - NP^{;i}\right]_{;i}.$$

$$\tag{6.22}$$

(4) Fundamental Poisson brackets

Non-vanishing fundamental Poisson brackets are the following:

$$\{h_{ij}(\mathbf{x}, t), \pi^{kl}(\mathbf{y}, t)\}_{PB} = \delta^i_{(k}\delta^j_{l)}\delta(\mathbf{x} - \mathbf{y}), \tag{6.23a}$$

and

$$\{Q_{ij}(\mathbf{x}, t), \Pi^{kl}(\mathbf{y}, t)\}_{PB} = \delta^i_{(k}\delta^j_{l)}\delta(\mathbf{x} - \mathbf{y}). \tag{6.23b}$$

(5) Wheeler-DeWitt equation

A primary application of the canonical formalism is the Wheeler-DeWitt (WDW) equation(DeWitt, 1967). Before writing down the WDW equation, we make a canonical transformation

$$(Q, P) \rightarrow (\bar{Q}, \bar{P}) \equiv (P, -Q), \tag{6.24}$$

which removes the negative powers of the momentum P. The resulting Hamiltonian is expressed as follows:

$$\mathcal{H} = N\mathcal{H}_0 + N^k\mathcal{H}_k + \text{divergent term}, \tag{6.25}$$

where

$$
\begin{cases}
\mathcal{H}_0 = \dfrac{2}{Q}\, \pi^{\dagger ij}\pi^{\dagger}_{ij} - \dfrac{2}{d}P\pi + \dfrac{1}{2}Q\psi(Q/2\sqrt{h}) - \dfrac{d-3}{2d}QP^2 - \dfrac{1}{2}RQ \\[2mm]
\quad - \sqrt{h}\, f\left(\psi(Q/2\sqrt{h})\right) + \Delta Q, \\[4mm]
\mathcal{H}_k = 2\pi^{\dagger}_{kj}{}^{;j} - \dfrac{2}{d}\pi_{;k} - QP_{;k} + \dfrac{2}{d}(QP)_{;k}.
\end{cases}
\tag{6.26}
$$

The WDW equation is written as

$$
\hat{\mathcal{H}}_0 \Psi = 0,
\tag{6.27}
$$

where $\hat{\mathcal{H}}_0$ is obtained from $\mathcal{H}_0$ by replacing π^{ij} and P with $-i\partial/\partial h_{ij}$ and $-i\partial/\partial Q$, respectively. However, in order to apply (6.27) to the observed universe after compactification, we first carry out the dimensional reduction and then we should take into account the cosmological principle. Such procedures were done using the formalism of Buchbinder and Lyakhovich which, although is generally different from the one described above, is very similar in the case of gravity(Ezawa et al., 1994). It was shown by the semiclassical approximation method that the internal space could be stabilized.

6.3 Compatibility of the two sets of fundamental Poisson brackets

(1) Compatibility conditions

The canonical variables in the Einstein frame can be expressed in terms of those in the Jordan frame. So we can calculate the left hand sides of (6.10) using (6.23a,b). The compatibility conditions are that the results are the right hand sides of (6.10), i.e. the following relations should be satisfied:

$$
\{\tilde{h}_{ij}(\mathbf{x},t), \tilde{\pi}^{kl}(\mathbf{y},t)\}_{PB} = \sum_{m,n} \int d^d\mathbf{z} \left[\left\{ \frac{\partial \tilde{h}_{ij}(\mathbf{x},t)}{\partial h_{mn}(\mathbf{z},t)}\frac{\partial \tilde{\pi}^{kl}(\mathbf{y},t)}{\partial \pi^{mn}(\mathbf{z},t)} - \frac{\partial \tilde{\pi}^{kl}(\mathbf{y},t)}{\partial h_{mn}(\mathbf{z},t)}\frac{\partial \tilde{h}_{ij}(\mathbf{x},t)}{\pi^{mn}(\mathbf{z},t)} \right\} \right.
$$

$$
\left. + \left\{ \frac{\partial \tilde{h}_{ij}(\mathbf{x},t)}{\partial Q_{mn}(\mathbf{z},t)}\frac{\partial \tilde{\pi}^{kl}(\mathbf{y},t)}{\partial \Pi^{mn}(\mathbf{z},t)} - \frac{\partial \tilde{\pi}^{kl}(\mathbf{y},t)}{\partial Q_{mn}(\mathbf{z},t)}\frac{\partial \tilde{h}_{ij}(\mathbf{x},t)}{\Pi^{mn}(\mathbf{z},t)} \right\} \right]
$$

$$
= \delta^i_{(k}\delta^j_{l)}\delta(\mathbf{x}-\mathbf{y}),
\tag{6.28}
$$

and

$$
\{\tilde{\phi}(\mathbf{x},t), \tilde{\pi}(\mathbf{y},t)\}_{PB} = \sum_{m,n} \int d^d\mathbf{z} \left[\left\{ \frac{\partial \tilde{\phi}(\mathbf{x},t)}{\partial h_{mn}(\mathbf{z},t)}\frac{\partial \tilde{\pi}(\mathbf{y},t)}{\partial \pi^{mn}(\mathbf{z},t)} - \frac{\partial \tilde{\pi}(\mathbf{y},t)}{\partial h_{mn}(\mathbf{z},t)}\frac{\partial \tilde{\phi}(\mathbf{x},t)}{\pi^{mn}(\mathbf{z},t)} \right\} \right.
$$

$$
\left. + \left\{ \frac{\partial \tilde{\phi}(\mathbf{x},t)}{\partial Q_{mn}(\mathbf{z},t)}\frac{\partial \tilde{\pi}(\mathbf{y},t)}{\partial \Pi^{mn}(\mathbf{z},t)} - \frac{\partial \tilde{\pi}(\mathbf{y},t)}{\partial Q_{mn}(\mathbf{z},t)}\frac{\partial \tilde{\phi}(\mathbf{x},t)}{\Pi^{mn}(\mathbf{z},t)} \right\} \right]
$$

$$
= \delta(\mathbf{x}-\mathbf{y}).
\tag{6.29}
$$

Other fundamental Poisson brackets should vanish. These conditions may lead to some restrictions on $f(R)$.

(2) Expression of the conformal transformation in terms of canonical variables

Using (3.4), (6.2), (6,3) and (6.8a,b), we obtain the following form of the conformal transformation expressing the canonical variables in the Einstein frame in terms of those in the Jordan frame:

$$
\begin{cases}
\tilde{h}_{ij} = f'(R)^{2/(d-1)} h_{ij} = \left(P/2\sqrt{h}\right)^{2/(d-1)} h_{ij}, \\[2ex]
\tilde{\phi} = \sqrt{d/(d-1)} \ln\left(P/2\sqrt{h}\right), \\[2ex]
\tilde{\pi} = \sqrt{d/2(d-1)} N^{-1} \left[\partial_0 P - P(NQ + N^i_{;i}) - N^i P_{;i}\right], \\[2ex]
\tilde{N} = \left(P/2\sqrt{h}\right)^{1/(d-1)} N, \quad \tilde{N}^i = N^i, \\[2ex]
\tilde{\pi}^{ij} = \left(P/2\sqrt{h}\right)^{(d-3)/(d-1)} \sqrt{h}\left[-\frac{2}{P}\pi^{tij} + h^{ij}\left\{\frac{1}{d}Q - (NP)^{-1}\left(\partial_0 P - N^k P_{;k}\right) - N^{-1} N^k_{;k}\right\}\right].
\end{cases}
$$

$$(6.30)$$

(3) Calculation of the Poisson brackets

It may seem that the calculations are carried out easily. However, the evaluations of the brackets involving the time derivatives of the momenta are difficult. It is noted that it is impossible to use the field equations. Since, in that case, changes of variables are restricted to those along the paths of motions, which does not fit to Poisson brackets which use changes in any direction. Nevertheless, we can show, using (6.30), that assumption that all of the equations (6.10), (6.23a,b) hold leads to contradiction(Ezawa et al., 2006, 2010). In other words, two frames are not related by a canonical transformation.

Therefore, in the framework of the canonical formalism used here, we cannot quantize the theory canonically in both frames. That is, if the $f(R)$-type theory is quantized canonically, corresponding Einstein gravity with a scalar field has to be quantized non-canonically, e.g. in the non-commutative geometric way.

7. Summary and discussions

In this work, we reviewed the equivalence theorem in the $f(R)$-type gravity by deriving it in a pedagogical and self-contained way. Equivalence of this theory with Einstein gravity with a scalar field, related by a conformal transformation, holds on the classical paths. Strictly speaking, description in the physical frame is equivalent to the description in the unphysical frame, since only one frame is physical. If the description in the unphysical frame is simpler, calculations could be done in the frame.

Concerning the surface term in the $f(R)$-type gravity, it is not necessary in the Jordan frame. Necessity of the surface term in the Einstein frame comes from the structure of the Lagrangian density that it contains the 2nd order derivatives linearly. A concrete example of the surface term is obtained that shows the dependence of it on the variational conditions. The usual variational conditions in the Einstein frame leads to the GH term. On the other hand, if the variational conditions are taken as in the Jordan frame, the surface term is different and is given by (5.6).

In the canonical formalism, the conformal transformation is not a canonical one. So the fundamental Poisson brackets are not equivalent in the sense that the sets of fundamental Poisson brackets in both frames are not compatible. Thus if the theory is quantized canonically in the Jordan frame, quantization in the Einstein frame has to be non-canonical, e.g. in the non-commutative geometric way(Kempf, 1994). It is pointed out that similar situation occurs in the inflation model in multidimensional Einstein gravity(Ezawa et al., 1996). In this model, the n-dimensional internal space continues to shrink during inflation and loses its gravitational potential energy which is transferred to the inflating space. The potential energy behaves as $-a_I^{-(n-2)}$, which is expected by the Gauss law in n-dimensional space, so that the shrinkage of the internal space leads classically to the collapse of the internal space similar to the situation in the case of atoms. However if $n > 3$, the canonical quantum theory cannot prevent the collapse of the internal space contrary to the case of atoms, so that non-canonical quantum theory is required. Recently, in the noncommutative geometric multidimensional cosmology, it is shown that stabilization of the internal space is possible(Khosravi et al., 2007).This suggests that in the multidimensional $f(R)$-type gravity, extra-dimensional space would be stable. This result is in conformity with that obtained by the semiclassical approximation to the WDW equation noted above.

Thus, considering the renormalizability of the graviton theory and stabilization of the internal space in the semiclassical approximation to WDW equation, it is plausible that $f(R)$-type gravity can be quantized canonically in the Jordan frame. In addition, similar stabilization is possible in Einstein gravity if the noncommutative geometry is used, so quantization in the Einstein frame would be non-canonical.

8. Appendix: Conformal transformations of geometrical quantities

We consider a conformal transformation given as

$$\tilde{g}_{\mu\nu} \equiv \Omega^2 g_{\mu\nu}. \tag{A.1}$$

Transformations of geometrical quantities are given as follows.

Christoffel symbols

$$\tilde{\Gamma}^\lambda_{\mu\nu} = \Gamma^\lambda_{\mu\nu} + \delta^\lambda_\mu \partial_\nu(\ln\Omega) + \delta^\lambda_\nu \partial_\mu(\ln\Omega) - g_{\mu\nu}\partial^\lambda(\ln\Omega). \tag{A.2}$$

Covariant derivatives

For a scalar field, we have

$$\tilde{\nabla}_\mu \tilde{\nabla}_\nu \phi = \nabla_\mu \nabla_\nu \phi - \left[\partial_\mu(\ln\Omega)\partial_\nu\phi + \partial_\nu(\ln\Omega)\partial_\mu\phi - g_{\mu\nu}\partial^\lambda(\ln\Omega)\partial_\lambda\phi \right], \tag{A.3a}$$

or

$$\tilde{\Box}\phi = \Omega^{-2}\left[\Box\phi + (D-2)\partial^\lambda(\ln\Omega)\partial_\lambda\phi\right], \tag{A.3b}$$

Ricci tensor

$$\tilde{R}_{\mu\nu} = R_{\mu\nu} - (D-2)[\nabla_\mu\nabla_\nu(\ln\Omega) - \partial_\mu(\ln\Omega)\partial_\nu(\Omega)] - g_{\mu\nu}[\square(\ln\Omega) + (D-2)\partial_\lambda(\ln\Omega)\partial^\lambda(\ln\Omega)]$$
(A.4)

scalar curvature

$$\tilde{R} = \Omega^{-2}\left[R - 2(D-1)\square(\ln\Omega) - (D-1)(D-2)\partial_\lambda(\ln\Omega)\partial^\lambda(\ln\Omega)\right]$$
(A.5)

Einstein tensor

$$\tilde{G}_{\mu\nu} = G_{\mu\nu} - (D-2)\left[\nabla_\mu\nabla_\nu(\ln\Omega) - g_{\mu\nu}\square(\ln\Omega) - \partial_\mu(\ln\Omega)\partial_\nu(\ln\Omega) - \frac{D-3}{2}g_{\mu\nu}\partial_\lambda\partial^\lambda(\ln\Omega)\right]$$
(A.6)

9. References

Alves, M.E.S., Miranda, O.D. & de Araujo, J.C.N.(2009).Probing the $f(R)$ formalism through gravitational wave polarizations, *arXiv:0908.0861[gr-qc]* (accepted for publication in *Phys. Lett. B*)

Alves, M.E.S., Miranda, O.D. & de Araujo, J.C.N.(2010).Extra polarization states of cosmological gravitational waves in alternative theories of gravity, *arXiv:10045580[gr-qc]*(accepted for publication in *Class. Quantum Grav.*)

Arnowitt, R., Deser, S. & Misner, C.(2004).The Dynamics of General Relativity, *arXiv: gr-qc/0405109*

Barrow, J. D. & Cotsakis, S.(1988).Inflation and the Conformal Structure of Higher-Order Gravity Theories, *Phys. Lett. B* 214: 515-518

Bennett, C.L., Hill, R.S., Hinshaw, G., Larson, D., Smith, K.M., Dunkley, J., Gold, B., Halpern, M., Jarosik, N., Kogut, A., Komatsu, E., Limon, M., Meyer, S.S., Nolta, M.R., Odegard, N., Page, L., Spergel, D.N., Tucker, G.S., Weiland, J.L., Wollack, E. & Wright, E.L., (2011).Seven-year Wilkinson Microwave Anisotropy Probe(WMAP) Observations: Are There Cosmic Microwave Background Anomalies? , *ApJS* 192 issue 2, article id 17

Buchbinder, I.L. & Lyakhovich, S.L.(1987). Canonical quantisation and local measure of R^2 gravity, *Class, Quantum Grav.* 4, 1487-1501

Carroll, S.M., Duvvuri, V., Trodden, M. & Turner, M.S.(2004).Is cosmic speed-up due to new gravitational physics?, *Phys. Rev. D* 70: 043528

Chiba, T.(2003).$1/R$ gravity and scalar-tensor gravity, *Phys. Lett. B* 575: 1-3

Deruelle, N., Sendouda, Y. & Youseff, A.(2009). Various Hamiltonian formulations of $f(\mathcal{R})$ gravity and their canonical relationship, *Phys. Rev. D* 80: 084032

DeWitt, B.S.(1967). Quantum Theory of Gravity. I. The Canonical Theory, *Phys. Rev. 160*: 1113-1148.

Dyer, E. & Hinterbichler, K.(2009).Boundary terms, variational principles, and higher-derivative modified gravity, *Phys. Rev. D* 79: 024028

Ezawa, Y, Kajihara, M., Kiminami, M., Soda, J. & Yano, T.(1996). A possible way of stabi- lizing a higher-dimensional universe, *Nuovo Cim.* 111B: 355-362

Ezawa, Y., Kajihara, M., Kiminami, M., Soda, J. & Yano, T.(1999). Semiclassical approach to stability of the extra-dimensional spaces in higher-curvature gravity theories, *Class. Quantum Grav.* 16 :1873-1888

Ezawa, Y., Iwasaki, H., Ohkuwa, Y., Watanabe, S., Yamada, N. & Yano, Y.(2006). A caninical formalism of $f(R)$-type gravity in terms of Lie derivatives, *Class. Quantum Grav.* 23: 3205-3214

Ezawa, Y., Iwasaki, H., Ohkuwa, Y., Watanabe, S., Yamada, N. & Yano, Y.(2010). On the equivalence theorem in $f(R)$-type generalized gravity in quantum theory, *Nuovo Cim.* 125B: 1039-1151

Fadely, R., Keeton, C.R., Nakajima, R. & Bernstein, G.M.(2010). Improved Constraints on the Gravitational lens Q0957+561 II. Strong Lensing, *ApJ* 711: 246-267

Faraoni, R. & Nadeau, S.(2007).(Pseudo)issue of the conformal frame revisited, *Phys. Rev. D* 75 : 023501

Fu, L., Semboloni, E., Hoekstra, H., Kilbinger, M., van Waerbeke, L., Tereno, I., Mellier, Y., Heymans, C., Coupon, J., Benabed, K., Benjamin, J., Bertin, E., Doré, O., Hudson, M.J., Ilbert, O., Maoli, R., Marmo, C., McCracken, H.J. & Ménard, B.(2009). Very weak lensing in the CFTLS Wide: Cosmology from cosmic shear in the linear regime, *Astron.Astrophys.* 479: 9-25

Gibbons, G.W. & Hawking, S.W.(1977). Action integral and partition functions in quantum gravity, *Phys. Rev. D* 15: 2752-2756

Hawking, S. & Ellis, G.F.R.(1973). Ag*Large scale structure of spacetime*Ah , Oxford University Press, London Hicken, M., Wood-Vasey, W.M., Blondin, S., Challis, P., Jha, S., Kelly, P.L., Rest, A. & Kirshner, R.P.(2009).Improved Dark Energy Constraints from A'100 New CfA Supernova Type Ia Light Curves, *ApJ* 700:1097-1140

Jarosik, N., Bennett, C.L., Dunkley, J., Gold, B., Greason, M.R., Halpern, M., Hill, R.S., Hinshaw, G., Kogut, A., Komatsu, E., Larson, D., Limon, M., Meyer, S.S., Nolta, M.R., Odegard, N., Page, L., Smith, K.M., Spergel, D.N., Tucker, G.S., Weiland, J.L., Wollock, E. & Wright, E.L.(2011). Seven-Year Wilkinson Microwave Probe(WMAP) Observations: Sky Maps, Systematic Errors, and Basic Results, *ApJS* 192: issue 2, article id 17

Kempf, A.(1994).Uncertaity Relation in Quantum Mechanics with Quantum Group Symmetry, *J. Math. Phys.* 35: 4483-4496

Kessler, R., Becker, A.C., Cinabra, D., Vanderplas, J., Frieman, J.A., Marriner, J., Davis, T.M., Dilday, B., Holtzman, J., Jha, S.W., Lampeitl, H., Sako, M., Smith, M., Zheng, T.M., C., Nichol, R.C., Bassett, B., Bender, R., Depoy, D.L., Doi, M., Elson, E., Filippenko, A.V., Folet, R.J., Garnavich, P.M., Hopp, U., Ihara, Y., Ketzeback, W., Kollatschny, W., Konishi, K., Marshall, L., McMillan, R.J., Miknaitis, G., Morokura, T., Mörtsell, E., Pan, K.,Prieto, J.L., Richmond, M.W., Riess, A.G., Romani, R., Schneider, DE.P., Sollerman, J., Takanashi, N., Tokita, K., van der Heyden, K., Wheeler, J.C., Yasuda, N. & York, D.(2009). First-year Sloan Digital Sky Survey- II(SDSS-II) Supernova Results: Hubble Diagram and Cosmological Parameters, *ApJS* 185: 32-84

Khosravi, N., Jalalzadeh, S. & Sepangi, R.R.(2006). Non-commutative multi-dimensional cosmology, *JHEP* 0601 :134(10 pages)

Khosravi, N., Jalalzadeh, S. & Sepangi, R.R.(2007). Stabilization of internal spaces in non-commutative multidimensional cosmology, *Int. J. Mod. Phys.* D16:1187-1196

Komatsu, E., Smith, K.M., Dunklet, J., Bennett, C.L., Gold, B., Hinshaw, G., Jarosik, N., Larson, D., Nolta, M.R., Page, L., Spergel, D.N., Halpern, M., Hill, R.S., Kogut, A., Limon, M., Meyer, S.S., Odegard, N., Tucker, G.S., Weiland, J.L., Wollock, E.

& Wright, E.L.,(2011). Seven-tear Wilkinson Microwave Anisotropy Probe(WMAP) Observations: Cosmological Interpretation, *ApJS* 192: issue 2, article id 18

Larson, D., Dunkley, J., Hinshaw, G., Komatsu, E., Nolta, M.R., Bennett, C.L., Gold, B., Halpern, M., Hill, R.S., Jarosik, N., Kogut, A., Limon, M., Meyer, S.S., Odegard, N., Page, L., Smith, K.M., Spergel, D.N., Tucker, G.S., Weiland, J.L., Wollack, E., & Wright, E.L.(2011).Seven-year Wilkinson Microwave Anisotropy Probe(WMAP) Obser- vations: Power Spectra and WMAP-Derived Parameters, *ApJS* 192: issur 2, article id 16

Maeda, K.(1989). Towards the Einstein-Hilbert action via conformal transformation, *Phys. Rev. D* 39: 3159-3162

Magnano, G. & Sokolowski, L.M.(1994).Pyhysical equivalence between non-linear gravity theories and a general-relativistic self-gravitating scalar field, *Phys. Rev. D* 50: 5039-5059

Mantz, A., Allen, S.W., Rapetti, D., & Ebeling, H.(2010).Observed Growth of Massive Galaxy Clusters I: Statistical Methods and Cosmological Constraints, *MNRAS* 406:1759-1772

Massey, R., Rhodes, J., Leauthand, A., Capak, P., Ellis, R., Koekemoer, A., Réfrégier, A., Scoville, N., Taylor, J.E., Albert, J., Bergé, J., Heymans, C., Johnston, D., Kneib, J-P., Mellier, Y., Mobasher, B., Semboloni, E., Shopbell, P., Tasca, L. & Van Waerbeke, L. (2007).COSMOS: Three-dimensional Weak Lensing and the Gowth of Structure, *ApJS* 172: 239-253

Nariai, H.(1971). On the Removal of Initial Singularity in a Big-Bang Unuverse in Terms of a Renormalizable Theory of Gravitation, *Prog. Theor. Phys* 46: 433-438 Nariai, H. & Tomita, K.(1971). On the Removal of Initial Singularity in a Big-Bang Unuverse in Terms of a Renormalizable Theory of Gravitation II, *Prog. Theor. Phys.* 46:776-786

Nojiri, S. & Odintsov, S.D.(2007).Introduction to Modified Gravity and Alternativefor Dark Energy,*Int.J.Geom.Met.Mod.Phys.* 4 115-146

Percival, W.J., Eisenstein, D.J., Bahcall, N.A., Budavan, T., Frieman, J.A., Fukugita, M., Gunn, J.E., Ivezić, Z., Knapp, G.R., Kron, R.G., Loveday, J., Lupton, R.H., McKay, T.A., Meiksin, A., Nichol, R.C., Pope, A.C., Schlegel,D.J., Schneider, D.P., Spergel, D.N., Stoughton, C., Straus, M.A., Szalay, A.S., Tegmark, M., Vogely, M.S., Weinberg, D.H., York, D.G. & Zehavi, I.(2009). Baryon Acoustic Oscillations in the Sloan Digital Sky Survey Data Release 7 Galaxy Sample, *MNRAS* 401 : 2148-2168

Reid, B.A., Percival, W.J., Eisenstein, D.J., Verde, L., Spergel, D.N., Skibba, R.A., Bahcall, N.A., Budavari, T., Fukugita, M., Gott, J.R., Gunn, J.E., Ivezić, Z., Knapp, G.R., Kron, R.G., Lupton, R.H., McKay, T.A., Meiksin, A., Nichol, R.C., Pope, A.C., Schlegel, D.J., Schneider, D.P., Strauss, M.A., Stoughton, C., Szalay, A.S., Tegmark, M., Weinberg, D,.H., York, D.G. & Zehavi, I.(2010). Cosmological Constraints from the Clustering of the Sloan Digital Sky Survey DR7 Luminous Red Galaxies, *MNRAS* 404: 60-85

Riess, A.G., Macri, L., Casertano, S., Sosey, M., Lampeitl, H., Ferguson, H.C., Filippenko, A.V., Jha, S. W., Li, W., Chornock, R. & Sarkar, D.(2009). A Redetermi- nation of the Hubble Constant with the Hubble Space Telescope from a Differential Distance Ladder, *ApJ* 699: 539-563

Schrabback, T., Hartlap, J., Joachimi, B., Kilbinger, M., Simon, P., Benabed, K., Bradać, M., Eifler, T., Erben, T., Fassnacht, C.D., High, F.W., Hilbert, S., Hildebrandt, H., Hoekstra, H., Kuijken, K., Marshall, P., Mellier, Y., Morganson, E., Schneider, P., Semboloni, E., Van Waerbeke, L. & Velander, M.(2009). Evidence for the accelerated expansion of the Universe from weak lensing tomography with COSMOS, *Astron. Astrophys.* 516: id A63

Sotiriou, T.P. & Faraoni, V.(2008).$f(R)$ theories of gravity, *Rev. Mod. Phys.* 82: 451-497. is a good review.

Starobinsky, A.A.(1980). A new type of isotropic cosmological model without singularity, *Phys. Lett.* 91: 99-102

Stelle, K.(1977). Renormalization of Higher-Derivative Quantum Gravity, *Phys. Rev.* D16: 953-969

Suyu, S.H., Marshall, P.J., Auger, M.W., Hilbert, S., Blandford, R.D., Koopmans, L.V.E., Fassnacht, C.D. & Treu, T.(2009). Dissecting the Gravitational Lens B1608+656. II. Precision Measurements of the Hubble Constant, Spacial Curvature, and the Dark Energy Equation of State, *ApJ* 711: 201-221

Teyssandier, P. & Tourrence, P_H.(1983). The Cauchy problem for the $R + R^2$ theories of gravity without torsion, *J. Math. Phys* 24: 2793-2799

Utiyama, R. & DeWitt, B.S.(1962).Renormalization of a Classical Gravitational Field Interacting with Quantized Matter Fields, *J. Math. Phys.* 3:608-618

Vikhlinin, A., Burenin, R.A., Ebeling, H., Forman, W.R., Hornstrup, A., Jones, C., Kravtsov, A., Murray, S.S., Nagai, D., Quintana, H., & Voevodkin, A.(2009). Chandra Cluster Cosmology Project II: Samples and X-Ray Data Reduction, *ApJ* 692:1033-1059

Wands, D.(1994).Extended Gravity Theories and the Einstein-Hilbert Action, *Class. Quantum Grav.* 11:269-280

Whitt, B.(1984).Fourth-Order Gravity as General Relativity Plus Matter, *Phys. Lett.* 145B:176-178

Permissions

The contributors of this book come from diverse backgrounds, making this book a truly international effort. This book will bring forth new frontiers with its revolutionizing research information and detailed analysis of the nascent developments around the world.

We would like to thank Prof. Ion Cotăescu, for lending his expertise to make the book truly unique. He has played a crucial role in the development of this book. Without his invaluable contribution this book wouldn't have been possible. He has made vital efforts to compile up to date information on the varied aspects of this subject to make this book a valuable addition to the collection of many professionals and students.

This book was conceptualized with the vision of imparting up-to-date information and advanced data in this field. To ensure the same, a matchless editorial board was set up. Every individual on the board went through rigorous rounds of assessment to prove their worth. After which they invested a large part of their time researching and compiling the most relevant data for our readers. Conferences and sessions were held from time to time between the editorial board and the contributing authors to present the data in the most comprehensible form. The editorial team has worked tirelessly to provide valuable and valid information to help people across the globe.

Every chapter published in this book has been scrutinized by our experts. Their significance has been extensively debated. The topics covered herein carry significant findings which will fuel the growth of the discipline. They may even be implemented as practical applications or may be referred to as a beginning point for another development. Chapters in this book were first published by InTech; hereby published with permission under the Creative Commons Attribution License or equivalent.

The editorial board has been involved in producing this book since its inception. They have spent rigorous hours researching and exploring the diverse topics which have resulted in the successful publishing of this book. They have passed on their knowledge of decades through this book. To expedite this challenging task, the publisher supported the team at every step. A small team of assistant editors was also appointed to further simplify the editing procedure and attain best results for the readers.

Our editorial team has been hand-picked from every corner of the world. Their multi-ethnicity adds dynamic inputs to the discussions which result in innovative outcomes. These outcomes are then further discussed with the researchers and contributors who give their valuable feedback and opinion regarding the same. The feedback is then collaborated with the researches and they are edited in a comprehensive manner to aid the understanding of the subject.

Apart from the editorial board, the designing team has also invested a significant amount of their time in understanding the subject and creating the most relevant covers. They scrutinized every image to scout for the most suitable representation of the subject and create an appropriate cover for the book.

The publishing team has been involved in this book since its early stages. They were actively engaged in every process, be it collecting the data, connecting with the contributors or procuring relevant information. The team has been an ardent support to the editorial, designing and production team. Their endless efforts to recruit the best for this project, has resulted in the accomplishment of this book. They are a veteran in the field of academics and their pool of knowledge is as vast as their experience in printing. Their expertise and guidance has proved useful at every step. Their uncompromising quality standards have made this book an exceptional effort. Their encouragement from time to time has been an inspiration for everyone.

The publisher and the editorial board hope that this book will prove to be a valuable piece of knowledge for researchers, students, practitioners and scholars across the globe.

List of Contributors

Takayuki Hori
Teikyo University, Japan

T. Görnitz
FB Physik, J. W. Goethe-Universität Frankfurt/Main, Germany

Paul Benioff
Physics division, Argonne National Laboratory, Argonne, IL, USA

Mayukh Lahiri
Department of Physics and Astronomy, University of Rochester, Rochester, NY, USA

Boaz Galdino de Oliveira
Instituto de Ciências Ambientais e Desenvolvimento Sustentável, Universidade Federal da Bahia, Barreiras – BA, Brazil

Regiane de Cássia Maritan Ugulino de Araújo
Departamento de Química, Universidade Federalda Paraíba, João Pessoa, PB, Brazil

Sture Nordholm
The University of Gothenburg, Sweden

George B. Bacskay
School of Chemistry, The University of Sydney, Australia

Ion I. Cotăescu
West University of Timisoara, Romania

Allan Ernest
Charles Sturt University, Australia

Y. Ezawa
Dept. of Physics, Ehime University, Matsuyama, Japan

Y. Ohkuwa
Section of Mathematical Science, Dept. of Social Medicine, Faculty of Medicine, University of Miyazaki, Japan

Printed in the USA
CPSIA information can be obtained
at www.ICGtesting.com
JSHW011438221024
72173JS00004B/853

9 781632 383846